AF682533

Volume • 2

Industry and Agriculture in India Since Independence

OTHER VOLUMES IN THIS SERIES

Volume 1 State and Nation in the Context of Social Change

Volume 3 Region, Religion, Caste, Gender, and Culture in Contemporary India

Volume 4 Class Formation and Political Transformation

Social Change and Political Discourse in India

Structures of Power, Movements of Resistance

Volume • 2

Industry and Agriculture in India Since Independence

Edited by

T.V. SATHYAMURTHY

OXFORD UNIVERSITY PRESS

DELHI CALCUTTA CHENNAI MUMBAI

1998

Oxford University Press, Great Clarendon Street, Oxford OX2 6DP
Oxford New York
Athens Auckland Bangkok Bombay
Calcutta Cape Town Dar es Salaam Delhi
Florence Hong Kong Istanbul Karachi
Kuala Lumpur Madras Madrid Melbourne
Mexico City Nairobi Paris Singapore
Taipei Tokyo Toronto
and associates in
Berlin Ibadan

First published 1995
First published in Oxford India paperbacks 1998

ISBN 0 19 564394 1

Typeset by All India Press, Kennedy Nagar, Pondicherry 605 001
Printed at Rajkamal Electric Press, New Delhi 110033
and published by Manzar Khan, Oxford University Press,
YMCA Library Building, Jai Singh Road, New Delhi 110 001

To the memory of

KRISHNA BHARDWAJ

b. Dharwad 21 August 1935
d. New Delhi 12 March 1992

and

PRAMIT CHAUDHURI

b. Calcutta 4 December 1935
d. Lewes (England) 19 July 1996

Preface to the Paperback Edition

The Project on 'Social Change and Political Discourse in India' was started in January 1989. The four volumes resulting from it have appeared during the two years between mid-1994 and mid-1996.

The last seven or eight years have witnessed the crystallization of new social, political and economic trends in India that had already been in the making at least since the Internal Emergency (1975–7). The remit of the project was to address the task of analysing the changes that have occurred in India from a variety of perspectives. A number of social science perspectives have been brought together in these volumes by contributors with a commitment to an inter-disciplinary treatment of India's social, economic, political and cultural development since Independence.

My aim was to produce a set of studies that would concentrate on the *political* dimension of the different facets of change in India that can be identified. Thus, it has been possible to provide a single focus for contributors whose interests and accomplishments range from academic expertise in one or more social sciences to activism in specific spheres of political and social life.

This project began during the last year of the Congress(I) government under Rajiv Gandhi's Prime Ministership. The paperback edition of this volume will appear during the first few months after the eleventh general election. The interregnum has been marked by an intensification of political change which constitutes an important strand of the subject matter explored in the four volumes of this project.

Indian politics oscillates between extremes–e.g. between incorruptible and immaculate personal values and rectitude of certain individual statesmen (e.g. V.P. Singh, A.B. Vajpayee and Jyoti Basu), and an ever-expanding body of *netas* with reputations sullied by persistent and mounting allegations of corruption (e.g. Chandra Shekhar, P.V. Narasimha Rao and J. Jayalalitha); between the institutional malaise engendered by a lugubrious and complacent bureaucracy, and the momentum provided by certain institutions of governance (e.g. the Election Commission since 1991, and the Supreme Court of India during the last few years) that can be whipped into shape as pro-active and dynamic agents capable of insisting that legal and constitutional norms be enforced on public servants and politicians alike; between the undemocratic and authoritarian tendencies of the ruling elements representing the interests of the dominant classes, and the irrepressible democratic urge of the mass of the Indian people; between a thrust towards centralization of power and the resistance of decentralizing tendencies at different levels of the polity, (e.g. the countervailing power of the State and the Centre within the dialectic of state power in India; and the tension between the State and the levels below); between political parties and organizations pursuing policies based on caste domination and communal calculus and those that follow a more or less consistent policy of affirmative action towards disadvantaged castes, *Adivasis* and minority communities as the centrepiece of their political agenda[1]; and, between the economic compulsions of liberalization, globalization, marketization and privatization (in order to cater to the needs of an expanding middle class of consumers), and the political economy of expanding poverty (to which increasing numbers of people are condemned by the very process by which the middle class becomes more affluent).

The contributors to the project have addressed the dialectics of the interaction of these pairs of opposites and the tension generated by them in the quotidian politics of India at various levels. Volume 1 and this volume look at political and social change, largely from an institutional and policy nexus. Volumes 3 and 4 focus on the impact of the political actions of different segments of the population on institutions and policies. They also deal with the pressures exerted by the latter on popular movements of resistance under conditions of rapid socio-economic change. This volume is centrally concerned with the nature of India's economic policy exploring the role that the Indian state has played in intensifying

[1]Singh, V.P., 1996: 'Affirmative Action in India', *Mainstream* 34(18 May): 24, pp. 7–13.

the demands of the dominant (producing and non-producing) classes in Indian society.

The essential methodological characteristics of the project which give it its unique quality can be summed up as follows:

1. It is primarily concerned with identifying and analysing the *political* dimension of change.

2. It is committed to an *inter-disciplinary* approach. It draws together (in respect of both the selection of topics and methodology) practitioners of different social science disciplines. They are called upon to discuss the political dimension of their specific topic(s)/remit(s) from the methodological perspective of their particular social science discipline(s). In practice, this has meant that only a small number of contributors have been drawn from the academic discipline of political science. Economists, lawyers, sociologists, social anthropologists, historians, students of literature and mass communication, journalists, and activists figure prominently in the choice of those invited to write. The aim has been to enhance explanatory effect by combining sound economic and historical scholarship on the one hand, and, on the other, sensitivity to the nuances of political change.[2]

3. It is a *co-operative* endeavour which has been richly influenced by informal bilateral discussions between individual contributors and myself, as well as among larger gatherings of social scientists at the three workshops (Mysore, 1990; Bhubaneswar, 1991; and Bhubaneswar, 1992).

4. Its scope is *comparative* because of the emphasis given to comparing different regions of India in the discussion of several topics and to the unevenness of the Indian experience of change.

5. In the selection of contributors, care was taken to include younger social scientists as well as activists (in several specific cases the same person embodied both roles), whilst ensuring that the discussion of each topic remained scientific and academic.

6. The editor defined each topic in relation to the whole of the project. Each contributor worked from a detailed brief containing a broad indication of the scope of the material to be covered and the relationship that it would bear to the project as a whole.

7. Considerable effort has gone into relating each chapter of each volume to the objectives of the project as a whole. In this sense, thematic variety has been made to serve the interests of unifying what would otherwise have remained a fragmented exercise.

I should like to express my gratitude to the (anonymous) reader whom Oxford University Press approached in 1990. The reader's sterling advice was an important factor in the shaping of the original aim of the project and in the decision to focus attention primarily on the content

[2]Byres, T.J., 1995: 'Political Economy, Agrarian Question and Comparative Method', *Economic and Political Weekly* 30(11 March): 10, pp. 507–13.

and dynamics of social change in the concrete as a guide to an understanding of the shifts in political discourse.

It is perhaps worth reporting that social scientists familiar with the work presented here have drawn attention to the fact that no other major society of the world has yet been analysed with the tools of such a comprehensive methodological approach as that followed in this project.

At the time of its completion contributors were of the view that the project as a whole was sufficiently comprehensive and did not reveal any significant gaps. I am sure that with a much wider dissemination in a paperback edition and with passing years, gaps will come into view. My hope is that the four volumes are so structured that gaps will be easily identified and can be filled without radically altering the basic foundation of the project.

York
November 1997

T.V. Sathyamurthy

Acknowledgements

This work originated in the project 'The Terms of Political Discourse in India' for which the Ford Foundation made the necessary funds available. I would like to record my thanks to the Ford Foundation, New Delhi and New York for the financial support that has made these volumes possible.

During the last three years, a vast debt of gratitude has been built up to a large number of individuals, not all of whom are named here, but whose help and advice have been invaluable for the project.

R. Sudarshan's inspiration and support, wise counsel, and unfailing vision immensely benefited the project.

Those who attended the three conferences, besides contributing to extremely stimulating and wide-ranging discussions on the spot, gave invaluable help not only through their own specific contributions but also through discussion and debate over the entire period.

The project would have been impoverished without the organizational skills and intellectual qualities provided by Valerie Atkinson who, as the Project Administrator, not only fulfilled the daunting remit of her initial appointment in setting up the administrative base, but, in addition, undertook considerable editorial responsibility. The form, style, and substance of a number of chapters of these volumes are the result of her literary expertise and clarity of expression.

Throughout the last four years, Krishna Raj has been a tower of strength, always supportive of the project. But for him, several of the crucial chapters would not have materialized.

Kalpana Wilson, Research Assistant of the project for two years (1989–91), worked closely in the execution of the project, the preparation of individual briefs for contributors, the commissioning of work, the content of the conferences, and the organization of the volumes. This work has undoubtedly been enriched by Kalpana Wilson's exceptional intellectual input, including a note that she prepared soon after leaving the project as her contribution to the General Introduction. She also worked on building a data base, including an annotated bibliography covering a number of topics included in these volumes.

A number of persons have provided intellectual, moral, technical, and organizational support. I benefited from editorial suggestions in respect of the introductory chapters given by R.A.W. Rhodes.

Suzanne Dekker and Susan Popel have provided invaluable support in the speedy and extremely efficient production of computerized typescripts. The Computing Service has given important technical help. Phil Rees of the J.B. Morrell Library has been a never-failing source of ready bibliographical assistance in ample measure.

In India, Snigdha Mitra, Radha Kanta Nayak, and S.K. Pradhan (ably supported by the staff and students of the National Institute of Social Work and Social Studies) were responsible for the excellent arrangements that guaranteed not only a good intellectual environment but also ensured a salubrious social setting for the two conferences that were held in Bhubaneswar. S. Chellani of the Ford Foundation, New Delhi, went beyond the call of duty in facilitating transits in India.

T.V. SATHYAMURTHY

York
November 1994

Contents

Contributors

Amiya Kumar Bagchi	Researcher, policy advisor, and economist (Calcutta)
D. Bandyopadhyaya	Civil servant and scholar (Calcutta)
Sanjaya Baru	Economist and journalist (Delhi)
Krishna Bharadwaj	Economist, activist both within the women's movement and on the general political scene (Delhi)—now deceased
Pramit Chaudhuri	Economist and teacher (Brighton)
Sucha Singh Gill	Economist, teacher, and political activist in human rights (Patiala)
Zoya Hasan	Teacher and researcher in political science, and activist in the women's movement (Delhi)
Utsa Patnaik	Economist, researcher, teacher, and activist in the women's movement (Delhi)
Sumit Roy	Teacher, researcher, and writer in economics (Brighton)
T.V. Sathyamurthy	Teacher and researcher in politics (York)
E. Sridharan	Researcher in economics and politics (Delhi)
Ravi Srivastava	Teacher and researcher in economics (Allahabad)
A. Vaidyanathan	Teacher, researcher, and policy advisor (Madras)

Note

Throughout this volume the term 'State' (upper case 'S') is used to refer to the States of the Indian Union or Federation, and the term 'state' (lower case 's') to the Indian state, the power of which at any given time is controlled by the government at the Centre.

General Introduction

T V SATHYAMURTHY

Over the past fifteen years there has been a serious academic neglect of political processes and their ramifications for Indian society and the post-independence Indian state. These four volumes on 'Social Change and Political Discourse in India: Structures of Power, Movements of Resistance' are the first fruit of an international collaboration exploring diverse aspects of political, economic, social, and cultural change in India.[1]

The aim of the present project is to strike out a new path and close the gap between empirical observation on the one hand, and theoretical analysis and understanding of dynamic political processes on the other. In setting out on such an expedition, we lay no claim to a grand general theory capable of explaining Indian political reality in all its aspects.

Even though the work presented in these volumes was shaped by the initial objectives of the project, its present form results from the interchange of views and perspectives between contributors who took part in the three conferences (Mysore, 1989; Bhubaneswar, 1990 and 1991). The central themes of the four volumes assumed their present

[1] These were originally defined in a project entitled 'Terms of Political Discourse in India' (1989–92).

shape as a result of the opportunity provided by these occasions for an exchange of ideas. This work does not claim to have emerged from a sealed chamber of academic objectivity; rather, it has been shaped by the specific political conjunctures—domestic and international—in which its content and balance had to be determined.

The first volume explores the political dynamic of the state structures of contemporary India as products of the evolution from the colonial to the post-colonial era. The nationalist movement, under the leadership of the Congress party, viewed the process of state formation as an aspect of the political process underlying the forging of an Indian nation.

Having failed to prevent the partition of the country and inherited a truncated version of the 'Indian nation' envisaged in the independence movement, the leadership of the ruling Congress opted for a federal structure which inclined more towards centralization of power than towards decentralization of authority. The Constitution of India bore the marks of the Congress party's 'revisionist' orientation to the notion of federal structure. After showing some interest in establishing a state that would lie somewhere between a confederal and federal configuration of political forces, the Congress after Partition showed a distinct preference for a political system that lay somewhere between a federation and a unitary structure. During the past four decades, numerous political movements have arisen in various regions and among different strata of the population, representing a challenge to the dominance of the Indian state and its increasingly authoritarian rule.

The second volume is centrally concerned with analysing the Indian state's economic policies in the spheres of industry, technology, and agriculture. The overall picture that emerges from the volume is one of acutely uneven development, development skewed in favour of certain regions and specific classes at the expense of others. In the process, contradictions of a vertical nature (e.g. between dominant and dominated classes, upper castes and middle castes, upper and middle castes, on the one hand, and lower castes on the other etc.) have been woven into a complex web of interaction with horizontal (i.e. intra-ruling class) contradictions which have gathered momentum during the last quarter of a century.

The third volume addresses the problems faced by civil society as a whole, and especially by different segments of the mass of the Indian people in raising basic economic and political demands. The divided and fragmented character of Indian society calls for a multi-layered and diachronic analysis, building up from the micro- to the meso- and macro-levels, of the tensions engulfing the relations between its different segments and regions, on the one hand, and on the other, the Indian state and the classes and groups controlling its power. The divisive role played by region, religion, caste, gender, and culture in the articulation of conflicting interests constitutes the main theme running through this largely empirically orientated volume.

The fourth volume attempts to bring together the larger themes of class formation and political transformation in India.

The theme of class formation is a difficult one to explore, especially in the context of social change involving numerous horizontally divided intra-class categories of deprived and oppressed people. Yet, political action from the grassroots points not only to the problems of different elements of society in a far-flung country, pursuing separate aims, but also to the potential contained in some mass movements for uniting in order to forge common demands on a widening scale. The phenomenon of political transformation is viewed from two angles—one of the actual cultural and social divisions that have come to the fore, and the other of potentialities contained in the present situation to overcome fragmentation and to struggle for common political and economic goals.

The division of this work into four volumes, each with its own distinctive subject, is not primarily intended to carve up the project into four discrete parts for two reasons. First, the project as a whole has steadily evolved through a number of discussions with a wide range of social scientists and political activists. The themes that emerged as a consequence of these interactions are sufficiently interconnected, varied, and complex to encompass a large variety of political processes and socio-political forces. Their examination and analysis from a number of different standpoints and perspectives constitutes the principal impetus for this work. The cohesiveness of the work as a whole is ensured by integrating it around a set of themes which weave in and out of the individual volumes.

The four volumes are thus presented as a coherent work which should be judged as a whole. The political discourse of those who control the power of the state and that of the mass of the population have diverged over the last five decades and have often stood in mutual contradiction. Social and economic changes have triggered off political movements against the increasingly unacceptable hegemony of the Indian state and its domination by the Congress party. The increasingly coercive character of the state structure has evoked popular resistance at a number of different levels in society, not all of which can be said to be progressive or democratic in their aims. The Emergency showed that even such a highly centralized and authoritarian state is vulnerable to pressures from a politicized and politically conscious civil society. The Indian state which inspired awe and fear in the minds of the people by its sheer fire-power and ruthless exercise of authority during the first half of Indira Gandhi's rule became—during the latter half of the '70s—more and more susceptible to domestic and international pressures. Throughout the post-independence era, however, it has neither lost its essentially élitist class character nor given up its tendency to compromise democratic principles.

The background

The experience of the internal Emergency (1975–7) stimulated a new kind of activism in the form of political parties opposing the ruling Congress party. There were also numerous organizations concerned with a variety of social, economic, cultural, and environmental problems, raising demands of broadly two kinds. The first set relates to safeguarding elementary civil liberties and democratic rights (especially those guaranteed in the Indian Constiutution) of the mass of the Indian people. The second set of demands was for greater participation of the people in the making and implementation of policies affecting their daily lives.

During the years leading up to the Emergency, the exercise of the coercive power of the state, through numerous civil–bureaucratic and police–para-military structures of government, both at the Centre and in the States, reached hitherto unprecedented levels of concentration and acquired a more and more monolithic character. The first phase of Indira Gandhi's Prime Ministership (1966–75) was marked by strenuous attempts on the part of the ruling party to shore up its power by a variety of means, including the propagation of populist slogans; short-circuiting democratic processes in the society as a whole and within the Congress organization; remoulding the bureaucracy into a tool of the Congress high command. Attempts were made to pervert the judiciary politically, to undermine opposition parties; and to topple State governments; and to appoint and dismiss Congress Chief Ministers in utter disregard of elementary democratic conventions.

Underlying these developments, culminating in the massive 'derailment' of constitutional democracy in 1975, were complex interweaving political, social, and economic processes. They were taking place not only in a specific international setting but also within a domestic political economy that was fraught with internal contradictions. These contradictions affected the capitalist and producing classes in ways unexpected by the government.

During the period since the end of the Emergency, a number of tensions gathered momentum. The '80s witnessed a clash between the tendency of the state to centralize power and militant opposition to it (especially in Punjab, Kashmir, and Assam) taking more and more violent forms. There was also an intensification of political contradictions and conflicts centred in the various nationalities and religions. At the same time, the efforts of successive governments to usher in an era of economic liberalization emphasized the worsening plight of the mass of the people in general and a fragmenting working class in particular. They also highlighted inter-segmental contradictions of a proliferating bourgeoisie in a context of rapidly deepening dependence of the Indian economy on international capital. Problems, deeply rooted in the past, were resurfacing in a rapidly changing domestic and international context.

The academic study of politics in India (as at 1975) was singularly ill-equipped for the task of unravelling the complexities underlying these phenomena. Its so-called mainstream (Morris-Jones, 1964; Rudolph and Rudolph, 1967) was exclusively concerned with formal structures and the prescribed ways in which they were supposed to function; and social categories (mainly, but not only 'castes') derived from 'traditional' Indian society. It ignored the dynamic ways in which castes were becoming transformed and the crucial role played by existing and rapidly emerging new classes to which economic change had given rise. Nor has it shown an awareness of the refraction of classes into liminal categories (i.e. categories which straddle different orientations—e.g. class–caste; caste–region; region–religion), partaking of the characteristics of old and new patterns of social and economic organization. At the same time, mainstream Indian political scientists (with the notable exception of Kothari) found themselves encumbered with the conceptual and methodological strait-jackets in which Western political science in general, and comparative politics and development studies in particular, sought to squeeze the extremely varied, subtly nuanced, richly-textured, and multi-layered empirical material which the Indian field offered.

The Indian scene in academic political science as at 1975 was still conditioned by the era of political stability under the Congress government. Kothari's (1970) energetic and distinguished early work, identifying the Congress system of power as the linchpin of Indian democracy, provided the necessary initial thrust to academic political science writing. However, the vast bulk of political scientists writing in India before 1975 were either poorly cloned versions of Kothari or insufficiently sophisticated disciples of the American school of political development. It is to Kothari's credit that he responded to the 1975–7 Emergency by radically altering his essentially structural–functional perspective and recasting his approach to Indian politics by centring it in discussions of the relationship between the state and democratic civil society (Sathyamurthy, 1991).

Moreover, a great premium was placed on theory derived from

highly restricted empiricist forays into the field of observation (e.g. Weiner, 1967; Morris-Jones and Dasgupta, 1975), whilst the task of drawing out the complex and qualitatively significant interactions embedded in empirically observed phenomena, extending far beyond the realm of quantitatively measurable characteristics, sadly remained neglected (with significant exceptions, e.g. Brass, 1965; more recently Kohli, 1990, and Srinivasavardan, 1992). By 1970, mainstream Indian political science had reached the stage at which it was finding out more and more about less and less (O'Brien, 1972; Sathyamurthy, 1971, 1973, 1986, 1987).

Among the most serious lacunae in the development of knowledge of Indian politics, according to the mainstream school, were

1. an understanding of the nature of the state;
2. the dynamic underlying class, caste, and other forms of politics and the interrelationship between these;
3. the interactions between the various—i.e. political, social, economic, ideological, cultural and international—vectors affecting change in post-independence Indian society; and
4. an insight into the factors underlying the tenacity and stubbornness shown by ordinary people in their continuous struggle against the persistent effort of the state to abridge, curtail, and even eliminate constitutionally guaranteed elementary civil liberties and democratic rights.

Yet, over the past fifteen years or so, we have witnessed an accumulation of a rich variety of political action, of differing range, depth, and coverage, originating in the grassroots. Even though not all of these struggles have been mounted on an all-India scale, such forms of political action have struck deep roots in the economy, ideology, and culture of the society (e.g. movements espousing the demands of peasants, farmers, women, Adivasis and Dalits). There is no book which provides a comprehensive and penetrating understanding of such political processes taking place in different parts of the country.

In contrast to mainstream or bourgeois social theory, Marxism offers analytic tools appropriate for an understanding of rates and degrees of change under the impact of dynamic forces. In carrying out the task of providing empirical substantiation of concrete phenomena, Marxist economists studying India have always taken 'economics' to mean 'political economy' in the strict interpenetrating sense of the term. Their studies generally follow up analyses of economic relationships with some indication of their political significance. This served as a good example for Marxist political scientists to follow in their own more centrally and explicitly politically orientated studies.

However, concentration on empirical study in this mode was handicapped by two related factors. First, there seems to have been some sort of unwritten understanding between the communist parties (i.e. practitioners of communist politics in a generally non- or anti-communist milieu) on the one hand, and Marxist intellectuals (with a few distinguished exceptions such as Kosambi and Randhir Singh), on the other. The communist parties were assumed to constitute the source of theory, whilst Marxist intellectuals were believed to be substantiators of theory (of which there was an abundance, authored by leading intellectuals in the CPI) with data collected in the course of fieldwork. Second, this derivative and flawed quality of the work of a number of Marxist political science academics in India had the twin effect of stifling a vigorous discourse between questions of theory and method, on the one hand, and empirical data, on the other. As a consequence, their work suffered from vulgar reductionism and 'formulaistic' jargonism and failed to generate new insights.

If the besetting sin of bourgeois (or mainstream) political science lay in its failure to recognize the ongoing dialectic between contradictory social and economic forces underlying political phenomena, the most debilitating shortcoming of Marxist political science arose out of its academic practitioners' general unwillingness to disentangle the dialectical method from the rigid orthodoxies of the different prevailing 'lines' within the fractured Indian communist movement (Sathyamurthy, 1986; Wilson, 1994).

As an academic discipline, political science has suffered greatly, caught, as it has been, between the Scylla of international compulsions (i.e. compulsions emanating from the West in general and America in particular) leading to the adoption of a positivist–functionalist sociological paradigm (of which Béteille, 1965; Bailey, 1960; and Hanson, 1969 are outstanding examples), and the Charybdis of domestic compulsions of mechanically adhering to party orthodoxies and 'lines' arising out of too close an identification with the communist movement (Sathyamurthy, 1971, 1987).

In attempting to strike out a new path, we have approached our tasks from two broad angles. The first involves focusing attention on the richness and diversity of the language used to refer to political processes, political action and aims by different actors involved in day-to-day affairs involving institutional structures of the society and the state. Starting with the semantic and semiotic dimensions of the terms of political discourse generated through the concrete experience of people, the research strategy quickly moved to substantive areas of material significance. As the project progressed, the analytic emphasis shifted to the evolution of the continuing relationship between changing terms of political discourse, on the one hand, and, on the other, the changing agenda of politics in the concrete. Thus, even from its very early stages, the project became involved in historical analysis and the examination of concrete social, economic, political, cultural, and ideological processes. The task facing the contributors was one of establishing links between the terms of political discourse and the dynamic of the political processes involved in social change, and of interweaving the two.

The second and equally important angle from which the project views its remit is concerned with a balance between the essentially *political* dimension of what is studied and the economic, social, cultural, ideological, and international underpinnings of politics. This balance has been secured by following an uncompromisingly interdisciplinary approach to the political questions that are selected for study. Whilst it is indeed a difficult balance to achieve in practice on a sustained basis, even a modest degree of success in such an endeavour can only result in an enhancement of the intellectual 'toughness' of political study and the analytical value of the conclusions reached. This project represents a serious attempt at providing a single coherent account of the Indian polity as a whole, and systematically analysing its evolution and principal tendencies.

From the outset we recognized that in order to realize this aim, not only must the project involve economists, sociologists, historians, and students of politics, but also political activists. The four volumes of this project address not only the question of how India's concrete experience of politics has contributed to enriching our understanding of theoretical concepts and to layering political vocabulary (according to its nuances), but also the central questions arising out of political practice in India.

Political concepts and terms of discourse

The various concepts involved in understanding the dynamic underlying change and continuity in Indian political life embrace diverse spheres of thought and action. Some concepts refer to political processes (e.g. elections, party politics, institutionalization of 'corruption', and the changing modalities of Centre–State relations). On a somewhat different level, there are concepts that address ideological developments—e.g. the various 'isms' of politics, the political economy of 'liberalization' policies, and the emergence of new mass-based constituencies in politics with special reference to, for example, community-based or environmentally-orientated issues.

A third category of concepts and terms refers to institutional features of politics (e.g. resource allocation mechanisms, and the relationship between trade unions, government, and employers). At yet another level, we encounter concepts which carry a constitutional import (e.g. the role of the state in maintaining national security and integrity; the relationship between the executive, legislature, and the judiciary; the role of the armed forces; and the elaboration of para-military police forces in the name of maintaining security). A whole gamut of concepts, analytically separable along several different theoretically identified spectral bands for purposes of discussion, overlap in actual practice, and this gives rise to explanatory complications which must be taken into account in clarifying our ideas of Indian politics.

The dynamic of complex political processes has been captured over the years through expressive phrases or terms imaginatively conceived by those active in politics as well as those who write about it. These have passed in and out of media parlance and informed discussion without any systematic attempt to assess their overall significance in a broader context or, for that matter, the different levels of significance that they acquire under different concrete conditions. Thus, for example, terms such as secularism, nationalism, communalism, regionalism, liberalism, revolution, liberation, and 'value-based' politics are commonly used. Yet, on the other hand, the application of these terms to specific circumstances[2] reveals the ambiguities and contradictions to which they are prone in ordinary use.

At the same time, a large number of economic and sociological terms are used to describe aspects of the political process which pertain to the subterranean reaches of politics rather than strictly to the sphere of formal institutional mechanisms.[3] These represent but a thin cross-section of the enormous range of concepts of differing vitality and import that have arisen in discussions of Indian politics.

Method of approach

In the process of defining the tasks facing the project, a number of methodological points have been addressed. Among these, five are of

[2] For example, 'communalism' in Punjab where a discussion of say *khalsa* or *quaum* has gained political relevance in recent years, as indeed the term Dravidistan (or, latterly Eelam) which has undergone a transformation of meaning during the past several decades.

[3] For example, *dalal, mamul,* and Permit-Quota-License Raj in discussions of corruption; loan *melas* in discussions of partisan favours in the economic sphere; and *chacha, dada, mastaan, Aaya Ram Gaya Ram,* and *Garibi Hatao* in discussions concerning the manipulation of political and economic processes.

crucial importance: the need to spell out the project's interdisciplinary character clearly; the need for social scientists from disciplines other than politics to focus on the political implications of their work; the cultural dimension of political change; the need to link the two levels of analysis relating to the 'terms of political discourse' and the 'substance of political action'; and, the need to link the qualitative and quantitative aspects of political change.

1. Interdisciplinary nature (vide *Figure 0.1*) *of the task of analysing political change*

The interdisciplinary nature of the task of analysing political change needs to be spelled out clearly. Over the decades, the institutional structures of the state have been substantially drawn into the development of the economy and into ensuring the stability and order that are necessary to accomplish the government's stated as well as hidden goals. These structures have a specific origin in constitutional arrangements that preceded and followed Independence. The fact that state structures are (directly or implicitly) rooted in the Constitution alongside a number of socio-economic and political values and civil rights, introduces a dynamic which has worked itself out in numerous predictable and unpredictable ways, but which almost invariably can be specified and interpreted. The different bodies involved in the interpretation of the Constitution (from the Supreme Court downwards, and including such important structures as the Finance Commission)—against a backdrop of conflicts of interest and changing power equations in the rural and urban social economy, as well as the central and regional political formations competing for power—have contributed to shifts in the overall political direction of the system as a whole.

The theoretical and substantive importance assumed by issues raised outside the milieux constituted by institutional structures, during the past fifteen years, has compelled some social scientists (e.g. Kothari, 1983, 1988) to increase the attention paid to popular movements (more specifically, movements of resistance to state power). The struggles which they wage do not originate in recognized institutions. Nor are recognized institutions of state power or political parties able to co-opt or absorb them. In fact, such popular movements often develop a dynamic of their own that needs to be located within the political arena in their specific relationships.

We are thus involved in a task that has to embrace a dual aim. We must trace the quintessentially *political* evolution of the Constitution, on the one hand, and, on the other, the *political* evolution of the society as a whole (of which the structures provided by and ideas

FIGURE 0.1

A Conceptual Guide to the Four Volumes

Level of Analysis	*Social Science Disciplines*	*Themes and Concepts Explored*	*Volume No.: Chapter No.*
1. Constitutional and Institutional	Politics; History; Sociology; Law	state structures (electoral, judicial, bureaucratic); independence and nationalism; patriarchy; secularism; civil, fundamental and democratic rights; equality; social justice; dependence	1:2; 1:6; 1:3; 1:4; 1:5; 3:3; 3:4; 3:6; 3:7; 3:8; 3:9; 3:10; 3:11; 3:12; 3:13; 3:14; 3:15; 3:16; 3:17; 3:18; 3:19; 4:1; 4:3; 4:4; 4:5;4:6;4:8; 4:11; 4:13; 4:14; 4:15; 4:16
2. Socio-economic and related questions	Economics; Economic History; Social Policy	economic development; planning; technological policy; balance and terms of trade between industry and agriculture; equality of access to education, health and social services; political dissent, dissidence and contradiction; positive discrimination; inter-regional differences	2:1; 2:2; 2:3; 2:4; 2:5; 2:6; 2:7; 2:8; 2:9; 2:10; 2:11; 3:1; 3:2; 4:1; 4:2; 4:3; 4:4; 4:5; 4:6; 4:9; 4:10; 4:12; 4:13; 4:16
3. Extra-constitutional avenues of change	Social and Political Anthropology; Politics; Sociology; Social Psychology	movements outside the purview of the Constitution; movements 'to empower the poor'; movements to mobilize lower and lowest castes, Adivasis and Dalits; women's movement and women's organizations; and other activist movements.	3:1; 3:2; 3:3; 3:5; 3:6; 3:11; 3:12; 3:13; 3:14; 3:15; 3:16; 3:18; 4:1; 4:2; 4:5; 4:7; 4:8; 4:12; 4:16

NOTE: Only Volume 2 deals with a single interdisciplinary field, namely, political economy. In all other volumes, there is an interplay between several social science disciplines. This is reflected not only in the selection of themes for individual chapters, but also in the choice of contributors, each giving their analysis an appropriate focus derived from the relevant social sciences.

embodied in the Constitution are but a part). It is vital that our analysis of this dual process of evolution should include popular movements which act from outside the institutional structures in staking their claim for a part in the shaping and implementation of policies affecting their life and environment. There is much that needs to be done here in order to reflect the impact of specific movements on the general direction of politics in India and to focus on the interrelationships between different arenas of political action. Volume 1, which focuses attention on the political evolution of the Indian state, and Volume 3, which deals with the impact of numerous social vectors (stemming from cultural and religious diversity, religious plurality, caste hierarchy, gender inequality, and Adivasi discrimination) that cut across the vast mass of the dominated and the oppressed, are crucially related to this methodological requirement.

In a society in which (thanks to Gandhi's influence), special significance continues to be attached to decentralized local action on a 'voluntary' (as opposed to a 'regimented') basis (often moving from one practical issue to another, rather than on the basis of a total or holistic ideology), the bases of struggles originate in the concrete experiences of communities. This is especially relevant since the economy and social bases of a number of communities (in Indira Gandhi's euphemistic language 'the weaker sections of society') are threatened, often as a direct consequence of government policies seeking to fortify the classes and interests that sustain the ruling parties in power. This is in fact attempted by focusing on specific topics, especially in Volumes 3 and 4, with some reference to simultaneous processes of fission and fusion between segmented identities and class identities with which popular struggles in a variety of socio-political contexts is suffused.

Volume 3 includes, among other themes, for example, Rao's characterization of the political dimension of the socio-economic struggle for survival of the Adivasis (Ch. 14), Nayak's treatment of the conditions of life of the 'indigenous peoples' (i.e. Adivasis) of Orissa (Ch. 13), Patel's analysis of the political dynamic underlying the discourse of communalism (Ch. 3), Chandra's examination of the relations between Hindus and Muslims in the rural north Indian context based on a critical reading of two novels (Ch. 4), Sheth's discussion of 'Other Backward Castes' in the context of the Mandal controversy (Ch.10). Suresh's account of the Dalit movement (Ch. 12), and Omvedt's treatment of the anti-caste movement (Ch. 11).

In Volume 4, a number of contributors deal with specific themes concerned with the interpenetration between class, on the one hand, and segmental identifications such as caste, region, religion, and gender, on the other. Examples of such discussions include Vidyasagar's discussion (Ch. 6) of the challenges posed by the new agrarianism to the left political forces; Tulpule's account (Ch. 4) of the fragmentation of trade unions; Hensman's examination (Ch. 2) of the impact of industrial re-structuring on men and women workers; D'Souza's analysis (Ch. 3) of the casualization of labour; Nauriya's essay (Ch. 9) on the struggles in the sphere of civil, human, and democratic rights; and Krishna's treatment (Ch. 14) of the tension between the state which tries to appropriate dissent and people's movements which attempt to keep dissent from being stifled by the state's partial or fragmentary efforts to meet popular demands.

2. *State policy and the politics of economic development*

A highly sophisticated body of literature (mainly but by no means entirely economic in its disciplinary origins) has developed on the impact on Indian society of the government's development policies in general, and of planning in particular. It has addressed the policies of the government in respect of industrialization: food production, accumulation, and security; anti-poverty drives; droughts, famines, and other 'Acts of God'; generation of employment opportunities; health, education, and social welfare; as well as long-term strategies involving the reorientation of markets, export, import, balance of payments, the terms of trade between agriculture and industry within the domestic economic sphere, and the controversial package combining massive doses of devaluation with far-reaching measures of decontrol and liberalization of trade. These themes are explored in several chapters, especially in Volumes 2 and 4.

Volume 2 specifically focuses on a number of discussions which include the relationship between the discourse of planning and the actual economic strategy followed by the government; continuity and change as represented by the fluctuating fortunes of planning in the history of independent India; the political and economic consequences of the government's policies of liberalization and technological development; the unevenness of agricultural development between different regions and its political consequences; and the government's approach to poverty alleviation. Volume 4 focuses on questions centrally involving the transformation of trade unions; fragmentation of labour; proliferation of the bourgeoisie; the impact of the educational policy of the state on the poorer sections of the population; and the political economy of the state's orientation towards environmental issues.

Very little analytic work has been carried out by political scientists about crucially important questions such as:

i. the impact of political forces on development;

ii. dynamic choices and strategies that have entered the sphere of policy since Independence;

iii. the impact of balancing the requirements of 'capital' and the demands of 'labour' in particular, and of society in general;

iv. the role of the state in setting the agenda (hidden and otherwise) of economic management and development; and

v. the experience of a vast proportion of the mass of the population which contributes more to the process of development than it receives from it.[4]

The increasing awareness shown by an expanding section of socially conscious economists relating to the political consequences, costs, options, and opportunities that choice of strategy in the economic sphere inevitably entails, is not matched by a similar awareness on the part of political scientists of the interaction between political and economic factors in the unfolding of a particular pattern of economic development. The chapters in Volume 2, exploring economic policies, consciously address the political processes entering this area, whilst the two views of liberalization regard the underlying dynamic from the different perspectives of technological progress, on the one hand, and political cost to the mass of the people, on the other.

3. Uneven and unequal development

Far-reaching social and cultural changes have followed in the wake of unequal and uneven economic and political development in India. These have been subjects of sociological and anthropological studies as well as interdisciplinary explorations by students of 'cultural change'. Some attempts have been made to provide accounts of change in India in this general sphere. As sociological research has a predilection in favour of meso-level investigations whilst anthropologists focus mainly on micro- (and occasionally meso-) level research, a general overview of social change (especially with particular reference to its underlying political dynamic) is hard to come by.[5]

[4] The seven volumes resulting from the American Social Science Research Council project on 'Development' (all published by Princeton University Press) focused attention primarily on 'governance', 'order', and formal 'democracy'. Their interest lay in comparing the 'democratic' potential of underdeveloped countries with the actual conditions prevailing in the industrialized bourgeois–democratic societies of the West. It did not include a serious attempt to view politics from below. Neither did it take into account the role played by political forces that mobilized opposition against government policies which resulted in uneven and dependent development (*see*, for example, O'Brien, 1972).

[5] For an excellent discussion of this general problem, *see* Cohn 1987. *See also* Srinivas 1969, Gough, 1981, and Saberwal, 1986, for contrasting orientations to the question.

Because India has a rich and varied past, much of the discussion of social change revolves around 'continuities' from and 'adaptations' of past practices, rather than concentrating on the impact of movements orientated towards safeguarding the vital interests of the poor. These are threatened, for example, by (a) modernization; (b) development; (c) emphasis being laid on 'private' property at the expense of 'communal' property in the process of aggrandizement of economic development by powerful classes: (d) new forms of penetration of 'primordial affiliations and loyalties' (e.g. caste) into socio-economic and political structures; (e) intensification of 'unequal exchange' in a context of acute forms of 'uneven development'; and, last but not least, (f) the emergence of mass-based movements with the specific aim of advancing the submerged interests of 'subaltern classes'.

A number of chapters brought together in these volumes (especially Vols. 2, 3, and 4) reflect an increasing awareness, especially on the part of younger and less conventionally academic students of politics, of the need to analyse and characterize the political underpinnings of the dynamic that has been unleashed since Independence within the country as a whole and in different parts of India by forces of social and cultural change. Two different balances have been kept in mind in the choice of topics and contributors—between empirical substantiation and analytic interpretation, and between different social sciences and substantive themes.

4. Levels of analysis

On the one hand, the work presented here is interdisciplinary (and in fact, cross-disciplinary) in character. On the other hand, our analytical objective is to come to grips with concrete substantive concerns that provide the context for examining underlying political, economic, social, and other forces. This has in some sense compelled contributors to be methodologically particularly aware of the implications of the two different ways of segmenting spheres of interpretation and analysis—namely, in accordance with the social science disciplines involved, and by the substantive topic addressed (*see*Fig. 0.1)—and to refrain from running them into each other, whilst taking care to differentiate between the two levels of analysis.

We have also attempted to be analytically conscious of two different levels of treatment of political processes which in practice overlap: (1) the level of analysis that stems from a linguistic semiological orientation (e.g. as reflected in new slogans, words, and the political loadings that they acquire over time); and (2) the level of analysis that stems from an understanding and exploration of substantively significant components of the Indian polity. This awareness has been sought to be complemented by the task of identifying (taking into account the historical dimension of all socio-political processes), significant phases of change, of analysing quantum leaps and transformations, and of characterizing qualitative and quantitative changes with a degree of precision. Thematic coherence has been striven for in order to make it possible to draw out analytic threads that are applicable across the board to different facets of Indian politics.

Two issues relating to the concern of this method with the link between terms of discourse and the substantive processes from which they are derived need to be made explicit. First, whilst a critical analysis of various terms and theories of discourse is essential, studies involving these must show an awareness of how the terms of discourse themselves change and do not remain frozen constant. It must be emphasized that a pure discourse theory cannot deal with a dynamic situation. Second, we ought to remember too that Western categories of description and analysis of indigenous political processes can have the effect of obscuring our vision. The linguistic philosophy underlying the analysis of these terms is based on concrete experience that is radically different from reality. Most political discourse actually takes place in the vernacular. We are aware of the need for work (not contained in these volumes) dealing with the question of how mediation takes place between terms encoded in law (for example) in English and political practices locally in vogue. This is an area which will bear much future exploration.

5. *Quantitative change and qualitative transformation*

As far as possible, an attempt has been made throughout this work to capture a sense of the qualitative aspects of change as well as to provide some comprehension of the orders of magnitude involved by carefully examining the transformations that the language of politics (and the terms of its discourse) has undergone in the last four decades—its slogans, concepts, categories of thought and analysis, its referents in the social, economic, cultural, and other spheres, and the terms used to describe different aspects of politics 'then and now'. It is this concern that has provided the specificity for this exercise and a set of themes for analysis.

In other words, we have been engaged in looking at politics (over time) through the reflecting mirror of the concepts, categories, and terms generated in a context provided by the interplay of dynamic factors affecting the interpenetrating spheres of political, economic, social, and cultural change, without prejudice to explorations of the historical roots of change (stretching back to the colonial and even pre-colonial era) as topics in their own right.

We do not underestimate the need for a degree of commitment on the part of the social scientists engaged in this work to the society in general and to the mass of the Indian people in particular. A scientific contribution arising out of a combination of academic commitment and social commitment can be expected to be distinctive in its own right; it can also be expected to be significantly different and stand out from the work of Western scholars who have the resources and time to fly in and out of India at will, but whose commitment to Indian society is (with distinguished exceptions of course), more often than not, largely subordinate to their commitment to their careers or their own particular specialized academic interests.

The strategy

The studies presented in these four volumes are oriented around a three part strategy (Fig. 0.1). On the first level, we have concentrated mainly on the tiers of government, including the state structure (arising out of the springboard of the Constitution), electoral processes, the judiciary and other watchdog institutions, the bureaucracy, and concepts of secularism, civil fundamental and democratic rights, equality, social justice (including aspects of it with no explicit economic content). Volumes 1 and 3 contain a number of chapters which pursue these and other related themes.

In overlap with these concepts, but constituting a separate category, are broad concepts of the role of the state in economic development, planning, the attainment of technological self-reliance, the adjustment of conflicts between the industrial and agrarian sectors of the economy, ensuring equality of access to different classes and segments of the population to state-provided education, health, and social services, the operation of the multi-party system with the forces of inter-party and intra-party oppositions ranged against one another along lines stressing dissent or dissidence or contradiction, the pursuit of policies aimed at positive discrimination of certain groups or categories of people. These are explored in specific thematic contexts in various chapters in Volumes 2 and 3.

The third category of concepts embracing movements outside the purview of the constitutional framework (e.g. movements 'to empower the poor', caste organizations, women's movement(s), and other kinds of activist politics) is analysed in various chapters in Volumes 3 and 4.

In these volumes a serious attempt has been made to knit together themes involving an analysis of, (1) political and moral notions; (2) the role of the state in augmenting pressures on resources and on distribution; and (3) the resilience and vitality of Indian society as a whole. The essays specifically addressing the theme of the terms of political discourse, especially in Volumes 1 and 3, are concerned to analyse the evolution of these broad concepts over the past forty years or so with the help of essential historical/empirical material around which their different operational variations gyrate.

Contributors have followed the methodological device of focusing on topics relating to politics, economics, social change, culture, and ideology from the particular vantage point of the circumstances surrounding the formulation and adoption of the Indian Constitution. This procedure has enabled several of them to explore the significance of the realm of politics which has given the Indian public a rich and varied accumulation of political slogans over the years. Not only have these slogans and bywords served as markers and as media for conveying messages and challenges; but they have also been used as codes with varying degrees of effectiveness by ruling parties, opposition parties, and mass movements alike in order to win public support and raise consciousness.

Throughout we have remained aware of the limitations inherent in confining our attention to an exploration of the political sociology of new usages and slogans within a framework of detailed linguistic and content analysis. Whilst this is a minimum necessary analytic activity, its usefulness is bound to be minimal unless it is brought within a more general framework of explanation and interpretation of the political dimension of economic, social, cultural, and ideological changes taking place in India.

These volumes have been written with the awareness that in offering an analysis of social and political change in India we have to take into account not only an incredibly rich and rapidly expanding field of academic writing that is in existence but also a constantly changing and complex political reality that is being shaped before our eyes by a whole range of forces. The interdisciplinary nature of the project is reflected in the choice of contributors who reflect a variety of intellectual and political backgrounds as well as theoretical positions and academic specializations, but who also include individuals whose work in the political field has directly involved attempts to alter existing structures and reshape political, social, and economic forces at work in contemporary India.

The main themes of this work

Our aim is to deepen understanding of formal and informal political processes, to set the context of the major developments affecting different segments of the population, and to predict the likely course of change and development in Indian society as a whole. Our concern is with the tension that has developed over the decades between the state structures or structures of political power, on the one hand, and, on the other, popular expressions of dissent ranging from protest arising out of the dissatisfactions of civil society to organized movements of resistance challenging the authority and power of the state. The nationalist movement, under the leadership of the Congress, sought to transform the colonial political legacy into a powerful state with the aim of firing the engines of capitalist economic development and establishing a just, socialistically inclined, civil society capable of overcoming poverty.

It is interesting that, in the event, India's development policies have, in fact, led to the emergence of underdeveloped and dependent capitalism (discussed in some detail in Ch. 4 of Vol. 1) on the one hand, and, on the other, a civil society in which absolute poverty has tended to grow in intensity. The systematic discussion of how political terms are generated and deployed by different political forces and sections of the population takes place within the structural framework of the post-independence Indian state in which the Indian nation of the independence struggle was embodied in 1947.

The Indian nation, to which the colonial regime transferred power, was the end product of a specific political discourse. It gave rise to political structures which have developed along certain lines, whilst at the same time they have atrophied in respect of some of the expectations initially associated with their emergence. These structures have their provenance in the Indian Constitution which constituted the main site for the elaboration of the political discourse of the Indian state. The Constitution thus becomes central to an analysis of the changing role of political and bureaucratic structures and an analytic mirror against which concrete experience and the gap that it reveals between discourse and reality must be held.

Volume 1 seeks to demonstrate that two broad perspectives have been deployed by almost all the contributors simultaneously—namely, the perspective of those on whom institutionalized power in general and state power in particular is exercised; and, a perspective of differentiation of the effect of structures of power on diverse segments of the population identified according to a variety of criteria (based, for example, on gender, caste, class, and region).

Central to an understanding of the impact of state power on different segments of the population is a discussion of its variegated economic policy. Starting from a general standpoint of furthering the interests of industrialization (and modernization) in the spheres of private and public enterprise, the government embarked on a planning exercise of a magnitude not exceeded by any capitalist country since the War. Both industry and agriculture came under the aegis of a planned economy, the expansion and modernization of the latter under the Green Revolution leading to a rapid growth of food production and far-reaching if uneven differentiation of the peasantry (namely, Chs. 7, 8, and 9 of Vol. 2 and Chs. 5 and 6 of Vol. 4).

After seven Five Year Plans, however, the Indian government was compelled to take cognizance of deep-seated changes in the world economy and, in particular, the crisis facing the socialist economies since the beginning of the '80s onwards. Its first step consisted of an experimental measure in 1981 when the Indira Gandhi (Mark II) government (1980–4) decided to climb down from the high ground of 'autonomous' and 'self-reliant' economic development and sue for the hand of the International Monetary Fund (IMF) for Special Drawing Rights (SDRs) to a maximum limit of 5 billion US dollars.

Within five years, however, under the stewardship of Rajiv Gandhi, India embarked on a policy of thoroughgoing 'liberalization', an essential ingredient of which was a turning away from planning, and an acceptance of a radically changed code of behaviour on the part of the government towards foreign capital. By 1991, when a minority Congress (i.e. Congress (Indira)) government took power after the tenth general election, India's economic policy had come full circle. Thenceforward, there was nothing to distinguish India from the other supplicants lined up before the doors of international capital and its inter-governmental instruments.

Volume 2 examines the political implications of the changing economic policy of the Indian government for different segments of the capitalist and producing classes. The consequences of planned industrialization and the Green Revolution for uneven development, on the one hand, and regional differentiation and differentiation of agriculture, on the other, throw light on the reality underlying the discourse of planning which dominated government thinking for nearly four decades. A sudden break with the strategy of planned development and a lurch in the direction of a policy of economic liberalization emphasizes that the government may have painted itself into a corner with no alternative policy available to it either from its own advisers or from those opposition parties that are against the shift away from autonomous development and self-reliance.

During the four decades since Independence, social categories in India have manifested a tendency towards refraction under the mediation of change. Categories such as 'region', 'religion' or 'community', 'gender', and 'caste' and 'class' in particular, have tended to become more and more interpenetrating. The pressures generated by an unevenly developed economy—under conditions of increasing economic dependence of the Indian state on the West, on the one hand, and of Indian capital on international capital, on the other, have militated against a policy even of minimum welfare for the vast number of people living below the poverty line.

Economic development in India has resulted in a number of overlapping imbalances: between the agrarian and industrial bourgeoisie; between politicians and bureaucrats at the central level and at the regional level; between the organized working class and marginalized workers; between a rapidly differentiating middle peasantry and the agrarian classes below it; between upper castes and middle/lower castes; between middle castes and Dalits; between the majority community and the minorities; between Adivasis and non-Adivasis; between men and women; and between the state and women.

A relationship of domination and subordination has become an enduring characteristic of the Indian polity. In this, the horizontal (and, to date, essentially non-antagonistic) contradictions between different segments of the dominating élite at different levels of the system (for example, central–local; industrial bourgeoisie–agricultural rich; politicians–bureaucrats) are reflected on the vertical (and fundamentally antagonistic though dormant) contradictions of the society (managers–workers; bureaucrats–mass of the people; police–women, poor peasants, Adivasis and Dalits; local Hindu communalists–poor Muslims, etc. etc.) in such a way as to intensify oppression and ignite violence.

Over the last four decades, the state has increasingly turned ever more sophisticated weapons of violence against a politically conscious and critical mass of deprived and oppressed people. Violence has spread from the structures of the state to the sinews of civil society. The rigours of the new economic policy, the sharp edge of which will be felt more intensely by the lower rungs of society, point in the direction of greater and more anomic urban violence, on the one hand, and ruthless and systematic upper and middle caste violence against lower castes, on the other (accompanied by some retaliation by the latter).

Volume 3 pursues this general theme of refraction of socio-economic (e.g. class and gender) and socio-political (e.g. caste; communal identity; regional location) analytic categories mediated by forces of change under a number of different rubrics chosen for their thematic relevance and empirical appropriateness.

India is perhaps the only poor non-socialist country in which the analytic category of 'class' has consistently exercised a presence on both sides of the line separating the dominant and the dominated. It has remained more or less indestructible though the major segments of each class—namely, the industrial bourgeoisie, the agrarian bourgeoisie, and the commercial petit bourgeoisie; the industrial working class, the rural proletariat, poor peasantry, and landless labour; and the vast under-class of displaced labour, marginalized workers, migrant labourers, and lumpen elements in urban and peri-urban areas have been subject to a process of differentiation, proliferation or fragmentation. The constant process of formation, transformation, and reformation of classes has posed enormous problems for class-based political organizations in general and the communist movement (which is the most important organization of the working class) in particular.

Changes in the configuration of classes have resulted mainly from:

1. severe domestic and international resource limitations placed on the role of the state (despite its undoubted class character and its repressive capacity) as the main promoter of the interests of the industrial and agrarian bourgeoisie;

2. the incompleteness of the 'market revolution' (in other words, the vast expansion and penetration of market forces has not been matched by production and purchasing capacity of large sections of the population);

3. the tendency of capital to squeeze disproportionate surplus from labour (resulting from uneven, distorted, and dependent development of capitalist production);

4. the persistence of pre-capitalist socio-economic relations (and agrarian structures inherited from the colonial period), and mode of production in agriculture side by side with capitalism;

5. the aggressive role played by the non-productive classes (e.g. the bureaucracy in the sphere of the exercise of state power, regional politicians and petty bureaucrats, as well as shopkeepers in a number of States on the northern, north-western, and north-eastern periphery);

6. the deliberate policy of lumpenization of politics practised mainly by certain regional parties (of which the AIADMK, DMK, and Shiva Sena are outstanding/pre-eminent and by no means the only examples); and

7. the large-scale displacement of workers in the rural economy and in certain trades in cities resulting in extensive marginalization, expansion of the migrant seasonal workforce, and casualization of large segments of (especially female) labour and the consequent enfeeblement of working class and trade union organizations.

The processes of class formation in the various sectors of the economy since Independence have been influenced by the political forces, contradictions, and forms of expression associated with the processes enumerated above.

Yet, despite these obstacles to the crystallization of class forces and to political action based on class interests, class organization, and class identity, the basic class structure of Indian society and political economy has remained more or less intact. The ability of different segments to launch political struggles with the aim of achieving their class interests has, however, been severely circumscribed by the domestic milieu in which political activities take place, and the international constraints that have become increasingly crucial to the development of the Indian economy over the years. The political transformation of India over the last two decades has adversely affected the growth of class power, as reflected in class-based political organizations. The interdependence between the processes of class formation and political transformation is a field that offers a number of themes for investigation.

Class struggle, an important form of political action during the '40s, '50s, and '60s, has become more and more muted. The CPI after the Telangana insurrection, the CPI(Marxist) consequent upon its electoral successes since the 1967 (fourth) general election, and even some of the CPI(Marxist–Leninist) groups since the Emergency ended in 1977, have distanced themselves from their long-term revolutionary programmes, concentrating their attention on short- and medium-term electoral politics. This change has been so far-reaching in character that it has had the effect of 'regionalizing' communism like a number of other political forces in the country (e.g. DMK and AIADMK, TDP, AGP, and even the Janata Dal in the post-National Front government period and especially during the Tenth General Election of 1991). In other words, the communist parties are content to nurture the constituencies in which they are already strong rather than seek new revolutionary political pastures in the hope of spreading their power in the future. The fact that these trends were followed by Indian communist parties long before recent developments in the Soviet Union and eastern Europe, so catastrophic to the future of communism, makes it extremely unlikely that they will be entrusted with the task of revolution during their political lifetimes.

Large segments of the Indian population have become more and more clearly aware of their civil rights, democratic liberties, fundamental rights (specified in the Indian Constitution), human rights (enshrined in such documents as the Unilateral Declaration of Human Rights), right to a sound environment, and related spheres. Increasingly, they have shown readiness to engage in collective action in order to resist various forces that are ranged against them. They struggle against the encroachments of the state (usually in the form of the bureaucracy, the police and paramilitary organizations), the upper castes, and local machineries of oppression that the structures of power only too readily mobilize against them. At the same time, they have to face the terrorism of the lumpen elements ordinarily mobilized at the behest of populist political forces, including certain ruling political parties at the Centre and in the States, and the polluters and robbers of the environment and community-based amenities. Impressive political struggles and campaigns in various parts of India have been waged in the last two decades. Some of them have been won against heavy odds.

The waning of class action *qua* class action has been paralleled, especially since the late '70s and early '80s, by the rise of other forms of protest directed towards safeguarding and claiming ordinary citizens' elementary but important rights as social beings.

In Volume 4, a number of themes relating to the contours of political transformation under way in India are considered and an assessment of the phenomena of class formation and the political factors affecting class polarization, class fragmentation, class proliferation, and class cohesion is provided.

The wide spectrum of topics chosen for the analysis of these themes should not be viewed as an attempt on our part to conceptualize the book as an exhaustive 'list' of topics which would attempt to cover every relevant force, event, or phenomenon. Far from it. The major focus has never been allowed to stray away from the general themes identified in the course of the project. To a certain extent, our overall strategy may have diminished the number of major gaps in the areas covered. Of course, the problem of gaps cannot be completely wished away in any enterprise of this kind, however ambitious its scope. However, our discussions with social scientists have convinced us that there are no large omissions, and the general method of approach pursued here has the advantage of easily and clearly bringing to light such gaps as there may be or may be revealed in the future with further unfolding of political processes.

Second, none of the four volumes is intended to be self-sufficient or *sui generis*. Each volume forms an integral part of a coherent work. The contributions were commissioned not with each individual volume in mind, but rather according to the major themes which run through the work in its entirety. The final decision to divide the work into these four volumes was made only after a large proportion of the contributions materialized. The distinctive focus of each volume resulted from the evolution of the project and was not determined in advance. We feel that such a procedure has the advantage of maintaining a focus on the identification of key concepts (see Fig. 0.1) which have been illuminated in a number of chapters written from a variety of perspectives and empirical bases. Rapid changes–changes that elude our analytic grasp–taking place in India, have compelled social scientists and political activists of all ideological and scientific persuasions to rethink their general understanding of the processes under way. This work is offered as a stimulant and as a beginning of such an effort. Our aim is to combine academic purpose with a degree of practical utility to those involved in bringing about change.

The motivation for this project lies in the attempt to try a different kind of experiment in generating political analysis from the usual discipline-bound perspectives in which social science literature abounds. We make no claims to completeness in any sense of the term, our aim being to communicate to readers some analytic impressions about dynamic political processes in a context of social change captured through a feel for texture, insights, and examples, some of which have been briefly alluded to above.

Introduction to Volume 2: Industry and Agriculture in India since Independence

This volume is centrally concerned with the character of India's economic policy. It explores the role that the Indian state has played in intensifying the demands of different segments of the dominant (producing and non-producing) classes in Indian society. Intra-élite competition among these classes for a share in resources has become increasingly fierce in the regions. At the same time, state policy has augmented pressures for more equitable distribution from different segments of the mass of the people. In the four decades following Independence, the government's overall economic policy—and its industrial, technological, and agricultural policies in particular—has exercised a differential impact on the capitalist and working classes.

Class differentiation in agriculture under the impact of the Green Revolution (in selected areas), the compulsions of serving an expanding market with industrial production largely conditioned by the uneven and dependent character of capitalist development, and the difficulties experienced by the state as well as the private sector in achieving a breakthrough in the sphere of technological development have contributed to several important changes in India's political economy. Of these, the most significant relate to the intensification of rural poverty side by side with the rise of a class of rich and middle peasantry whose demand for a share in state power has become irresistible. Other major influences include the proliferation of the industrial bourgeoisie (and the concomitant intensification of conflicts and contradictions between its different segments), and the fragmentation/marginalization/casualization/large-scale migration of the poor peasantry, landless agricultural workers, and (particularly the semi-skilled and unskilled segments of) the industrial working class.

Until the recent policy thrust in favour of wholesale liberalization, planning remained an integral part of the vocabulary of discourse on political and economic development. Protagonists subscribing to conflicting orientations and manifesting different ideological and cultural tendencies have tended to appropriate 'planning' to the language that they use to refer to development. From different parts of the mass base of the society, critiques have been advanced centring around those representing divergent interests in different regions. These have often drawn attention to the imbalances generated by the uneven development of productive forces in the country under an umbrella concept of 'planning'.

Far-reaching socio-economic changes have taken place in India in an international setting which, during the last two decades, has encouraged a market-friendly approach to development. At the same time, creditor nations (and especially the USA) have become increasingly hostile to the governments of those poor countries which have pursued policies aimed at redistribution of wealth between different segments of society or that divert greater resources to the uplift of 'weaker sections' (Birtill, 1992; Seabrook, 1992).

During the first four post-Independence decades, a number of contradictory elements have surfaced in the orientation of state power in India. Included among these are, (1) the unfolding of its class interests; (2) the progressive dilution of the neo-Keynesian active welfare-based aims of state policy; and (3) the breakdown of an earlier consensus within both government and parliamentary opposition circles in favour of a more and more aggressive pro-market and profit-orientated approach to economic policy.

The detailed discussions, centred on the Constitution in Volume 1, can be directly linked to the concerns of this volume by raising certain specific questions, such as:

(i) What is the relevance of the concept of 'socialism', embodied in the Preamble to the Constitution, to the main body of the document?

(ii) what are the gender assumptions of the Constitution?

(iii) what are the assumptions in the Constitution about market economy, natural resources, societal economic assets, society, profit, interest, wages, and rent?

The ambivalence of the Constitution would point to certain fundamental contradictions. Thus, for example, the fundamental rights guaranteed in the Constitution (in this particular discussion, the right to property) would clearly militate against the poor Adivasis holding on to their land in the face of the aggressive encroachments of powerful rural elements. Some fundamental rights restrict the domain of state interference or intervention, whilst others actively promote the enlargement of the economic space. By the same token, some provisions in the Constitution would appear to favour socialist principles, non-discrimination, equality of opportunity, and the right to democratic participation, whilst others would appear to undermine or even repudiate these values.

There is thus a theory of distribution that is *implicit* in the Constitution and does not meet the eye. Ambedkar was keen to give an *explicit* content to the Constitution, but it was deliberately kept out by the Constituent Assembly.[1] A major consequence of this omission was the contradictory character of, say, offering compensation to landlords (in lieu of the land that they would be compelled to give up under land reform laws embodied in the land reform legislation) which resulted in lopsided distribution between better-off peasantry and poor peasantry. One of the main platforms of the anti-colonial struggle was provided by widespread opposition to the colonial policy of fleecing the Indian peasantry. Therefore, the idea of compensating a parasitic class of landlords ran counter to the thesis of the exploitation of the peasantry. In the actual working of the concepts embodied in the Constitution, however, the balance of forces is invariably tilted in favour of the economically advantaged segments of society.

Implied in these ambivalences are different kinds of social and political tensions. These should be brought out systematically and methodically with empirical substantiation. In the Indian polity, the state itself is tilted by acting as a vehicle in the restructuring of the economy, in a direction which in practice runs counter to the theoretical assumptions underlying the Constitution. Thus, by way of illustration, on the specific example considered above of payment of compensation to landlords under the land reform legislation, it is worth

[1] As an unelected body, the Constituent Assembly reflected, more or less exclusively, the interests of the dominant classes. The interests of the dominated—which included a wide spectrum of Indian society—were represented only by proxy by those amongst the Congress leadership who claimed to represent the interests of the poor.

noting that landlords not only obtain compensation from the state under the law, but, by sophisticated legal representation the article enshrining the sanctity of property (Article 32 of the Constitution) prevents resort to the Supreme Court for a rectification of the injustices that might occur in the course of the application of the land reform legislation. Thus, lawyers have been able to work wonders with the Constitution! Our aim is to throw light on the specific dynamic of the state that was envisaged under the Constitution and examine the tension in the Constitution aimed at providing political legitimacy for a state of a certain specific character. It is not our intention to suggest that there was a calculated balance, but rather a balance of tendency which it is the purpose of a number of discussions in this volume to clarify in their specific contexts.

We ought therefore to give greater importance to concepts than to physical content of policy. Thus, it is essential, for example, that we look carefully at debates on property in the Lok Sabha and ascertain who said what about it, and compare/contrast it with the actual conduct of policy. Thus, returning to our example of land legislation, precisely at the time when *solatium* for land acquisition was increased from 15 per cent to 30 per cent, the term 'property' was removed from the Constitution (to a chorus of protests from a number of Members who claimed that property represented identity and dignity). Yet another aspect of the state is reflected in its ownership of more than half of the capital stock in the country; at the same time, it is under considerable pressure to divest—not only from various groups within the country but also from such agencies and institutions as the World Bank and the International Monetary Fund (IMF), dependence upon which has increased manifold over the last decade.

The ambivalence in the Constitution was reflected, during the immediate aftermath of Independence, in the economic policy of the Indian government. Despite the enormous popularity enjoyed by the Congress party, Nehru entrusted all the economics (i.e. finance, trade and commerce, and industry) portfolios to non-Congress 'experts'—R.K. Shanmukham Chetty, C.H. Bhabha, John Mathai, Chintaman Deshmukh, and Shyama Prasad Mookerji. Indeed, it could be said of the team that it would have won the enthusiastic approval and unqualified trust of a latter-day World Bank or International Monetary Fund! From 1957 onwards, however, the Prime Minister appointed 'experts' to these ministries from within the Congress party itself. These largely consisted of men who were themselves industrialists or traders, or who enjoyed the confidence of the Indian national bourgeoisie. They included T.T. Krishnamachari, C. Subramanian, Manubhai Shah, Morarji Desai, and in more recent times, such figures as R. Venkataraman and Vishwanath Pratap Singh.

Thus, a long and uninterrupted line of ministers have remained in charge of the overall economic, financial, commercial, and industrial orientation of India that can be legitimately described as capitalist and pro-Western in character. More recently, since 1977 when the first non-Congress government took power at the Centre, the socialist shibboleths of the Nehru era and Indira Gandhi's populist slogan-mongering have given way to central government policy, which has become more and more explicitly technocratic, market-friendly, and pro-capitalist in orientation. In all the central ministries that have been in power since 1977, economic portfolios have been held by ministers openly committed to free market, capitalist, pro-liberalization, and denationalization policies. They come from diverse political backgrounds and include George Fernandes, Rajiv Gandhi and his friends, Manmohan Singh, P. Chidambaram, and Narasimha Rao.

The pro-poor or left-wing elements in the Congress and left wing parties, which have been elected to power from time to time in a few States, have always remained on the defensive. They have been compelled by circumstances—not least by the clout exercised by the Congress party—to make low-key demands for a better distribution policy, investment in socially necessary projects, projects designed to expand employment opportunities for the poor, and universal education. Even in their heyday, when they appeared to exercise some influence on Nehru (1957–62) and, briefly, under Indira Gandhi (1969–73), their role consisted of fighting a rearguard action in the Congress party organization against a Congress government which never followed pro-poor policies and often pursued priorities that went against the interests of the poor.

Thus, Indian economic policy has enjoyed a continuity which the left wing has not been able to interrupt at any time since Independence. In other words, the difficulties in which Indian economic policy has become embroiled over the decades have been due to its inability to adjudicate between the conflicting and competing demands of the different segments of the bourgeoisie, and not because it has given priority to aims embodied in grandiose slogans such as *Garibi hatao* or, for that matter, in the propagandistic 'Twenty' points or their progeny embodied in Sanjay Gandhi's 'five' points.

Ministers presiding over the economic destiny of the Indian Republic have thus consistently viewed the main problem facing the national economy not so much as one of distributing what is produced more or less equitably between the different classes of a vertically divided society (down from a small minority of the most well off to the vast majority of the least well off, a large proportion of whom live below the official poverty line), but rather as one of adjusting the claims and counter-claims of competing segments of a horizontally proliferating capitalist class.

During the last four decades, the Indian capitalist class has vastly expanded. Besides the historically distinctive national industrial big bourgeoisie, it now includes the agricultural bourgeoisie (consisting of the rich and middle peasantry engaged in cultivating foodgrains and commercial crops). Moreover, a new class of industrialists has grown from largely provincial (if not parochial) roots and expanded nationally (e.g. the Ambanis of Reliance). At the same time, there are those belonging to a rising class of regional bourgeoisie (largely consisting of entrepreneurs and captains of home-grown industries in certain States—e.g. Andhra Pradesh, Gujarat, Maharashtra, and Tamil Nadu), and sections of the bourgeoisie, including non-resident Indians (NRIs) with capital to invest in Indian industry and the Indian market.

Even though the original remit of the Indian state included a specific role for it in safeguarding the interests of capital, the different segments of capital no longer believe that their interests are in its safe-keeping. The Indian state, for its part, has come under increasing pressure to beat a retreat from the economic scene. Different segments of capital in the country, as well as foreign capital and creditors, have become more and more impatient with the Indian government's slow pace of economic liberalization. So much so that, since July 1991, the Indian Government has set itself the goal of making room for market forces to take over the economic space vacated by the state by initiating a radical policy of liberalization. In the management of the Indian economy, and in the development of industry, agriculture, and commerce, government policy has thus come full circle.

The chequered career of economic planning in independent India is intelligible in the broad context of the general economic policies of successive governments. Two mutually contradictory reasons are commonly offered to explain the fact that Nehru excluded any mention of the Planning Commission from the Constitution. The first is that a vast majority of those nominated to the Constituent Assembly would be hostile to egalitarianism of any kind, and certainly to the idea of a planned economy which gave importance to the interests of the poorer sections of society. It would have been in their interest to place obstacles in the way of a national statutory planning body being established.

The second reason given for the origin of the Planning Commission outside the ambit of the Constitution lays stress on Nehru's deliberate intention of releasing its working dynamic from bureaucratic and other restraints imposed by constitutional discipline and orthodoxy.

In actual fact, however, with the partial exception of Prasanta Chandra Mahalanobis and Damodar Ramachandra Gadgil,[2] none of the executive heads of the Planning Commission (designated Deputy Chair)—from Vangal Thiruvenkata Krishnamachari to Pranab Kumar Mukherji—had a vision of India's economic development that could be deemed to be a genuine alternative to that projected by the successive ministers of the central government in charge of economic portfolios.

Also, during the last two decades, the place of an earlier generation of higher echelon bureaucrats and economists (e.g. Pitambar Pant, K.N. Raj, and C.M. Trivedi), tinged with a degree of social idealism, has been taken by a new breed of hard-headed technocrats (e.g. L.K. Jha, Arjun Sen Gupta, Montek Ahluwalia, Manmohan Singh, M. Narasimham, and Abid Husain) whose impatience to chart an entirely new course for the Indian economy influenced government policy more and more during and since the Emergency.

It is thus not at all surprising that, after 40 years' experience of planning, a difficult situation has arisen in respect of the unequal and uneven development of the Indian economy, and the impact of these two aspects on different segments of Indian society. As an important organ of state power, the Planning Commission has been a structure of crucial significance for the development of the Indian economy. For its part, the Constitution, which has defined the broad nature of the state and its relationship with the people, has remained silent on a number of aspects of the internal structure of the state (which includes the Planning Commission) and the way it operates (Sudarshan in Volume 1; Nauriya in Volume 4). It is clear that as an agent of government which spends a large proportion of the total appropriation, the Planning Commission wields enormous power and influences the course of events, often decisively.

Yet, in the Planning Commission itself, how are decisions made, and how is its internal structure operated? What are the checks and balances by which accountability and continuing vigilance can be maintained on its activities or inactivity, as the case may be? Answers to these questions would be vital to an understanding of the health of public enterprises. There is an area of uncertainty and dubiety in the relations between the executive and the bureaucracy, where the working of public enterprises are concerned. The important point to note is that we are as yet far from generating sufficiently rigorous and

[2] The general approach adopted by Mahalanobis, a statistician and physicist, was generally pro-people in character, but his methodology was far too academic and lacked practical thrust or imaginative goals. Gadgil, an economist, was also an academic committed to liberal and humanitarian values, from whom no daring or far-reaching pro-poor initiatives could be reasonably expected.

clear concepts capable of capturing the essence of the internal structure of the state and its working.

A number of related questions in the general area of economic development and growth from the perspective of this work suggest themselves for consideration, not least for their relevance to the political dimension of social change. Some examples of such questions which are taken up for discussion by contributors in the other volumes, against the background of political economy provided in this volume, are given below:

1. the historical antecedents of planning and the convergence of thinking between the emerging national bourgeoisie, the nationalist leadership, and professional economists;

2. explaining the paradoxes of development—e.g. why is there a secular decrease in per capita production output side by side with a steady uninterrupted increase in economic activity;

3. the social roots of economic conflicts;

4. the impact of economic development policy on labour in general and on the marginalization of labour in particular;

5. the impact of the economic development policy on private enterprise;

6. analysis, with special reference to development planning, of the concrete experiences of different minorities and other disadvantaged segments of society;

7. the impact, if any, on the planning process of pressures in the direction of decentralization of power and decision-making;

8. the identification and sociological analysis of the role played by different groups (e.g., women, students, Adivasis, child labourers, marginalized and/or migrant labour, etc.);

9. an enquiry into the specific characteristics as well as the changing political, economic and social configurations of the Scheduled Castes (SCs), Scheduled Tribes (STs), Economically Backward Castes (EBCs), Most Backward Classes (MBCs), Backward Classes (BCs), Other Backward Classes (OBCs), etc.

10. the impact of consciousness-arousing movements on economic demands and government policies for development, especially in respect of distribution;

11. the problem of corruption in the polity and bureaucracy as perceived by different segments of society;

12. the role played by the lobby wielding political power in rural areas on state policy;

13. the consequences for planning of widening inter-regional and intra-regional disparities in agriculture;

14. the impact of land reform legislation in different States and the Green Revolution on the differentiation of the peasantry;

15. identification of the factors underlying the resilience of the political system and exploring the links between the development of the state, the evolution of popular forces in different segments of society, and the process of adaptation of different organized political forces to changing conditions;

16. evaluation of newly emergent forces in specific spheres such as 'ecology', 'environment', 'human rights', 'bonded labour', etc.;

17. examination of economic and political questions specifically relating to such crucial sectors of development as small industries, multinational capital, the big bourgeoisie, and the rural rich classes.

Discussions involving such questions seek to illustrate the complex nature of interaction between the social, cultural, political, and economic aspects of India's development strategy during the past four decades.

Analysis of the political dimension of India's economic development in this volume begins with a two-pronged proposition advanced in Chapter 1. Emphasizing that in India the discourse of planning has been heavily implicated in the statist paradigm, Bagchi draws attention to the widespread assumption that the state is the autonomous active agent which impinges on a passive society, galvanizing it into appropriate responses as desired by the planners and policy-makers. At the same time, the major economic project in which the Indian state has been engaged—that is, of creating conditions for capitalist development (through the adoption of 'economic nationalist' policies, including agrarian transformation)—has failed. The failure of the capitalist class in India to achieve a modicum of homogeneity is reflected in the different groups of which it consists, acting to protect their own specific arenas.

The shift, over the last decade or so, from this paradigm to its diametric opposite, which is based on an equally *simplistic* model of the free market, has been accompanied by the assumption that an all-powerful state has been the root of all economic evil and the progenitor and promoter of all unproductive ('rent-seeking') activities. Now as before, however, the state is viewed as a structure fundamentally unconnected with society, or, at best, linked to the latter by a semi-permeable membrane. This presupposition ignores the numerous devices at the disposal of dominant elements in the ruling classes (including the bureaucratic apparatus of the state itself) to manipulate the state apparatus, and by the same token, the strategic responses evoked by state action. It is also negated by the extreme regional and ethnic diversity of the country and the distinct ways in which various nationalities relate to the Indian state.

Bagchi locates this problematic in a wide ranging and comparative discussion. He points out that, under the present circumstances, progressive elements in India must face the political task of guiding the process of internal liberalization along a clearly understood path. They should advocate policies designed to strengthen those parts of the public sector which are beyond the means of the private sector to take over. These would include, for example, demands for more stringent economic nationalism and the enforcement of 'bourgeois laws' in order to discipline capital.

Against the backcloth of Bagchi's general discussion of the issues confronting India's political economy, Chaudhuri examines the political context of economic planning in India in Chapter 2. His main conclusion that planning has not affected poverty is considered against the question of whether the government's failure to tackle poverty effectively was caused by the fact that its overall strategy was neither 'support led' nor 'growth led'. Combining the two in a single strategy inevitably causes it to fall between two stools. Chaudhury points out that planning in India has not been marked by efficient management and allocation of resources. For reasons already discussed in this chapter, major areas of economic policy have lain outside the planning exercise. Both in agriculture and in industry, the benefits of growth have accrued to those who control land and industrial capital, at the expense of the rural and urban poor. He concludes the chapter with a sceptical note as to whether a liberalized free enterprise economy could deliver faster or better results in India. Nevertheless, the task ahead would require that high and productive investment must be directed towards particular sectors and scarce resources should be husbanded from other uses.

Changes in India's industrial and technological policy and their implications for the future constitute the subject matter of Chapters 3, 4 and 5. Of these, Chapter 3 focuses on the past and the others focus largely on the present and the future. Baru's analysis of continuity and change in India's industrial policy (Chapter 3) is based on the view that in the immediate aftermath of Independence, there was wide agreement in the country on the economic policy that the government should pursue. Both the government and the left parties saw eye to eye on the major imperatives of India's developmental aims, and the methods that should be pursued to achieve them.

The evolution of this consensus took place over several decades, stretching back to the last phase of colonial rule. The proliferation of the bourgeoisie during the last three decades has led to the appearance of new segments of the capitalist class which are less committed to 'economic nationalism' and positively attracted to far-reaching policies of liberalization. Baru documents the erosion and eventual breakdown of the consensus underlying the economic policies which the Indian government followed during the first 25 years of independence. The breakdown represents not only a division between the government and the left forces but also a divergence of views among the different segments of the national and regional bourgeoisie as well as the agrarian rich.

Chapters 4 and 5 deal with the dynamic underlying the recent lurch of the government towards complete liberalization of the Indian economy. They, however, differ in their primary focus and in their method of approach. In Chapter 5, Sridharan examines the policy of liberalization exclusively from the point of view of India's failure to establish a self-sufficient technological base under the aegis of its immediate post-Independence policies. His aim is to consider the ramifications of the long drawn out debate on 'self-reliance' in technology which, during the '80s and since, has come to be dominated by what is referred to in the chapter as the 'revisionist critique'. Sridharan traces, from this vantage point, the ongoing shift in the terms of discourse in India's technology policy which started in the early '80s, gathered fresh momentum with the initiation by Rajiv Gandhi of a policy of deregulation of industry, and received a forward thrust of immense magnitude with the coming to power of the Congress(I) minority government under the leadership of P.V. Narasimha Rao.

Roy's discussion of the liberalization of the Indian economy in Chapter 4 is explicitly political in character. It focuses attention on the impact of recent changes on the poorer sections of society. Analysing the key explanations in vogue of stagnation in the Indian economy, Roy emphasizes that they ought to be placed in the context of the Indian state, its class character, the interests of the dominant classes, and the extent to which government policies have been shaped by them. Roy traces the impetus for liberalization back to the '70s, identifies its underlying motivations, distinguishes between internal and external factors shaping its course, and points to the differentiated response of different segments of rural and industrial classes to it. The concluding part of the discussion advances the argument that a superficial understanding of the politics of liberalization will, more likely than not, lead to an exacerbation of the already parlous state of the poor in India, whilst adding to the inter-societal tensions gripping the country. Liberalization, he concludes, cannot be separated from the discourse on development and politics in India. Roy's method of approach and his arguments are thus congruent with the general line pursued in Chapter 1.

Since the mid-'60s, agricultural development has significantly contributed to the sharpening of horizontal contradiction *within* or amongst the dominant classes in India. As the impact of the Green Revolution came to be felt in different parts of the country, the terms of trade between industry and agriculture were radically altered. As the effects of this change registered themselves in different parts of the country, powerful new class interests made their appearance in the agrarian economy of different regions. These were reflected in the emergence of rich and middle peasantry, generally though by no means universally, congruent with the middle castes (and some of the lower castes).

The predominance of these new agrarian classes in the States led to

1. a political displacement of the old upper caste (= landlord) élite;

2. a sharpening of the socio-economic contradiction between the newly emergent dominant agrarian classes, on the one hand, and the increasingly politically conscious poor peasantry and landless labour, on the other;

3. eventually, an intensification of Centre–State tensions, mirroring the rising antagonism between the Congress party (which ruled at the Centre continuously from 1947 to 1989, barring a brief Janata interregnum) and the parties of the opposition with their bases in the regions, and

4. an exacerbation of intra-ruling class conflict, i.e. conflict of interest between different segments of the industrial bourgeoisie, on the one hand, and the agricultural bourgeoisie, on the other.

Chapters 6, 7, 8, 9 and 10 are concerned with outlining the nature of agricultural development in India under the aegis of central government policy during the first four decades of independence and analysing its economic dynamic and political repercussions.

In Chapter 6, Bharadwaj sets the scene for the second half of this volume with a long-term view of the particular path followed by policy-makers in the sphere of agricultural development, and its impact on regional differentiation in India. She addresses the regional differentiation of industry and agriculture in the context of productive accumulation of surplus in agriculture needed for industrialization, which she analyses under the headings of:

1. Conditions of production in agriculture;

2. conditions of reinvestment of surplus into productive channels; and

3. interrelation between agriculture, household industry, non-household industry, and the role of public investment.

Two broad generalizations are offered on the basis of an analysis of data for the periods from 1947 to the mid-'60s, and from the mid- '60s to the end of the decade of the '70s. First, the States which manifested some industrial vitality were also the ones in which agricultural growth was promising, thereby underlining the importance of agricultural surpluses in adequate quanta in order to sustain industrial expansion. Second, the States in which the general level of living was reasonably high were also the ones in which an effective demand for products of industry was created.

During the '80s, however, the attention of policy-makers was dominated by the dramatic intensification of the balance of payments crisis. This shift, which resulted from the fiscal and development strategies of the state, has been underlined by a 'dichotomous' path of development. India's economic development is characterized by a profound asymmetry between rapid industrial growth within the economy, on the one hand, and on the other, a vast expansion of the 'dependent segment', i.e. a multiplication of the numbers of people who have no regular or assured means of livelihood resulting from, among other factors, the disappearance of rural industries and radical changes in the pattern of consumption of the people as a whole.

Srivastava's analysis of data on India's uneven development in agriculture and its implications for political processes (Chapter 7) logically follows from, amplifies, and brings up-to-date Bharadwaj's long look at the phenomenon of regional differentiation in India. His aim is to throw light on two significant trends in the Indian political process in recent decades—the growing strength and importance of regional politics, and the rise of new social groups (defined regionally and caste-wise) as wielders of political power in the regions.

Building up his analysis from the district to the State level with the aid of relevant sources, Srivastava has provided a picture in which the main features of agricultural growth, trends in labour productivity (e.g. trends towards increased proletarianization of rural labour) and labour use, patterns of commercialization, marketed surplus and terms of trade, and the disparities in investment performance between regions are identified. The socio-political scene that emerges from such a development of the rural economy is marked by an acutely uneven development of capitalism in agriculture. Inter-regional differences have been intensified and the differentiation of rural classes varies from region to region. Furthermore, even some of the hitherto agriculturally stagnant regions which in the past have been resilient to change, are currently undergoing a transition in which 'it is possible to discern some differences both within the dominant proprietary classes, and between these classes and the intermediate groups (the new rich and middle peasantry) in matters relating to agricultural performance and investment behaviour'.

These changes in the agrarian scene have influenced politics in two principal ways. First, the influence of rich and middle peasantry and other rural classes with a relatively well-developed capitalist orientation has, in general, been strengthened. Second, the demands of the rural capitalists in developed regions (e.g. western U.P., western and southern coastal Maharashtra, southern Gujarat, and Haryana) have been incorporated in the national agenda. The rich peasantry has thus consolidated its position in Indian politics both at the national and at the State level.

By the same token, Central and State policies of industrial development have also been more effective where the regional capitalist class has emerged as a strong force and where it has traditionally played a more important role alongside the rich peasantry. Srivastava has clearly demonstrated that there is a reciprocal relationship between the state and the capitalist class, buttressed by historical and economic reasons, which is notably weak in the economically backward regions. These features of Indian development raise problems not only in the sphere of conventional fiscal relations between the Centre and the States, but also in the far more delicate and complex political sphere of new demands raised by regional movements (e.g. Punjab, Assam, and Jharkand), especially since the late '70s.

The origins of the political tensions gripping the relations between the people of Punjab, on the one hand, and the central government, on the other, can be traced to the failure of the Centre to give Punjab a measure of regional political autonomy commensurate with the important economic role that it has played in food production during the last two decades. The political grievances which were originally focused on the Anandpur Sahib Resolution during the early '70s spilled over, in the face of the Centre's prevarications and general intransigence, into the arena of religious and ethnic conflict with strong cultural and communal overtones revolving around the state of the 'Sikh psyche'.

Taking the case of Punjab for detailed scrutiny, Sucha Singh Gill (Chapter 8) has drawn attention to the contradictions created by agrarian capitalism in the State, the political changes that have taken place in Punjab, and the specificities of the political situation prevailing in that State. He draws a distinction between the internal contradictions of the State itself (which are reinforced by communal division), and a second set of contradictions between Punjab and the Centre which has become sharper with the rising political consciousness of the class of capitalist farmers in the State.

The resulting political changes are analysed under the three headings of changes in the support bases of the various political parties (especially the Congress(I), the Akali Dal, CPI, and CPI (M)), the class base of those holding elective offices at various levels (e.g. *panchayat*s, *taluka*s, *zilla*s, the Vidhan Sabha, and the S.G.P.C.), and the emerging pattern of class organizations in rural areas. More recently, the interests and demands of the peasantry have been taken up by the Bharatiya Kisan Union (BKU)—of which Punjab's Khetibari Zamindara Union (KZU) was a forerunner. The BKU, along with militant unions representing the interests of agricultural workers, has altered the tone of the internal politics of Punjab. The discussion of the case of Punjab is set in the context of the struggles that have been waged against the state by different segments and classes of the peasantry and urban traders in the State.

The BKU constitutes the central focus of Chapter 9, in which Patnaik and Hasan have provided an analysis of the farmers' movement with special reference to Uttar Pradesh. It seeks to link the economic significance of the rich and middle peasantry of western U.P. with their rising political importance. Starting with the observation that, in relation to the all-India trends in agricultural production during the last three decades, the growth of capitalist production in Indian agriculture (involving investment embodying technical change and productivity rise) has been concentrated in north India to a degree greater than is usually appreciated, the authors document the rise in importance and bargaining strength of the rural capitalist producers of the region.

The beneficiaries of the Green Revolution in the 19 districts of western U.P. are identified on the basis of an analysis of the social impact of the *Zamindari Abolition and Land Reforms Act* (1952) and the economic consequences of the government's initiative in raising food production. The farmers' agitation in U.P. has spread with greater rapidity in the more advanced areas than in the less advanced parts of the State.

An important indicator of the capitalist nature of the investment taking place is the concentration of not only physical productive asset formation but also eventually increasing class-concentration of land itself. The net losers in the changes in landholding accompanying capitalist transformation are high caste Hindus and Sikhs, whilst the net gainers are drawn from the middle castes.

One of the political consequences of adverse movements of output price–cost and output–price consumer goods has been that small farmers support agitations for reducing the rate of rise of input costs or for raising final output prices. The fact that these struggles are led by well-to-do farmers, who stand to gain a great deal more than the small farmers, in no way deters the latter from taking part in them in the interest of their survival.

The only losers are the poor peasantry and agricultural labour which are substantial net purchasers of foodgrains and, understandably, do not constitute a part of the support base of such agitations. A detailed account of the struggles waged by the BKU in western U.P. and the different ingredients of its base empirically substantiates the conclusions to which the general analysis points.

Land reform is a State subject under the Indian Constitution. As such, one of the major factors underlying the differences in the agrarian political economy of the various States is the content of the land reform measures and their implementation by different State governments. It is a truism that in India land reform has, by and large, led to variable results in different States, depending on the extent to which it has proved to be capable of meeting the basic criteria laid down by the Congress party even before Independence was granted.

A major divergence of viewpoint between the Congress party and the two major communist parties has provided us with two conflicting paradigms of land reform—one which has been instituted in the three States of Kerala, Tripura, and West Bengal (all of which share a strong political presence of the different ingredients of the communist movement), and the other which is prevalent in different variations in other parts of the country.

In Chapter 10, Bandyopadhyay has documented the experience of land reform in West Bengal from the perspective of an enthusiastic scholar–administrator (belonging to the I.A.S.), who was given the responsibility of directing the implementation of Operation Barga which, to date, remains the only comprehensive package of land reform that addresses the interests and needs of the poor peasantry (and to a less successful degree) landless labour.

The political background of the Left Front government's Operation Barga goes back to the significant role played by the two United Front (UF) governments (1967–70) in the sphere of land reform and implementation of land ceiling laws under the leadership of Hare Krishna Konar. Konar drew a distinction between the act of vesting surplus lands (in which he strictly followed the legal process) and the distribution of vested land (in which he allowed the political process to assume precedence over bureaucratic procedures).

By following radically different procedures in respect of these two aspects of land reform, Konar succeeded in politicizing the poor peasantry. He was also able to increase the support for the CPI(M) in the West Bengal countryside. The level of support held firm in spite of the repressive policies of the Congress government which held power for six years (1971–7)—during the interval between the dismissal of the second UF government and the resounding electoral victory of the first LF government.

The experience of Operation Barga points to the possibilities and limitations of land reform under the aegis of the Indian Constitution. While it shows the State of West Bengal to advantage by comparison with most other States in India, it also points to the fact that even the most progressive land reform legislation reaches a dead end sooner or later if it cannot be further extended along the line of socio-economic progress. Under conditions of uneven dependent and truncated development, it is difficult to sustain for long even the momentum provided by socially progressive and politically radical change.

Chapters 1–5, concentrating on different aspects of India's industrial strategy and technological policy, focus on the politics of the Centre. Chapters 6–10, which are principally concerned with the political economy of agriculture, are thematically rooted in the States. An important dimension of the Centre's economic policy towards the poor was provided by Prime Minister Indira Gandhi's resort to populist rhetoric (which reached its high point in the *Garibi hatao* slogan of 1972) as a means of vanquishing opposition forces within the Congress party. An element of political opportunism did enter the Centre's policy calculations from 1972 onwards. It took the form of according a central place to the government's poverty alleviation programme under the aegis of which a number of new initiatives was undertaken.

In Chapter 11, Vaidyanathan describes these policy initiatives and their social, economic, and political results. The populism of the Centre was copied in some of the States (e.g. Tamil Nadu and Andhra Pradesh). This led to a duplication or reproduction of similar programmes in some of the States also. The tensions to which these rival populist initiatives gave rise were amply reflected in political struggles over decentralization of power and improvement of administrative efficiency.

In some States (e.g. West Bengal and Karnataka, and latterly Kerala) opposition parties and coalitions, which came to power after 1977, have introduced new local institutions (*panchayati raj*) to which power has been radically devolved from the State capitals. The Centre, under Rajiv Gandhi's administration, made an unsuccessful attempt to hijack the progressive measures adopted by some of the States (governed by parties of the Opposition) by introducing a constitutional amendment Bill, according *panchayati raj* a statutory status in the state structure. Even though a general election had to be called before this initiative could fructify, subsequent governments at the Centre have taken on board the need to institutionalize *panchayati raj* so that the resources needed and the authority required to carry out various projects involved in poverty alleviation are given to appropriate institutions at the base rather than handed down from an omnipotent or omnicompetent Centre.

The topics chosen for discussion in this volume relate the economic compulsions of India's development to the political consequences of choosing a particular set of (unitary) structures under the Constitution. The strategy of industrial and agricultural growth adopted by the Indian government has been viewed from a variety of perspectives. A fairly coherent account of the problems to which the government's economic policy has given rise and the political forces that have been unleashed emerges from this volume.

The discussions in Volume 1 concerning the nature of the state pointed to the conclusion that the Indian state has, over the years (and especially under Indira Gandhi), increased its power formidably, and vastly multiplied its police and paramilitary structures of coercion and repression. At the same time, their concrete experiences of the effects of the government's economic policy has sharpened social conflicts of various kinds involving the mass of the people—chiefly those between dominant and dominated classes, and among the different segments of the dominant classes.

Over the years, the clash between the repressive state apparatus and a population which has become increasingly aware of its civil and democratic rights, has exposed the Achilles' heel of the Indian state. Over the decades, state power has become less and less effective in dealing with intra-ruling class (or intra-élite) conflicts, regional/ethnic/communal clashes, and more and more dependent on its creditors abroad. Any pro-people pretensions, to which it might have aspired during the '50s and early '60s, have largely vanished. The Indian state, as a consequence, has experienced a loss of hegemony against the background of its ever-increasing coercive power.

Civil society in India has been subject to fragmentation and partial polarization along a number of different axes of contradiction and conflict. As a consequence, economic or class grievances and conflicts have been refracted along lines of cleavage which pit communal group against communal group, castes against each other, and different regions and nationalities against the Centre. The problems faced by women, children, marginalized labour, Scheduled Castes (SCs), Adivasis, and Dalits have been compounded over the years as Indian society as a whole may well be poised on the brink of possible disaster in the newly commenced era of liberalization. Volumes 1 and 2 will thus make way for Volume 3 which will deal with themes drawn from the concrete problems facing different segments of the mass of the Indian people.

REFERENCES

ALAM, J., 1991: 'State and the Making of Communist Politics in India, 1947–57', *Economic and Political Weekly* 26 (9 Nov.): 45, pp. 2574–83.

BAILEY, F.G., 1960: *Tribe, Caste and Nation* (Manchester: Manchester University Press).

BÉTEILLE, A., 1965: *Caste, Class and Power* (Berkeley: University of California Press).

BIRTILL, P., 1992: 'A note on foreign investment in India's "new" economy'. *Race & Class* 34 (July–Sept. 1992):* 1, pp. 17–22.

*Special issue entitled *The New Conquistadors*.

BRASS, P.R., 1965: *Factional Politics in an Indian State: The Congress Party in Uttar Pradesh* (Berkeley: University of California Press).

COHN, B.S., 1987: *An Anthropologist among the Historians and Other Essays* (Delhi: Oxford University Press).

GOUGH, K., 1981: *Rural Society in Southeast India* (Cambridge: Cambridge University Press).

HANSON, A.H., 1969: *Planning and the Politicians and other Essays* (London: Routledge & Kegan Paul).

KOHLI, A., 1990: *Democracy and Discontent: India's Growing Crisis of Governability* (Princeton: Princeton University Press).

KOTHARI, R., 1988: *State against Democracy: In Search of Humane Governance* (Delhi: Ajanta).

————, 1983: 'The Crisis of the Moderate State and the Decline of Democracy', in LYON and MANOR (Eds.), 1983, pp. 29–47.

LYON, P., and J. MANOR, (Eds.), 1983: *Transfer and Transformation: Political Institutions in the New Commonwealth: Essays in Honour of W.H. Morris-Jones* (Leicester: Leicester University Press).

MORRIS-JONES, W.H., 1964: *Politics and Government of India* (London: Hutchinson).

————, and B. DASGUPTA, 1975: *Patterns and Trends in Indian Politics: An Ecological Analysis of Aggregate Data on Society and Elections* (Bombay: Allied).

O'BRIEN, D.C., 1972: 'Modernization, Order and the Erosion of the Democratic Ideal: American Political Science 1960–1970', *Journal of Development Studies* 8 (July): 4, pp. 351–77.

RUDOLPH, L.I. and S.H. RUDOLPH, 1967: *The Modernity of Tradition: Political Development in India* (Chicago: University of Chicago Press).

SABERWAL, S., 1986: *India: The Roots of Crisis* (Delhi: Oxford University Press).

SATHYAMURTHY, T.V., 1971: 'American Science of Indian Politics: An Essay in the Sociology of Knowledge', *Economic and Political Weekly* 6 (5 June): 23, pp. 1133–6.

————, 1973: 'Social Anthropology in the Political Study of New Nation–States', *Current Anthropology* 14 (Dec.): 5, pp. 557–79.

————, 1986: 'India since Independence: A Research Note on the Development of the Power of the Indian State', *South Asia Research* 6 (May): 1, pp. 39–51.

————, 1987: 'Contemporary European Scholarship on Political and Social Change in South Asia: An Essay in the Sociology of Knowledge', *Economic and Political Weekly* 22 (14 March): 11, pp. 459–65.

————, 1991: 'A Unique Academic Understanding of Politics', *Economic and Political Weekly* 26 (7 Sept.): 36, pp. 2091–2100.

SEABROOK, J., 1992: 'The re-conquest of India: The Victory of International Monetary Fundamentalism', *Race & Class* 34* (July–Sept. 1992): 1, pp. 1–16.

* Special issue entitled *The New Conquistadors*.

SRINIVAS, M.N., 1969: *Social Change in Modern India* (Berkeley: University of California Press).

SRINIVASAVARADAN, T.C.A., 1992:* *Federal Concept: The Indian Experience* (New Delhi: Allied).

*Posthumously published (edited and compiled by C. CHARI). The author was a highly distinguished scholar/civil servant who retired as Home Secretary to the Government of India in 1980 after a career as a member of the Indian Administrative Service, stretching back over a period longer than three decades.

WEINER, M., 1967: *Party Building in a New Nation: The Indian National Congress* (Chicago: University of Chicago Press).

Dialectics of Indian Planning: From Compromise to Democratic Decentralization and Threat of Disarray • 1

AMIYA KUMAR BAGCHI

The patchwork consensus

As in the case of almost every other kind of discourse in post-colonial India, the discourse of planning also had its moorings in the colonial past. Deliberate policy-making with state help for economic development was part of the early economic nationalism of such thinkers as R. C. Dutt, G. V. Joshi, Dadabhai Naoroji, M. G. Ranade, and G. K. Gokhale (Chandra, 1969). But planning in the stricter sense of chalking out a programme of expenditure and administrative (or political) action for national economic development in the light of the availability of resources and assessment of needs seems to have gained currency only from the '20s onwards. M. Visvesvarayya can probably be regarded as the pioneer in talking about planning in India in this sense (Visvesvarayya, 1920, 1934; Chattopadhyaya, 1985).

As Visvesvarayya's case clearly underlines, economic planning was not the hobby-horse only of some economists or some nationalists with a bias towards public enterprise and socialism. As far as I can make out, at least three streams of thought fed the movement for economic planning from the '30s onwards. One was a basically conservative nationalist trend which looked upon private enterprise as the major instrument of development but which wanted the state to step in wherever private enterprise was inadequate for the purpose or wherever private enterprise had to be promoted through government subsidies or protection of various kinds. Visvesvarayya and the framers of the Bombay Plan clearly fit this pattern.

The second was an official or technocratic stream which regarded economic planning as an essential adjunct to the effort at administering and improving the poverty-stricken land that India was. For such advocates of planning, deciding between state and private action was purely a matter of administrative convenience or instrumental effectiveness. From Sir George Schuster, the retrenching Finance Member of the '30s with a private hankering after planning (Chattopadhyaya, 1985), down to Sir Ardeshir Dalal, many officials and policy-makers of the Government of India would fit into this mould. The third strand was represented by those economists, politicians, and publicists for whom economic planning was an essential part of a strategy to construct a socialist society in India. For many of them, indeed, economic planning and the expansion of public enterprise became synonymous with an advance towards socialism in India.

The ability of the young Soviet Union to withstand the effects of the global capitalist recession and to industrialize rapidly demonstrated to many—and they included professed conservatives as well as socialists—that national economic planning would be an effective instrument of economic advance. However, planning sustained by economic development was not part of the legitimating discourse for colonial regimes. A discourse of planning and a discourse of upliftment of the poor were part of the process of legitimation of the demand for a state which would shake off colonial control. The post-colonial Indian state did not come into being as the result of a successful armed struggle conducted by the native bourgeoisie. Kosambi (1946) might have heralded the adulthood of the Indian bourgeoisie with the publication of Nehru's *Discovery of India*, but this was an adulthood which hid behind the patriarchy of the traditional order and the squirearchy of British officialdom.

Side by side with the rhetoric of planning ran the rhetoric of Harijan improvement, revival of village industries, and restoration of a traditional *Ramrajya*, based on rural self-sufficiency and social harmony—a rhetoric which was personified by Gandhi's dominating and charismatic figure. While Gandhi's magic roused the popular masses, the take-over of an existing state apparatus with the least disturbance to the powers and class relations in society could not be effected without a strategy for incorporation of the most thrusting elements of the ruling classes in the emerging enterprise of transfer of power. Power was to be transferred rather than conquered: power was to be appropriated rather than destroyed and transformed. This overriding goal determined the character of planning that came to be theorized about and institutionalized in policy documents, administrative measures, and formal bodies such as the Planning Commission and the National Development Council.

The Gandhian perspective on rural development proved inadequate in providing a framework of administrative guidelines and a rationale for an activist state. One indication of this was the elaboration of a 'Gandhian plan' by Shriman Narayan. The feeble bureaucratic impulse towards planning in the Government of India circles was suppressed during the '30s but resurfaced with some vigour when top officials wrestled with the task of delivering much needed military supplies by squeezing a very poor agrarian economy. This impulse was given expression in the exercises carried out at the end of the World War by the Advisory Planning Board set up by the Government of India and by the various departments of the central government, with a view to reconstructing and developing major sectors of the economy (Government of India, 1947; United Nations, 1947; Bagchi, 1963).

The consensus about the rhetoric of planning was based not only on the vision of an activist state. It was based also on an agreement that the state would not seriously interfere in the processes of money-making and power-wielding within the society. Nehru's rhetoric of socialism had already been assessed at its true worth by G. D. Birla and big business in India in 1936 (Chandra, 1979; Markovits, 1985). Nehru emphasized his willingness to go along with the designs of big business—who were the framers of the Bombay Plan of 1946—when he publicly opposed the direct tax provisions of the budget of Liaquat Ali Khan, in spite of prior agreement to the contrary at a confidential cabinet meeting (Chattopadhyaya, 1988). The consensus involved not only Indian big business but also foreign capital with which Indian big capital was eager to collaborate. Again, in 1949 the central government brought out a special statement assuring foreign capital that its existing bastions would not be assaulted by the new government and that it would be allowed to expand in most fields of economic activity (with the exception of a few sectors reserved for the government).

The élite consensus began to look frayed at the edges when it came to the rural sector. Already in the '30s the Congress, in most major provinces of India, had come out in favour of abolition of *zamindari* and intermediary rights in land: the Bengal Congress, for narrow and sectional reasons, had dragged its feet over the issue and had paid dearly for it. The Congress pronouncements around the year of Independence seemed to promise more than the legal abolition of intermediary rights in land.[1] The *Report of the Congress Economic Programme Committee* (presided over by Nehru), submitted in January 1948, proposed the abolition of intermediary rights in land, the definition of a maximum size of holdings, the acquisition of the surplus over that size, and distribution of the surplus land to village cooperatives entrusted with the management of that land. The Congress Agrarian Reforms Committee, presided over by J. C. Kumarappa, went in detail into practically all the issues relating to agrarian relations in India. Besides recommending the abolition of intermediary rights and making concrete recommendations about the way the size of the maximum permissible holding should be fixed, the Committee also recommended that any tenant who had cultivated a piece of land for at least six years should be accorded full occupancy rights.

However, the rural oligarchs, with the support of the legal machinery, forced the central Congress leadership to effectively shed most of the land reform proposals except the abolition of intermediary rights.[2] The activist state at the centre became enmeshed in a patchwork of absentee landlordism, resident landlordism, rich and middle peasant farming and sharecropping on minuscule plots. The constitutional provision for the concurrent State rights to regulate agriculture underlined the nature of the consensus under which the central government could manage the visible commanding heights in the cities provided they left the mud fortresses of the rural oligarchs alone.

But, of course, the extending tentacles of a burgeoning administration, finance, and trade could not but reach out also to those rural fortresses. While the reality of oligarchic power was recognized, the nature of the exercise of that power was sought to be transformed through cooperatives, *panchayats*, and community development schemes. The imperative to reach out to rural areas, by circumventing rather than directly challenging the rural social order, partly explains the bizarre influence that Albert Mayer, a US architect, seems to have exercised on Nehru's community development initiative (Thorner, 1981).

[1] For a history of the Congress resolutions and policy pronouncements from the '30s up to the beginning of the '50s, *see* Malaviya (1954).

[2] The best account of the legal devices used by the rural oligarchs to defeat even the diluted land reform proposals in most Indian States, is to be found in Thorner (1976).

The filibustering of the rural oligarchs succeeded in slowing down the pace of land reforms and eventually stalling them altogether, despite vociferous and sometimes violent protests by the landless, the uprooted autochthonous groups, and by organized political groups on the Left. The Ladejinsky report on the Intensive Agricultural Development Programme (IADP) areas urging peasant-oriented land reforms as a condition for its success was discreetly buried in virtually all the States. The world-view of the rural magnates was reinforced by the technology and the ideology of the United States and World Bank lobby which proclaimed that, in the rationally ordained countryside of India, what was needed was not land reform but fertilizers and higher-yielding varieties (HYV) of crops.

If there was an 'urban bias' in Indian planning, it took the form of the refusal of the urban élite to disturb the rural order as supervised and lorded over by the Thakurs, the Rajputs, the Bhumihars, the Reddys, the Gounders, or the sugar and groundnut barons. The crisis in agricultural production in the middle years of the '60s had two contradictory effects. On the one hand, the patched-up consensus at the top had to be replaced by a patchwork of coalitions extending from New Delhi to various challenging networks of power in the States. The explicit assumption of authority by vocal representatives of local power networks was helped by the enrichment of landlords and rich peasants through a turn in the terms of trade in favour of agriculture, and the doling out of large subsidies for the use of irrigation facilities, fertilizers, and farm equipment.

However, the enrichment of the farm lobby and the local magnates was not always a smooth operation of 'transfer of power'. Many of the State governments along, of course, with the central apparatus of paramilitary forces and surveillance, were faced by peasant insurgencies. These insurgencies generally had deep historical roots of resistance against economic and political invasion by outsiders; but they were also newly empowered by various shades of Marxist and Maoist ideology.

The contentious élite and the discourses of incorporation, dissent, and resistance

The consensus at the top, even while it lasted, did not always look like an achieved fact. It often looked like contentiousness: there were, of course, issues to be fought over, resources to be contended for. The area of operation was the whole reach of the state and society. The Constitution and the laws defining the limits of state power, of central power as against the power of the regional satraps, of the private domain of custom, usage, and religion were all objects of contention as well as of attempts at transformation.

The state was forced to be minimalist as far as civil relations—of marriage, inheritance, and partition—were concerned, but in intention at least, bureaucratically interventionist when it came to the sharing of commercializable relations and resources. Investment licensing which was supposed to ration resources for attaining socially desirable ends became an instrument of patronage in the hands of bureaucrats and politicians. The division of the economy between the private and public domain shifted with shifting perceptions of the élite about the constraints and opportunities. When private investors felt strong enough to take over some domain which had earlier been reserved for public enterprises, a demand for liberalization surfaced in the journalistic literature. When on the other side, businessmen wanted the state to generate markets through its own actions at public risk, the demand for planning by the government became more strident.[3]

The coexistence of a public and a private sector had been accepted as part of the fabric of the Indian polity from the very beginning. Through all the mutations of the Congress slogan of 'a socialistic pattern of society', 'socialism', or 'a mixed economy', that design was never seriously questioned. The design required, however, that the central government's discourse should be statist in enunciation, however soft that state might be on the ground. The discourse on planning was part and parcel of that statist enunciation.[4] The government had at its command several resources and regulatory functions related to those resources. These included the domain of external economic relations, the allocation of several scarce commodities, the taxing and spending functions, and the delimitation of the domain of the open market. A rationalization had to be provided for these functions, and some principles had to be found for allocating the scarce resources among contending parties. Moreover, the outcome of the economic process would be determined by the actions of myriad agents in the society. A summary description had to be provided for that outcome—a description that would also serve as a conditional prediction—if the agents had to act 'rationally' (rather than blindly) in response to the actions of others, including the policies of the government. The discourse on planning served the dual purpose of providing the desired rationalization and the needed description.

[3] The shifting public–private divide and the reasons for the shift have been analysed in greater detail in Bagchi (1990).

[4] I have borrowed the particular meaning of the word 'enunciation' from Foucault's usage (Foucault, 1972). However, I do not claim for it the richness, complexity, or contradictoriness of Foucault's meaning.

The consequence of planning, however, was not what the technicians of the Planning Commission had intended. Indeed if one takes into account either an atomistic response to the enunciation of programme and regulations, or the strategic response of organized groups, cartels, or oligopolistic firms, then only under very special circumstances could it ever be that the intention of the technicians—even had those intentions been internally consistent in the first place—would be actualized in practice.[5] Those special circumstances would have to consist of a world which is somehow conjured up so as to provide a fixed point at which the 'rational' (in fact, often highly irrational, judged by either individualistic or by social normative criteria) expectations of all the agents converge.

The way the private agents react to various aspects of planning depends on whether they are producers or consumers, whether they are producers with substantial holdings or means of production or producers continually on the verge of bankruptcy, whether they have some control over prices or other non-price variables that they can manipulate, and so on. One area of interaction between public plans and private decisions that has received a substantial degree of attention has been the arena of farm policies. The analysis has progressed from simple-minded analyses of supply responses to changes in relative prices of particular products to analyses of changes in terms of trade of agricultural producers in general to differential changes in terms of trade of farmers producing surplus and those who are only in the farm business because the alternative is sinking into the mass of the proletariat.

In other words, willy-nilly the class composition of the rural population has entered into the discourse of population economy. Such a discourse has been further complicated by the recognition that different groups of farmers are differentially situated to act as organized farm lobbies, that farm lobbies are not always—in terms of class positions and family and clan connections—discretely separable from lobbies of traders, transporters, and industrial entrepreneurs poised to bestraddle both town and country with their class interests. The straight lines of the discourse have been broken off by the further recognition that, nonetheless, from time to time the interests of urban

[5] I distinguished between atomistic reactions and strategic interactions in the context of government planning in a mixed economy in an earlier analysis (Bagchi, 1960). I explored some issues relating to strategic responses to tariff policy, protection, etc. in two others (Bagchi, 1964: 1966). But, while I would claim some descriptive realism for the situation portrayed, I would not claim that any results were rigorously proved in any of them.

consumers have to be recognized as being in conflict with the interests of producers of foodgrains, and that global changes in terms of trade of primary products and manufactured goods can overwhelm domestic designs aimed at placating farm lobbies.

I have cited these examples only to indicate that the world away from the corridors of the Planning Commission has always been pitted against conflicts raging around the passive and the strategic responses to enunciations of particular kinds of government policy and its general directions. But my purpose is not just to indicate how complicated the discourse has always been, or to indicate how long and how labyrinthine the road has been, say, from Krishna's study of supply responses of Indian farmers, to Mitra's class-based analysis of the probable causes and consequences of changes in terms of trade of agriculture and industry, to Patnaik's search for credible models of inflation and consequential income dispersion, and to the elaboration of numerous models of dynamic disequilibrium to understand the behaviour of farm prices and the effects of farm policies in Indian agriculture (Krishna, 1963; Mitra, 1967; Patnaik, 1988; and Rakshit, 1989). The discourse of planning or economic policy-making has become entangled with wider political discourses of workers' and peasants' movements just as the latter have motivated discourses of political economy.

The discussions among the élite or their attempt to contain resistance within society have not, of course, remained confined within the domain of economic policy and property relations in a narrow sense. They have often taken the form of discussions about the power of the central government as against that of regional governments, the language policy of the Union of India, the role of traditional precepts and precedents in the formulation of law and the enunciation of legal discourse (Baxi, 1986; Dhawan, 1989; and Sudarshan, 1983). In a democratic society with contending parties, some of the differences have been marked off to define party positions. This applies to regional parties, to parties and organizations which claim to speak on behalf of farmers, as well as to parties whose class bases and related ideologies are professedly different from those of the Congress, Congress(I), or other dominant national parties.

The ideology of central planning required for its implementation a society and an effective legal framework in which claims of religion, ethnic, linguistic, or caste identity would be subordinated to those of legal equality, national citizenship, and secularization of public life. Such expectations were in fact belied on the ground: the struggle for equality often took the form of a struggle to be more equal than others. Affirmative action was often interpreted as nominal or quite simply as ascriptive discrimination rather than real discrimination.

Landlords appropriated cheap loans which were nominally granted to marginal or small farmers. The protection of minority rights often meant the right given to the privileged few to exploit the oppressed majority within a group that was corralled off as a minority interest.

Planners often saw these problems as only defects of implementation rather than as integral, constitutional birthmarks of a society in which landlords, usurious moneylenders, privileged bureaucrats, and policemen thriving on criminality remained in control of change. Planners also failed so see that in a world dominated by international capitalism, collaborationist landlordism, and mercantilism remained an enemy of the national self-reliance they were seeking to build in.[6]

It is possible to go on from here and to detail the various particularistic challenges that centralized planning has faced, the peoples' protests that *haute administrateur* social engineering has encountered, and the numerous ways in which the instruments of regulation have been utilized in order to defeat the public or social aims of regulation. Some of those issues are explored elsewhere (Bagchi, 1984), and other social scientists have analysed these aspects for a number of years. The contradictions created by class division, by pressure groups operating in the name of 'backward castes', particular communities, 'farmers', 'small industries', 'backward regions', have all come in for analysis and criticism.

The Approach to the Eighth Five Year Plan (Government of India, 1990) has managed to incorporate the rhetoric of many of the pressure groups, and this has endowed it with an air of inconsistency. But perhaps that is better than sweeping all the contradictions under the carpet and presenting a bland image of political wisdom and bureaucratic competence. The rhetoric and the reality are all the time intertwined, since as I have suggested, the rhetoric influences the action of various groups and performs as a signalling device to trigger off certain types of action which often act at cross purposes with the stated intentions of government policies.

However, one has still to ask: are there any conditions under which the rhetoric and the reality could ever mesh and produce an outcome which in some sense will be an improvement for most of the people much of the time?[7] One such scenario could be that of democratic socialism. Writing the script for that scenario is, however, at the moment a very difficult task to undertake. Since so many similar

[6] Chakravarty's last book (1987) will remain a testament to the faith and vision of a planner who refused to be daunted by the multiple obstacles that that vision encountered in practice.

[7] The *locus classicus* of discussion of the problems of constructing socialism in one country remains Carr (1970–2); the viability of social democracy in an interdependent world has been questioned in Bhaduri and Marglin (1987).

sounding scripts are being drawn up and jettisoned all over the world, we need to reflect long and reassess much of the received wisdom before venturing on even a tentative formulation.

The other alternative is to explore the different facets of capitalism and figure out whether something resembling viable capitalism can be conjured up in the Indian setting. Viability in this context requires the assurance of a reasonably stable growth with the prospect of bringing down the proportion of unemployed people and improving the standard of living of most of the people on a sustained basis. Such viability will still not mean that we will have 'capitalism with a human face' because most people will still be subjected to exploitation and insecurity, and will have to earn their bread with the sweat of their brow under the orders of a handful of owners and managers. But at least the size of their bread basket will grow larger over time, their standard of education will go up year after year, and their life-expectation will move up to the range of 65–75 years. We will also require that they live under a working democracy where people can change their rulers through a system of universal adult suffrage. Universal adult suffrage and a multi-party system are not sufficient but are necessary conditions for the operation of a democracy. Socialism has to be accepted culturally, politically, and socially before it can strike permanent root in a society: otherwise it will undergo the fate that has overtaken it in the Soviet Union and most countries of eastern Europe.

Economic nationalism and capitalism in a single country

The acceptance and eventual viability of socialist values will require long preparation in any country. India starts with one major advantage and some serious disadvantages in this respect. The major advantage is that formal democratic institutions have survived in this country in the 40 plus years since Independence, despite many assaults on such institutions by vested interests, and by sectarian and communal political movements. Any contemplated change has to work towards deepening the foundations of such institutions rather than overturning them in search of an instant realization of a millenarian vision.

The major disadvantages consist in the survival and growth of many pre-capitalist institutions, colonial cultural values, the symbiotic co-existence of highly exploitative capitalist modes of surplus appropriation, and the prevalence of high degrees of poverty, illiteracy, and non-market modes of exercise of power by private persons and institutions. Under the circumstances, the legitimate fear may arise as to whether it is at all possible to eradicate the exercise of non-market modes of power without a violent overthrow of all existing institutions. However, movements claiming to arrogate to themselves the right to exercise violence within the framework of a formally democratic but well-armed state apparatus are either doomed to failure or tend to produce their own pathologies which defeat the very purposes for which such movements were launched in the first place. I would not therefore like to elaborate on the possibility of such alternative choice in this chapter.

Instead, we can ask whether or not there are certain major commonalities between the foundation of a socialist society and those of a less overtly polluted or criminal capitalist order. On the basis of the experience of advanced capitalist countries in the phase of development in which social welfare programmes provided both the base of an expanding home market and a growing productivity of labour, I would claim that such commonalities can be discerned between the foundations of, shall we say, a welfare-oriented capitalist state and those of a democratically governed socialist state.

There are at least three major similarities between the foundational structures of the two societies which need stressing in the Indian context. The first is the total abolition of landlordism. By landlordism, I mean a state of affairs under which owners or controllers of land can exercise non-market power over the people who live or labour on that land or derive their livelihood from labour connected with the produce of the land. This non-market power of landlords could take the form of use of private armies, or a privileged use of local government and social institutions, or a privileged and largely illegal access to the use of the state apparatus in their own interests.[8]

The second major foundational similarity between welfare-oriented capitalism and democratic socialism would be the abolition of illiteracy and the empowerment of all human beings through access to one of the major instruments invented by *Homo sapiens* to control the environment and accumulate the skills for such control over generations.

The third similarity in the same genre which deserves to be stressed is the eradication of the kind of insecurity that is bred among people through the continual threat of starvation. Eradication of poverty in a country like India may take time. But eradication of desperate insecurity need not. A basic social insurance can be provided to everybody at a cost which is well within the means of the current

[8] In many parts of Africa, where the state apparently reserves the right to allocate the land as it pleases, the landlords belong to the ruling or chiefly oligarchy controlling the state apparatus even though the land may not be formally owned by them.

budgets of the Central and State governments. The continual threat of starvation not only saps the productivity of human beings. It also makes them prey to the illegal exercise of power by the privileged strata and thus undermines the foundations of a democratic, let alone an egalitarian, order (Chandra, 1988; Mohanty, 1989; Prasad, 1989).

However, the question may be raised as to whether capitalism can be at all viable during the period in which the foundations for a welfare-oriented capitalist state are being laid. We can appeal to history, and point out that in major capitalist countries the foundation of a capitalist order was laid by a national effort, however much the timing and the nature of that effort was conditioned by the development of international capitalism.

Capitalist relations began to develop in many countries of Europe in the later Middle Ages and especially in the aftermath of the plague epidemics of the fourteenth century. But it was by no means a story of uninterrupted progress or development. Capitalist social relations received a setback in most of the Italian city states when they came under the direct or indirect domination of either the Holy Roman Empire, the Papal State, or imperial Spain. Within Spain itself, incipient Catalonian capitalism was overwhelmed by the rule of the Castilian grandees.[9]

It took a series of revolutions in England from the sixteenth century down to the eighteenth for capitalist relations to be firmly established on the ground under the supervision of a group of triumphant profit-oriented landlords, merchants, and later on, industrialists. Adam Smith may have preached *laissez-faire* in 1776, but England pursued nationalist economic policies down to the 1840s when free trade was introduced in the interest of the manufacturers and wage earners, and that interest was firmly based on a global supremacy in manufacturing, trade, and finance.

The story of the spread of capitalism in the other countries of western Europe and in North America cannot be told only as a kind of passive adaptation of capitalist institutions borrowed from England. In France, Germany, or the USA, or even the small Scandinavian countries, it was a story of political revolutions leading to the eventual victory of interests championing the cause of capitalism; revolutions which were often reversed in the short run only to break out afresh in the setting of a new national and international conjuncture.

The champions of capitalism generally fashioned the new institutions in accordance with their perception of their national interests. It was only in rare circumstances that those interests were seen to

[9] I have argued about this possibility of underdevelopment within western Europe in Bagchi (1971).

coincide with those of the dominant country of the epoch, namely Britain. It can be argued that the ruling classes of the major groups of nominally independent countries, namely the Latin American countries, which embraced a free trade policy, condemned their own countries to industrial backwardness through that very act. Of course, there were other policies that sustained the social basis of that 'voluntary colonialism', such as the shoring up and strengthening of landlordism, an ideology that also included elements of racism, and the resultant condemnation of most citizens to a life of helotry (Bagchi, 1982, 1987, and 1987a).

The growth of capitalism in Japan since the late nineteenth century and in the four dragons of East Asia (Singapore, South Korea, Taiwan, and Hong Kong) since the end of the Second World War indicates that the origin of the capitalist system can be regarded as at least a bipolar affair. No simple theory of diffusion will account for the set of policies adopted by the Japanese ruling classes: they sought to introduce western applied science and technology while resisting the liberal values of free trade, freedom of expression, and democracy. A very similar rejection characterizes the policies adopted by Singapore, South Korea, and Taiwan. The success of the latter two regions and the quickening of Japanese industrialization after the World War—the so-called 'trend acceleration' referred to by Kazushi Ohkawa, Rosovsky, and other students of Japanese economic growth (Ohkawa and Rosovsky, 1973; Kaneda, 1989)—owes also a great deal to thoroughgoing pro-peasant land reforms and the consequent abolition of landlordism as a political and economic force.

With the collapse of the 'really existing socialism' in eastern Europe and with its economies coming under the aegis of the market, it is very likely that the origin of capitalism will come to be seen as a multipolar affair. Such countries will start with several enormous advantages in comparison with most of the less developed countries (LDCs). Landlordism has been destroyed in those countries, literacy rates are very high, there is a pool of highly skilled scientists, technologists, and industrial workers, and the occupational structure has changed in most countries so that industry employs more people than agriculture. It is not easy to predict whether these advantages will be swamped by irredentist ethnic and communal conflicts, or will ultimately triumph over the latter. But if the latter possibility is realized, it will be yet another nail in the coffin of a passively diffusionist theory of the spread of capitalism.

This brief and admittedly cursory glance at history is necessary because an ideology seems to be gaining dominance according to which either no national effort at economic development can succeed in an era of transnational capital, or if there is any hope for economic development it can happen only by blindly following the dictates of transnational capital. Some new work on theories of international trade and economic growth—work that overthrows the traditional verities of pure competition and free trade—is being adduced in support of this very old ideology.

The history of capitalism shows precisely that a country or region that passively adapts to the dictates of the dominant fraction of international or transnational capital generally finds itself condemned to underdevelopment. A new theory has to account for this signal fact; if it does not, we should be very careful before we accept it as a basis of discourse for planning, development, or political action. Moreover, I will argue in the last part of this section that these new theories strengthen the case for a nationally or a regionally oriented economic policy and for the operation of a semi-open system with regard to the international economic flows rather than for the kind of *laissez-faire* policies espoused by the World Bank–IMF lobby.

As of now, the Organization of European Commerce and Development (OECD) and more narrowly, the European Union (EU), can be regarded as a block of countries between which capital and trade flows face few barriers. Moreover, the direction of these flows can change, from time to time, depending on exchange rate differentials, differentials in productivity growth, and so on. The contrast we need to emphasize is with the vast majority of the LDCs which have historically received little net inflows of capital from the advanced market economies for any sustained period of time. Japan and the USA are also joined together by massive trade and financial flows, Japanese and German surpluses being used to finance the huge US budget and trade deficits. The ease of commercial and financial transactions is, however, rarely matched by a similar degree of labour mobility—except *within* the EU, strictly defined. The ideology of free trade, of course, never embraces the freedom of migration of labour from the LDCs to the advanced market economies. In the following discussion we shall take these stylized facts to be given.

For those private enterprise economies which are not already conjoined to the advanced market economies through a system of free mobility of capital, goods, and labour, the usual fate is that of sluggish growth periodically interrupted by short-lived booms, caused usually by external factors such as a rise in primary commodity prices, or longer-lasting recessions caused both by internal and external factors. During the '80s and since, recessions have become deeper and more prolonged for those countries which are already caught in the debt trap, and which have painfully to adjust their economies and societies to the requirements of repayment of debt contracted at usurious rates of interest.[10]

The four dragons of East Asia have not only managed to sustain a high rate of growth of national income since about the the mid-sixties. Unlike the countries of Latin America their growth has not been interrupted or turned into long period stagnation. In fact, it may also be claimed that the four dragons have experienced a 'trend acceleration' in GNP growth since the '70s.

The general conditions for such 'trend acceleration' have not been fully investigated. The intensification of inter-capitalist competition, the loosening of the constraint on growth posed by landlordism and the facilitation of technology absorption through appropriate policies promoting worker literacy and continuous growth have all contributed to this acceleration. One consequence has been that we can no longer take, following Arthur Lewis, any rate of national saving exceeding 5 per cent as adequate for sustaining a positive rate of economic growth in the long run. A principal requirement for accelerating the rate of growth of most of the economically successful economies of East Asia has been raising the rate of investment to a figure of at least 25 per cent of national income over a sustained period of time. The figures in Table 2.1 indicate the recent rates of investment in some major countries of Asia.

In Table 1.1, Taiwan (Taipei, China) is the only region in East Asia with a rate of investment of less than 25 per cent. But Taiwan had already emerged in the early '80s as a net exporter of capital. The rates of gross national savings of Taiwan, as the figures indicate, were far higher than 25 per cent. A high rate of investment and saving is by no means a sufficient condition for sustained economic development. But it is possible to argue that if the other conditions for raising the rate of growth cannot be satisfied, then the 25 per cent figure also cannot be sustained over a long period. Thus, this figure should not be seen simply as a multiplicand of the incremental–capital ratio (since $\frac{\Delta Y}{Y} = \frac{\Delta Y}{\Delta K} \cdot \frac{\Delta K}{Y}$ where K and Y are capital stock and national income respectively, and is the incremental operator), but also as a magnitude which reflects the influence of a host of other factors affecting economic growth.

[10] It is an irony of history that legal rates of interest were subject to a ceiling of 12 per cent in 'the first industrial nation' (England) during the phase of its graduation to industrial capitalism. The effective rates of interest for the external debt contracted by most Third World countries (after taking into account the fact that debts are denominated in hard currencies whereas their own currencies have undergone drastic rates of devaluation) generally exceed 20 per cent per annum.

Table 1.1

Gross Investment as a percentage of GNP in selected Asian countries (1987–9)

Year	Bangla-desh	India	People's Republic of China (PRC)	Hong Kong	Republic of Korea	Singa-pore	Taipei
1987	13.8	22.1	37.2	26.9	29.1	39.1	20.5 (39.9)
1988	12.8	22.6	37.6	28.0	29.8	36.6	23.3 (36.9)
1989	11.6	23.1	37.0	27.1	34.6	38.3	23.4 (33.3)

NOTE: Figures within brackets are gross national savings as percentage of GNP.
SOURCE: Asian Development Bank, 1990

The rate of gross investment over the long period reflects the propensity of the business community to invest, and that, in its turn, very much depends on what Keynes referred to as 'long-term expectation' (Keynes, 1936, 46–7). The only concrete factor affecting the state of long-period expectation that Keynes referred to was the long-term or 'normal' rate of interest; and, following him, Kahn (1954) linked the normal rate of interest to expectations about long-term monetary policy. Following Keynes again, Joan Robinson used the phrase 'animal spirits' to characterize the vigorousness of investment and competition within an economy, but did not try further to pinpoint their approximate determinants. Kaldor came closest to analysing the determinants of long-term investment in a capitalist economy, and his insights into the role of static and dynamic economies of scale, product differentiation, and competition as between producers with a certain degree of command on their business environment (such as location) and business parameters (such as advertising, price, and product differentiation) are yet to be fully developed.[11] However, Kaldor's attempt to put his insights into formal (and sometimes mutually inconsistent) models, and his preoccupation with the characteristics of a mature capitalist economy such as Britain, tend to render his analysis of use only at a second remove for our purposes.

I will instead try to develop my analysis along a different path.[12] I

[11] For attempts to develop some of the arguments arising out of Kaldor's theory, *see* Eaton and Lipsey (1978, 1979, and 1989); cf. also Romer (1986).

[12] The contours of this analysis emerged out of several previous efforts of mine (Bagchi, 1982, 1988, 1989).

assume that in the particular country we are concerned with there is a distinct group of capitalists who make a living (and derive power) by investing their funds in productive enterprises which yield their revenues over a long period of time (say 10 years). Profit-maximizing capitalists would be expected to invest their funds wherever their returns are highest. Moreover, assuming that the returns on productive investment increase sufficiently to more than compensate for any fall in consumption in the near future, the *ex ante* propensity to investment of individual capitalists should rise. However, it is more realistic to assume that the proportion of investment to GNP would rise if the share of investors (as against those of *rentiers* or wage-earners) rises in national income, without encountering any problems of shortage of demand or, in Marxist terminology, any realization crisis. This would in turn happen if investment rises strongly enough and is spread across the economy in such a manner as to more than make up for any loss of markets because of the relative (or even absolute) contraction of purchasing power of *rentiers* and the proletariat.

Post-Keynesian or post-Kaleckian economic theory elaborates many of the sub-themes or traces out many of the complications that invalidate the simple inferences suggested by the above scenario. I will not try to sketch those particular sub-themes or those complications that assume the existence of a basic two-class economy, consisting of investors (capitalists) and workers, and a closed economy at that, where it is further assumed that investors will necessarily invest their funds within the national economy. Instead, I examine the plausibility of the two basic assumptions in turn, namely, that the investors will regard the national economy as their principal field of investment, and secondly, that the society can be broken down into investors and workers, and all economic and political activities can be understood on that basis.

At any given moment, rates of return on invested money vary as between different sectors and as between different countries. A well-informed investor, scrutinizing the horizon, would then choose to invest in that sector and in that country where the return promises to be the highest. If this situation were to prevail, we would observe any number of investors crossing frontiers and investing in many different parts of the globe. Especially in periods of volatile exchange rates, we would expect investors in Third World countries, whether domestic firms or multinationals, to move their money to those other countries where long-term prospects of making profits are brighter.

This is in fact what tends to happen in the large majority of Third World countries today. Even apart from the charges for servicing debt and profits remitted by multinationals, a large amount of capital flows out of the Third World under various guises. But the flight of capital does not take place uniformly: nor is it responsive only to interest rate differentials or differentials in real rates of return (after exchange rate variations are taken into account).

To take a few striking contrasts, Japan, South Korea, and Taiwan were investing virtually every yen won or dollar earned, by investors within those countries even when they were emerging from war devastation, experiencing large doses of devaluation, or still working with relatively low levels of skill compared with most countries of the Atlantic seaboard. In contrast to Japan, South Korea, and Taiwan, Mexico and Argentina were sending large amounts of capital abroad even in their phases of high growth. These differences have to be related to differences in long-term expectations—expectations that are grounded in reality through government policies and business strategies.

To use a much-derided term, 'economic nationalism'—under which a particular territory and people are regarded by the domestic investors as peculiarly their own sphere of operation—has to inform the policies of the government of that country and guide the strategies of the business firms there. Such policies can be inward-oriented or outward-oriented, import-substituting or export-expanding; the strategies can be both monopolistic and competitive, narrowly targeted or diversified regionally and sector-wise. But it should normally be the case that for purposes of planning and strategic behaviour the nationals of those countries find it much easier to get together with one another than with foreigners, that many of the contracts and agreements can be informal or tacit rather than formal, and that foreigners would find it difficult to penetrate the *sanctum sanctorum* of businessmen's clubs and council chambers of the government where domestic businessmen, politicians, and bureaucrats get their act together. In the case of east Asian countries, the rules of the game were protected also by explicit barriers against the entry of foreign goods, capital, and people—although in different degrees at different times. But it has been repeatedly argued that Japanese barriers against the entry of foreign goods and firms are often non-transparent and are hard to overcome even when the formal barriers have been pulled down (Bagchi, 1987; Van Olferen, 1990).

Effective economic nationalism and the formation of an industrial bourgeoisie can be badly hampered by the pursuit of policies aimed at achieving the dominance of an ethnic (often defined both in ethnic and religious terms) majority. This is exemplified by the case of Ottoman Turkey in its last phase and of Republican Turkey under Kemal Ataturk and his successors (Keyder, 1990). Similar pursuit of economic dominance by the ethnic or communal majority in Malaysia,

Uganda, and other countries of Africa and Asia (often in the name of 'socialism' or equality) has hampered the formation of the bourgeoisie in those countries. Of course, this kind of political reaction has often been provoked by the attitudes and activities of an alien mercantile community. But there are states such as Switzerland which achieved a national homogeneity while abjuring any attempt to establish the supremacy of a particular linguistic or denominational community.

The consciousness of investors as belonging to a particular national space may spring from a feeling of achieved dominance, as in the case of imperial Britain, or from an aspiration to realize a subjectively perceived superiority to actual dominance, as in the case of Meiji Japan. When a number of countries are tied together by similarity in levels of income, education, or a joint *feeling* among its upper classes of superiority to other peoples because of 'civilization', superiority of religion or superiority of 'race', then capital and labour mobility may be relatively easy between those countries even if there are formal barriers against trade.

This is what obviously happened on the north Atlantic seaboard in the nineteenth and twentieth centuries. Here, nationalism can be transcended by a clubby or limited supranationality for the purposes of investment. This does not by any means imply that these countries do not feel threatened from time to time or even for a decade or so at a time by the unpredictability of capital movements (Nurkse, 1944; and Brown, 1987). But in the larger perspective of the growth of capitalism and its unevenness, such shifts can be treated as cyclical, or at their most dramatic turn, long cycle phenomena to which adjustments can be made without slipping out of the club.

Just as feelings of racial, national, and cultural superiority can power the nationalism of investors, these same barriers can hem in other potential investors within their chosen spaces. Thus, for example, the lack of access of overseas Chinese or Indian investors, up to the '70s, to the wider spaces of western Europe and the USA, has been one of the factors inducing them to invest in Hong Kong, Singapore, Thailand, and India. Of course, if such investors feel politically insecure (the reasons for which are generally to be sought in their own racialist attitudes and the political discrimination practised by host governments), their investment record will be correspondingly flawed. The paradoxical effect of the inability of potential investors to define a national space for themselves may perhaps be found in a heightened form in certain countries of Latin America such as Argentina and Mexico. The upper classes of these countries have behaved like exiled Europeans or potential American citizens and have accordingly wanted their investments to pay off within short time-frames. The track record of most sections of the upper classes of

Latin America being what it is, they also had objective reasons to fear the repercussions of their repressive policies.

If the investors are unable, for social, cultural, and political reasons, to take their own country as their space for investment, then, naturally, the prospects of indigenous growth along capitalist lines dim. Several consequences follow from a low rate of investment in the national economy. First, since in the era of international demonstration effects of various kinds, intended imports would increase and since a corresponding growth of exports, especially of manufacture, cannot take place without a healthy dose of investment powering it, the balance of payments would deteriorate. The country is very likely then to be strapped in an exchange-constrained regime (thus, saving-constrained growth and trade-constrained growth would be indistinguishable). This in turn would produce stop–go cycles and create further instability in expected returns (Bagchi, 1982).

Second, a low-growth, low-investment economy almost always starts with local or sectoral monopolies with considerable influence on trade flows and prices. When the economy is poised for investment in industry, these monopoly groups would try to extend their monopoly control over new fields, and there would also be a proliferation of monopolistic control and restrictive practices in newer industries.

Third, in industries which are characterized by strong economies of scale, monopolistic power would allow existing industrialists to reap high profits with plants which are below their optimum capacities; meaning that they have associated costs that are higher than for best-practice plants (that are larger in scale). It has been shown that with uncertainty about product demands, the size of plants that an incumbent monopolist has to operate in order to deter the entry of intending raiders can be smaller than in the case of certainty in product markets (Perrakis and Warskett, 1986). A low-growth economy subject to stop–go cycles will be subjected to a high degree of uncertainty of demand, especially for capital and intermediate products, and is therefore likely to help entrench this kind of pre-emptive investment behaviour.

Fourth, with current rates of population growth in most underdeveloped countries, sluggish investment and sluggish capacity-creation are likely to be associated with low rates of employment growth, and problems of shortage of effective demand will become endemic in the economy because of a high incidence of poverty (and low rate of investment) among the people.

Fifth, sluggish 'animal spirits' will also be reflected in low motivation to absorb imported technology, let alone innovate, and to expand markets abroad with aggressive marketing and financial strategies. Dependence on foreigners to bridge the balance of payments gap and supply (or rather lease) of technology will in any case restrict the freedom of choice of domestic investors and policy-makers in these areas.

Sixth, in an economy of endemic poverty and shortages, speculative behaviour on the part of traders and investors will become self-justifying.

Let us now look at the consequences of dropping the assumption that the national economy we are looking at only contains a group of investors and a group of proletarians. In particular, let us assume that an underdeveloped country also contains a class of landlords who exercise local non-market power, and who can influence the policy-making apparatus of the government. The existence of such a class would mean that power would often be exercised for purposes other than monetary gain, and that the source of greater power need not be a larger volume of wealth or command over the means of production. The non-market power of landlords would be extended over agricultural labourers, small farmers, and the local machinery for enforcing law and order. The landlords might tighten their hold over the poor by engaging in usurious moneylending themselves; or moneylending might be carried out by a separate group of people who would, however, normally be subordinate to, or in alliance with, the landlords.

Under these circumstances, the potential investors and their allies have to reckon with the intentions and political power of the landlords. How that political power would be exercised as far as the economic policy of the state *vis-à-vis* the rest of the world is concerned would depend on the kind of political and economic position of the country concerned. When the country is still relatively insulated from the rest of the world, it is conceivable that many landlords would act in a patriotic or at least xenophobic fashion. This is what happened in China before the 1911 revolution, when landlords opposed the plans of invading imperialist forces and comprador businessmen.

However, if the country is already penetrated by foreign capitalists and the national capitalists have failed to come out with an effective strategy of industrialization, then landlords would be in alliance with foreign capitalists and those domestic ones who have survived in alliance with foreign capitalists or have managed to create niches for themselves in a regime of free trade. This is roughly the situation which prevailed in most Latin American countries up to the great depression of the '30s.

Finally, we have a situation that is observed in most underdeveloped countries today: the landlords control agriculture and power in the countryside; some of them have moved into trade, transport, and industry; the landlords and domestic capitalists operate the state in collaboration, and in periodic conflict with multinationals and internationalized domestic capital.

Under none of these scenarios are land reforms and abolition of the non-market power of the landlords serious agenda items for the government. Popular resistance and struggle can, of course, force the governments to concede some types of land reform, but they remain ineffectual or marginal so long as the landlord–capitalist domination in the states endures (cf. the recent experience in the Philippines). The persistence of degrading poverty, illiteracy, and use of non-market power for structuring hierarchical relations impede investment decisions and depress the returns on investment, except where it can be guarded by monopolistic barriers. The problems that we instanced in the case of lack of effective economic nationalism on the part of investors become compounded severalfold by the persistence of landlordism. In multi-ethnic or multi-religious societies, culturally, landlordism tends to promote revivalism, inventions of tradition, and so on, which fragment the vision of a unified society, and investors cannot effectively anchor their investment to a uni-focal perspective.

In India and in most other LDCs, the persistence of landlordism and the failure to develop an effective investment-oriented economic nationalism have led to sluggish growth, repeated foreign exchange crises, inordinately delayed processes of absorption of technology, low rates of productivity growth, and a high degree of incidence of poverty and unemployment. Some analysts and policy-makers associated with the World Bank, IMF, transnational banks, and US and other OECD governments have used these failures as the grounds for advocating free trade in commodities and technology, but not men and women. In actual fact, free trade in technology has always been restricted officially and unofficially by restrictions placed in the name of protection of strategic secrets, patents, or other interests of the owners of advanced technology. In recent times, the onslaught of OECD governments has taken the form of a demand for unilateral imposition of US laws (such as *Super 301*) and practices irrespective of the laws, objectives, and interests of other countries.

Other theoretical analyses had demonstrated the hollowness of the discourse of internationalism in economic policy in a world divided into rich or dominant, and poor or economically and technologically dominated, countries, especially where the dominant countries restrict the mobility of labour and practise various kinds of protectionism in defence of their industries threatened by competition from the less-developed countries. Recently, new theories of international trade have been evolved which recognize the key role of economies of scale, learning by doing, and of strategic behaviour in international trade (Dixit and Kyle, 1985; Krugman and Helpman, 1985, 1989).

New theories of international trade, learning by doing, and strategies of self-reliance.

The ideology of free trade was long powered by a static theory of comparative costs or comparative advantage in which few people really believed. The attraction of the theory was its great simplicity and its ability to rationalize a situation that was originally brought about through differential relations of power in international affairs, but one that seemed to fit the colonial style trade flows that characterized the domination exercised by industrialized economies over the global economy.

Apart from Marxists and liberal theorists of imperialism, there were others, such as Alexander Hamilton, Friedrich List, and Mihail Manoilesco who had treated the free trade argument as a special case and one for government intervention in international trade as generally valid for most countries other than the global leaders of manufactures and trades. However, most of the arguments of such dissidents were dismissed by mainstream theorists as special pleading, and in most textbooks and journal articles the theory of static comparative advantage held sway and came nearest to the position of a piece of empty ideology masquerading as economic theory.

The theory proved increasingly inappropriate for explaining the major flows of world trade. In today's world, trade in manufactures and trade between the advanced market economies account for the major part of international commerce. The trade patterns in manufactures could not be explained by appealing to comparative advantage. There were too many different types of goods whose trade flows had to be accounted for; the varieties of goods produced and traded also increased rapidly, and finally their prices were obviously determined by decisions of price-setting firms.

These characteristics had already induced theorists such as Allyn Young, Nicholas Kaldor, S.B. Linder, Michael Posner, and Raymond Vernon to postulate the prevalence of economies of scale, continual product differentiation, the advance of technical knowledge as a normal product of deliberate R & D expenditures, increases in productivity through learning by doing, and regular changes in patterns of consumption brought about by transition of advanced market economies to a state of affluence. Such regular departures from the conditions of pure competition, beloved of protagonists of static comparative advantage, have also induced mainstream theorists to construct models incorporating strategic behaviour on the part of producing and trading firms and of governments formulating economic policies.

Much of this work is still confined to professional journals and surveys directed to fellow professionals (e.g. Jones and Kenen, 1984, 1985; Schmalensee and Willig, 1989). But attempts are already afoot to gloss over those implications of this work which are disturbing from the point of view of the World Bank–IMF lobby and their associates in top policy-making positions in the LDCs, including India. For example, in a recent semi-popular survey article pointing out that there is considerable empirical support for the new theory of international trade, Helpman (1989: 213) writes as follows:

> Proper evaluation of outcomes [of policy] requires detailed information about conduct, market structure, entry constraints, intersectoral links, and the like; we need more empirical studies designed to reveal this information ... since (a) the information needed for a successful policy design is not available; (b) the policy recommendations are very sensitive to the information; and (c) the 'calibration' studies indicate that good policies improve welfare only slightly, free trade remains a good rule of thumb ...

Lest this warning be thought inadequate, another mainstream trade theorist, Srinivasan (1989: 217) proceeds to warn the unwary who might be inclined to support 'an inward-oriented development strategy' because of the solace the new theory seems to provide for it. He proceeds then to place 'the inward-oriented-based theory in the perspective of traditional trade theory; to emphasize the extreme fragility of its conclusions and their lack of robustness....'

Reading Srinivasan's warning, I was continually reminded of the opening of Galbraith's *American Capitalism* (1952) where he discusses why the bumble bee should be an insecure insect and how provoking it must be to an *aficionado* of the then prevalent theory of aerodynamics that it was not. In defiance of the latter, the bumble-bee continued to fly and to thrive. Srinivasan lumps practically everything that goes by the name of development or planning as inward-oriented strategy, and then shows how theoretically and econometrically fragile such efforts must be, and refuses to even raise the question of how all the countries, ranging from the USA to Japan, successfully practised these arts of development in different phases of their history.

At the risk of being considered facile or superficial, let me pick out some of the strands of the package of theories emphasizing economies of scale, imperfect competition, strategic behaviour, learning by doing, and idiosyncratic product and process development. I will emphasize those aspects of the theories which incorporate decisions to invest in fixed capital, human beings, market development, or any decisions which take some time to implement and whose effects are also felt over a relatively long period of, say, one to five years.

First, suppose there is an industry which is characterized by economies of scale that are not exhausted by the size of a particular economy (say, Country A). It is also assumed that there are other firms in the world economy which are already fully exploiting the economies of scale. If the government of A does not offer some protection against external competition, then no investment will occur in that industry in that country. For, at the prevailing world prices, taking into account the risk of the investment not being as profitable *ex post* as it appeared *ex ante*, the potential investor's expected rate of return will be less than the prevailing rate of return in the economy (in the prevalent fully risk-free situation, the investor will just about be able to recover costs, including the cost of capital). The government may decide to offer protection or a subsidy to a potential investor: the latter may then be required to start exporting abroad as a test of the efficiency of investment. Of course, with the growth of the internal market, the investing firm may find it possible to sell a large fraction of its output at home. In such a sequence it is difficult to separate inward- and outward-directed strategies. But what is underlined is a case for public intervention in support of domestic investment (cf. Graham, 1923; Ethier, 1982).

The case for protection against external competition need not be dependent only on the existence of static economies of scale. One situation in which the government may offer tariff protection is when there is a potential for development of a particular industry to eventually become internationally competitive, which has not been developed because foreign competition—often of a predatory and pre-emptive nature—prevails in the domestic market. An import tariff or quota restriction may then induce investors to invest in the industry: the investment response may be so vigorous that the capacity of the industry may soon exceed the size of the domestic market and the price of the commodity then come down to international levels or even lower. Such a policy may be especially justified when investment is expected to come out of the potential savings of the upper classes instead of crowding out other, more socially profitable, investment, pushing up effective demand and reducing unemployment. Raising the import tariff on sugar by the Indian government in the '30s had such an effect (Bagchi, 1963), although that was hardly the intention behind the move.

In yet other situations, only some kind of public support or cooperative action can stimulate the development of an industry, and allows even a country trying to instal the particular technology for the first time to reap the full benefits of the transfer. The industrial advances made by Japan since the Second World War are studded with successful examples of instances in which it first imitated and then improved the technology. This is true of the Linz–Donawitz process for making steel (Lynn, 1984), and for the use of the semi-conductor in the manufacture of radio sets and other electronic equipment (Morita, Reingold, and Shimomura, 1986)—successes which allowed Japan to assume global leadership in those sectors from the '60s onwards.

There is an assumption in most of the conventional theories of international technology transfer that when an economically more backward country or a lower-wage country imports sophisticated technology from an advanced economy it has to follow the specification prevailing in the latter. Again, there are numerous examples to the contrary. A case in point is that of the adaptation of a component of Cummins diesel engine by Tsuzuki, a small Japanese firm. The latter succeeded in producing it in small quantities, but with a quality that was comparable with that of the US firm. 'The task was accomplished by substituting 22 labour intensive processing stages for an automated transfer line' (Odaka, Ono, and Adachi, 1988: 13).

Japanese success in importing technology and improving it and then beating the original developers in international competition have been replicated by other East Asian countries. As late as 1984, two major books on the semiconductor industry, one concerned with the general contours of development of the technology (Dosi, 1984) and the other devoted to the nature of competition between the USA and Japan (Okimoto, Sugano, and Weinstein, 1984), the two global leaders of the industry did not even mention South Korea and Taiwan. By 1987, both these regions were emerging as challengers of the two global leaders (Saghafi and Davidson, 1989).

In contrast to the mere assemblers of semiconductors, such as Malaysia, Philippines, and Mexico, South Korea followed by Taiwan, started full-time production of semiconductor chips which were competitive in the international market. The strategies followed by South Korea and Taiwan were different, the former depending on conglomerates for commercializing production, the latter developing a joint venture company, with the government as the chief shareholder for producing semiconductor chips and supplying private sector companies. But the success of both these regions has again shown how quickly patterns of comparative advantage, even in trade in sophisticated manufactures, can change.

Even though the governments of South Korea and Taiwan have pursued very similar strategies with respect to the import of foreign technology, the development of indigenous capability, and the simultaneous or phased exploitation of domestic and export markets, there are also some major differences with regard to the structure of control of private industry. On the whole, Taiwan's industrial plants have been more regionally dispersed than those of South Korea. Moreover, the South Korean government has actively aided the growth of huge conglomerates (the Chaebol), whereas Taiwan has relied far more on medium and small-sized firms. These differences have also been reflected in the different patterns of development of the electronics industry in the two regions (Mody, 1990).

The cases of Japan, South Korea, and Taiwan illustrate that in suitable socio-economic settings and with appropriate government and business strategies, even latecomers can make rapid advances in technology absorption and quickly overtake some of their erstwhile mentors. On the other side, the general impression often conveyed in many parables on the growth of modern industry and international trade, that the pioneers of the industrial revolution in western Europe were model playgrounds of competitive games and development or absorption of technologies, is also totally false. For example, Britain had highly imperfect capital and labour markets throughout the late eighteenth and most of the nineteenth century (Kennedy, 1987; Williamson, 1989). The upper classes of Britain also displayed a great deal of 'rent-seeking behaviour', much derided in recent literature, besides, of course, deriving an enormous amount of rent from land during its entire period of transition to an industrialized economy (Rubinstein, 1983). To cap it all, throughout the great age of technological innovation, British businessmen made heavy weather of absorbing the silk-spinning and silk-weaving technologies originally imported from Italy (Jones, 1987).

For a long time economists sought their analogies in classical mechanics and modelled their notions of equilibrium on those of the physicists. In this world, markets performed perfectly, and wondrously rational individuals ('rational fools' in the apt characterization of Sen, 1977) adapted instantaneously to any change in the state of the world. The introduction of notions of 'bounded rationality' (Simon, 1982), of behaviour in the presence of risk and uncertainty (Diamond and Rothschild, 1978), or of strategic behaviour, has not made a major dent in the mainstream economists' vision of an economy consisting of neurotically rational individuals operating in a social void.

If we have to absorb properly the facts of human experience as captured in economic decisions and outcomes, and draw lessons for behaviour aimed at creating better societies, we should turn to the philosophy of natural history rather than of classical mechanics for a guide to modes of reasoning. Students of evolution have found that nature has thrown up many different types of organisms which are wonderfully adapted to their different environments. They have also found that similar organisms, starting with apparently similar environments, have ended up with very different types of adaptation, and variations in some conditions during the course of evolution have led to very different outcomes in terms of modes of genetic organization of individuals and their behaviour in a group (for many marvellous examples, *see* Gould, 1977, 1984). For example, haplodiploid organisms[13] are much more likely to evolve tightly organized societies (such as those of bees or termites) than diploids.[14] But there are diploids who live in tightly-knit societies and there are haplodiploids who fail to evolve such societies. The varieties in the patterns of organization of human groups such as nations, nationalities, or multinational states far more resemble the varieties found in nature than the uniformity beloved of mainstream trade theorists.

Of course, analogical reasoning is not a safe guide: we must know how to discard it and when. Organisms in nature evolved through many millennia: their adaptation, selection, and survival was achieved through the blind working of myriads of natural forces. In cases of human organizations, we do have conscious agencies at work. Where our perspective differs from that of the conventional trade theorists is in treating the market as just another organization created by human beings, and the construction of the market and adaptation of human institutions as conscious acts of states, firms, and particular social classes in partial or complete control of the state apparatus. The experience of the development of industrial capitalism can teach us something about how to go about designing institutions and policies.

One basic set of strategies that distinguishes industrial capitalism from situations in which commerce is conducted under the dominance of feudal or other pre-capitalist relations is that human beings are treated as the major source of profit and power. Conscious investment in the upkeep of human beings in the shape first of poor laws, and then social insurance, has characterized the growth of industrial capitalism in England, France, and Germany. Reproducing a reserve army of labour in order to keep the workforce in line has ever been part of the *modus operandi* of capitalism. However, that reserve army has not been treated as an unlimited reservoir to be exhausted at will: control of labour has required its delimitation.

Furthermore, capitalists realized the value of a better-educated labour force. Universalizing education has again been an objective pursued by all successful capitalist states. Under conditions of imperfect information about the quality of products and services, the working of market forces tends often to throw on the market only low-

[13] Meaning, organisms in which one of the two genders, generally the male, has an endowment of chromosomes only from one parent but the other gender has chromosomal endowments from both the parents.

[14] Meaning, organisms in which members of either gender start with contributions from both the parents—as in the case of all mammals, including human beings.

quality products and services (Akerlof, 1970). Universalizing education and getting the labour market segmented along lines of education and skill are ways of checking such tendencies of the market. Such strategies are pursued not because capitalists are philanthropists but because they have to compete in markets where quality sells.

Along with such universal strategies of successful industrial capitalism, others grow up which are more particularistic in nature. Different societies evolve different strategies for attaining the aim of perfecting the social division of labour so as to gain and retain a competitive edge. In some cases, private enterprise and associations of private firms are the principal instruments for achieving the best social division of labour. The cotton textile industry of England in the nineteenth century seems to have fitted this pattern. In other countries and other industries, such as the metallurgical and power equipment industries of Germany and the USA, the growth of large-scale firms with the internalization of most production operation seems to have provided the greatest scope for a profitable social division of labour. In several French industrial cities in the nineteenth century, municipal or other local public bodies provided the framework for quality control, training of labour, and upgradation of technologies (Piore and Sabel, 1984). In Japan, a group of conglomerate firms using large numbers of subcontractors became the focus of social division of labour. But subcontracting is not a peculiarly Japanese phenomenon. It can grow up in other contexts too, such as the Lyons conurbation in France in recent times, where firms in various branches of engineering have resorted to subcontracting on a large scale (Lorenz, 1988).

The point I would like to emphasize is that, in designing appropriate trade policies and strategies of international competition, governments, firms, and technological research laboratories must cooperate so as to evolve operating principles suitable to each national setting. In socialist China, a publicly-guided system of cooperation between local governments, state and collective enterprises, universities or colleges, and research laboratories was the chosen instrument for development, transfer, and absorption of technology. In South Korea, it is private firms contracting out research needs to state-controlled laboratories and collaborating with them in order to adapt imported technology to local needs which became a highly efficient model of upgradation and absorption of technology (Bagchi, 1987b; D'Costa, 1989).

In India, in my view, we have failed, on a national scale, to evolve a durable system of subcontracting or technology transfer. India's international competitiveness has also suffered because of inefficient absorption and diffusion of both imported and indigenous technology. But, in a world which is characterized by monopoly or oligopoly, surrender to international capital could be the worst way of choosing a strategy for improving the internal productivity and international competitiveness of large-scale industry in India. As it is, it is the small and medium-scale firms which provide the major part of Indian exports. Any reorientation of economic policies and political strategies must take that fact into account, as it must reckon with the employment-generating potential of small firms. Simple-minded solutions for the sluggishness of investment in the shape of a mere loosening of investment licencing regulations have not worked so far in India, as the stagnation of rates of investment as a proportion of GNP during the last 10 years or more testifies. Some recent theoretical work has also borne out that, in the presence of monopoly and restrictive practices, simple auctioning of investment licences or import quotas is unlikely to be the answer (cf. Krishna, 1990; Krugman, 1989: sect. 3]).

Our earlier discussion should make it abundantly clear that a redirection of the discourse of planning cannot be confined to the realm of industrial or commercial policies. Redesigning some basic social arrangements must be put on the agenda if sensible industrial policies are to succeed. We have already adduced the success of several East Asian economies in the area of industrial and trade policies. We have emphasized that universalizing education and practising effective economic nationalism were necessary conditions for the success of their growth-oriented policies.

The East Asian economies have also done well in two other areas. One is that of land reform which abolished landlordism in Japan, South Korea, Taiwan, and, of course, China. These reforms abolished the non-market power of landlords in the countryside, released for purposes of investment, the rent that had earlier been consumed by the landlords for investments, and redirected the incentive structures of businessmen towards productive investment. The drastic reduction in the proportion of the poor through positive support policies and through the growth of incomes (Drèze and Sen, 1989) also had positive feedback effects on the productivity of the ordinary people and on the size of the internal market for industrial goods.

It is important to realize that East Asian success stories are based on a whole cluster of social and political institutions, and not to be misled by the isolated trumpeting of their so-called export-oriented strategies by the IMF–World Bank lobby. It is also to be stressed that if any country is excessively dependent on foreign aid or loans from commercial banks and gets into an external debt trap, it has to say goodbye not only to policies that seek to discriminate between domestic and foreign investors but also to policies that seek to redistribute incomes or social benefits in favour of the poor. The 'adjustment policies' forced on indebted nations by the IMF and the transnational banks have taken a heavy toll of the education, health, and longevity of the poor in most of the countries undergoing the treatment (Cornia, Jolly, and Stewart, 1987, 1988).

Although the growth of industrial capitalism (socialism) in Europe and East Asia were poles apart, there were major similarities in the strategies pursued by the countries in both regions in order to foster economic growth and industrialization. These strategies included effective economic nationalism, abolition of landlordism, universalization of literacy, and provision of some kind of social security for every citizen. The discourse of mainstream economics has ignored these specific social and political strategies and movements, and thereby provided a thoroughly unreal picture of how improvements in standards of living took place. The discourse of the political economy of planning has also sometimes confined itself only to action at the level of the state or an abstract market mechanism: an enriching of that discourse by bringing in actions by classes and groups below, outside, but often *in tandem* with or in opposition to the state has been overdue in the Indian context.[15]

There is a mistaken impression—deliberately propagated by some economists and business groups—that the troubles India has experienced in the shape of loss of internal or international competitiveness in large-scale industry are entirely due to the pursuit of a policy of self-reliance. By a state of self-reliance achieved by a country I would mean the ability of that country to freely *choose* macro–social policies, techniques of production, and management practices without having to obey the dictates of foreign firms, transnational banks or organizations such as the IMF and the World Bank. A country which is continually plagued by endemic foreign exchange shortages and does not have a backer to underwrite all such shortages (as the USA did for South Korea and Taiwan during the '60s and '70s) is unlikely to be able to pursue a policy of self-reliance. Hence the Indian government was not able to pursue a policy of self-reliance for most of the period since Independence—notwithstanding the official rhetoric to the contrary. But there have, of course, been fluctuations in the degree of infringement of the policy of self-reliance, depending on the strength of the externalization lobby.

[15] The same comment would apply to the discourse of political economy anywhere else in the world.

Can the discourse of planning and the discourse of people's choice be united?

As I have already noted, the discourse of planning in India has been heavily implicated in the statist paradigm. On a false analogy with the post-revolutionary situation in socialist countries, the state has been seen as a virtually autonomous entity acting on a passive society—or rather a society that could be galvanized into appropriate responses as desired by the planners and policy-makers. To this oversimplified paradigm has been opposed the equally simplistic paradigm of the free market—which also posits an all-powerful state, but a state which is the source of all evil and all unproductive ('rent-seeking') activities. As I have pointed out elsewhere (Bagchi, 1989a), these opposites share a basic presupposition, namely that the state in India can be regarded as an entity which is virtually disjointed from society.

This presupposition is violated not only by the numerous devices used by the dominant elements in the ruling classes to manipulate the state apparatus, and by the strategic responses evoked by state action. It is also violated by two other major features of Indian society, namely its permeation by various nationalities with their distinct ways of relating to the Indian state as a whole.[16] The centralist paradigm of planning has for a long time been confronted by locally-centred paradigms of planning: this has spawned a large literature on Centre–State relations.

But at a deeper level, in many communities, planning initiated by political and bureaucratic structures located in the national capital, State capital, and even district headquarters has been seen as an intrusion of external forces—bent on the same old goal, the exploitation of some of the most disadvantaged communities in the country.[17] The reality and perception of such exploitation can only be mitigated (but, of course, not removed altogether so long as a system dominated by private property rights endures) by involving the local people in the planning process.

Such a model of planning is being tried out in West Bengal and in some other States. The *Approach to the Eighth Five Year Plan* issued by the Planning Commission (Government of India, 1990) also gave its blessings to decentralized planning. But, of course, such planning cannot be effectively implemented without breaking the power of landlords all over the countryside. There is a discourse on decen-

[16] For a discussion of some of the economic issues involved in the nationality question, *see* Alam (1985 and 1985a).

[17] For a vivid account of some of the depredations wrought by bureaucrats serving consciously or unconsciously as tools of forest contractors and mining companies, *see* Anon. (1989).

tralized planning—a discourse which often claims a Gandhian pedigree—that attempts to sidetrack such issues of power and class, and seeks only to minimize the power of bureaucracies. But retaining old community structures with their own hierarchical means of control and involving the so-called non-governmental organizations or NGOs (which are often given lavish foreign funding and are rarely self-supporting) to provide new ideas for management will not convert the economic policy-making process into a democratic exercise. Moreover, communities with particularistic world-views—world-views that necessarily provide for domination over other groups or communities with different ones—will find it difficult to communicate with one another, and come to democratic solutions.

The short-lived (if not stillborn) *Approach to the Eighth Five Year Plan* attempted to ensconce many of the unresolved contradictions between democratic decentralization and authoritarian community structures. Between the aspiration to release market forces from the fetters of regulation and non-market constraints, and the refusal to put land reforms at the centre of a programme of social change. Between support for cottage and small-scale industries and encouragement of unfettered expansion of large-scale industrial units. And between endorsement of a further move towards external liberalization and achievement of a higher degree of self-reliance in science and technology, and a higher rate of employment growth. In fact, the masthead of the *Approach* document might have appropriately read: 'Let a hundred contradictions bloom!' This would not be a bad slogan if, in fact, the Planning Commission and the Government of India were prepared for the social turmoil that would result from the resolution of the contradictions through class struggle on the ground.

But let me get back to the theme of decentralized planning and promotion of production ventures geared to the use of local resources and local talents. One of the most daunting challenges against all these efforts has been the continued development of technologies which seem to generate further and further economies of scale, so that the reach of centralized enterprises seems to be ever-expanding. But in actual fact, side by side with further development of economies of scale, especially in process technologies, there is a continual differentiation of products which makes room for ever newer enterprises catering to new needs. The encouragement to smaller scale enterprises provided by the development of electronically-based information and control technologies has been widely discussed in the literature. But the same comment would apply to the catering and fast food industries, to clothing, and indeed to modes of transport where bicycles compete with sports cars as the favoured means of locomotion of the very rich.

In a country as variegated and as poor as India, there are two major avenues for the development of such local product differentiation: the first is that of empowerment of the poor by giving them a minimum income and food security, and providing them with access to land, water, and other resources. This would generate purchasing power and increased productivity to match it. Second, the development of existing skills to provide idiosyncratic bases for many modern industries to take up as a specific strategy. Within modern industries, skills still often remain firm—and location-specific. In India, many of the older locations of engineering industries, for example, function as pools of amazingly skilled labour. The need is to upgrade those skills and evolve a mechanism to keep them in continual contact with changing market profiles. A nationwide proliferation of extension agencies and arrangements for linking small input-producing firms with large finishing and marketing enterprises could be appropriate strategies in this area. But all this would have to be worked out within a framework of democratic decision-making with as unimpeded a flow of information, vertically and laterally, as possible.

The empowerment of local communities, which have been rid of traditional power structures, for planning and decision-making, could also be a forward step for better environmental planning. Poverty is a destroyer of environments; but, historically, predatory commercialization has been an even greater enemy. If local communities are given better control over their own resources and are made responsible for the consequences of their own decisions, and if they are linked with other local communities to permit interdependent decision-making, the outcome should be better than is achieved through bureaucratic planning imposed from above. In any case, it should be recognized that environmental planning is not sustainable in a densely populated country like India if the people refuse to cooperate. For example, if pastoralists are not provided with stable sources of fodder, they will graze their cattle on hill slopes, causing deforestation, avalanches of rocks, and siltation of rivers, wreaking devastation throughout the valley. If inequalities that cause persistent poverty endure, then the rich will always be in a position to utilize the poor to subvert the ends of well-intentioned planners. A two-way planning system based on the elimination of extreme inequalities should mitigate such possibilities for the opportunistic utilization of disequilibria in order to subvert the ends of planning.

The *etatist* approach to planning sought and still seeks to impose an *ex ante* consistency on the different components of the desired outcome through a combination of *fiat*, regulation, and market forces. There is little theorization on either the market responses—strategic or otherwise—to government action, and little recognition of possible reactions or even resistance on the part of ordinary people. An interactive, decentralized, democratic mode of planning would seek to ensure *ex post* consistency by taking the people into confidence at every stage of planning and altering the plan as it evolves, and it would only try to preserve certain external fortifications through *fiat* and macroeconomic management. This, however, must be distinguished from a mode of planning which takes existing power structures in the villages and towns to be given, and allows the market to play on the moats and inlets of these fortifications. The practice of the latter mode is only likely to produce results which are already familiar in various descriptions of semi-feudal formations and processes. One lesson in terms of discourse theory is that a post-Enlightenment logic cannot be grafted on to pre-Enlightenment structures of power, social interaction, and communication.

Epilogue

This chapter was originally written in early 1990. I had known for some time that there were powerful forces in India deliberately pushing the country into a debt trap with an attendant loss of economic and political sovereignty (Bagchi, 1985). These forces were internationalized domestic capital, transnational corporations, economic policy-makers in the government impatient with what they regarded as undue government interference in the market, large monopoly capital hoping to gobble up a huge chunk of the public sector (when the IMF–World Bank authorities enforced wide-ranging privatization as part of their nostrum), and politicians hoping to gain financially and otherwise from side-payments made by the major beneficiaries. Getting India firmly into a debt trap would put paid to the dreams of those people who wanted India to pursue an economic and social policy in the interest of the majority of the Indian people rather than for the benefit of international capital (including internationalized domestic capital and big business in India). The portents were very clear to all economists watching what was happening all over Latin America, and all over sub-Saharan Africa (Sachs, 1989; Bagchi, 1990a).

Even in early 1990 it seemed that India could have avoided a debt trap had a policy of austerity with Indian priorities in place been adopted by the Janata Dal-led National Front coalition government that was ruling at the Centre. This would have led to a decline in the measureed rate of economic growth for much of the economic exhilaration during the period of Rajiv Gandhi's Prime Ministership

(1984–9) was powered by increasing imports financed by fresh doses of commercial loans and by fuelling the greed of the upper middle classes for better cars, better colour television sets, more luxurious hotel accommodation, and other public and private extravagances. But the austerity measures could have been combined with a comprehensive, though basic, social insurance scheme for the poor, with land reforms in the heartland of India and with a new drive for universalizing elementary education (Bagchi, 1990b, 1990c).

However, this opportunity was missed by the National Front government, and, of course, by the 'rump' government of Chandra Shekhar which usurped its power when the Bharatiya Janata Party withdrew its support from V.P. Singh, the Janata Dal Prime Minister. The minority Congress(I) government, led by P.V. Narasimha Rao, decided to toe the IMF–World Bank line of totally dismantling measures of public regulation and pampering the delinquent rich even further. The July 1991 budget set about dismantling the apparatus of controls on external trade, slashing subsidies (especially those supporting the public distribution system), devaluing the rupee, and raising resources by further increasing administered prices and the tax burden on the poor rather than taxing the rich. Indulgence was shown to the rich who had been extremely lax in meeting their legal obligations. Amnesty for past tax evasions was granted almost indefinitely.

Even before the 1992–3 budget was presented, 'liberalization' of the economy along the lines recommended by the IMF–World Bank lobby had proceeded very far. An 'exit policy'—a euphemism for retrenchment of public sector employees—was put in place. Plans were announced for de-investment of large chunks of public enterprise equity. Maruti Udyog, for example, which had been set up as a joint venture with Suzuki of Japan, was effectively privatized with the Japanese holding the controlling interest. Most of the controls on monopolies, on capital issues by private companies, and on investments by the private sector were relaxed, but no measures were taken to discipline the private sector—measures considered essential for the smooth functioning of the market economy in advanced industrial countries such as America, Japan, France, or Germany (Bagchi, 1992).

A scheme for financial 'liberalization' was launched, following the recommendations of the Narasimham Committee, which in turn, and predictably, followed the usual IMF–World Bank package for structural readjustment. Rates of interest were raised all round, loans to agriculture and small-scale industries on preferential terms that had been in place for a number of years were drastically reduced.[18] The

[18] *See* Ghosh* (1992) and Narayana (1992) for a critique of the Narasimham report.

* D.N. Ghose is the former Chairman of the State Bank of India, the oldest and largest commercial bank of the Indian subcontinent.

warning that high interest rate policies and free-wheeling banking are major culprits in bringing about a decline in the British and American manufacturing industry was not heeded by the government.

There were, of course, drastic changes in Indian external economic arrangements. Rupee devaluation was accompanied by an across-the-board removal or abatement of quota restrictions on imports and drastic cuts in customs tariffs. Despite the fact that the devaluation of July 1991 failed to give a boost to exports and fuelled the forces of inflation, the government persisted with the policy during the budget session of 1992–3. The rupee was made partially convertible, with exporters being allowed to convert 60 per cent of their export earnings into rupees at the free market rate; they were also allowed to keep 15 per cent of their export earnings in foreign accounts. This arrangement replaced the earlier scheme of Exim scrips, which had allowed exporters to use up to 30 (in some cases 40) per cent of their export earnings for imports, or to sell the certificates of export to the highest bidders. Even this scheme had not proved sufficiently attractive to exporters.[19]

The rapid movement from a spate of external commercial borrowing into a debt trap and thence to an IMF–World Bank regime of structural adjustment is an all too familiar experience. Few of the countries experiencing such structural readjustment have again resumed a path of sustained growth and development (Branford and Kuchinski, 1988; Sachs, 1989). The signals provided by the IMF–style structural adjustment process tend to discourage investment all round: domestic markets shrink, the external value of the currency becomes highly unstable, and R & D expenditure appears to be infructuous in view of the unrestricted advent of foreign goods; high-technology industries are particularly hit since their learning and adaptation process, perhaps begun at the time of frantic external borrowing, remains incomplete when the structural adjustment wave overwhelms them.

I had suggested (Bagchi, 1990b) that India did have the alternative of choosing a domestically controlled adjustment process with a reordering of the economy towards mass consumption and producer goods-based growth with support being provided by land reform, universal education, and a social insurance system to guard against pauperization. Such an alternative would lead to a dip in the growth rate but ensure subsequent access to a higher growth trajectory. The choice of an externally orientated structural adjustment process by the current Indian government against a scenario of worldwide recession has made it almost certain that India has entered a prolonged period of recession accompanied by inflation, mass impoverishment, and pervasive social dislocation.

[19] *See* Sen and Das (1992) for a critical analysis of the experience of India with various schemes of import liberalization.

My view is that there is a fundamental conflict between the role of capital as finance seeking the highest rate of return at every moment of time, and the role of capital as a social factor of production allowing skills, learning, and productivity to be built up through a long-term commitment on the part of the controller of that social factor (the capitalist, the state, or a cooperative organization of workers) to a particular locale and/or a particular social network. Virtually all the advanced capitalist countries of today provided that commitment within the framework of a nation–state: in some cases, this was complemented by labour and capital mobility within narrow channels of race and territoriality. Finance was mobile to a limited extent throughout the period of the rise of capitalism (Braudel, 1984).

But whenever finance tended to dominate industry, the productive sectors of such an economy regressed in relation to those of the competing economies. This has been the case with the Netherlands in the eighteenth century, and Britain from the late nineteenth century onwards, and the USA since the 1970s. However, there is no law identifying the stage at which finance will dominate industry. If manufacturing industry remains underdeveloped, the influence of international forces and the mounting surplus in the hands of landlords, traders, and bankers, finance may begin to dominate industry before the latter has had a chance to be internationally competitive or to provide a base for sustained economic growth. Such premature development of finance has been the fate of most Third World countries which have been unable effectively to fight the forces of landlordism and predatory commercial capital at home and international capital abroad. India is also now experiencing such premature development of finance in addition to the endemic process of predatory commercialization that had created social tensions and over a long period weakened the forces of cohesiveness.

The determination of the USA and most other OECD governments not to allow the development of new, relatively autonomous, centres of capitalist growth becomes particularly clear when we scrutinize the provision relating to intellectual property rights, trade-related investment measures, and trade in services embodied in the draft prepared by Arthur Dunkel, the former Director-General of GATT. Their acceptance would deliver the less developed countries, including India, into a state of low-productivity helotry to the major OECD countries (Dhar and Rao, 1992; Sen, 1992). There has nonetheless been strong pressure from the IMF–World Bank lobby for India to accept the provisions of the draft. It was reported on 16 April 1992 that 40 per cent of the funds of the CIA would be devoted henceforth to economic surveillance. This is a clear sign, if such a sign were needed, that the power élite in the USA is about to impose a more fearsome order of economic imperialism on the rest of the world, including India.

Predatory commercialization with no safety nets for displaced artisans and pauperized peasants had been among the major economic forces rendering the vulnerable poor completely insecure in their aspirations. Such tensions were made use of by groups and interests seeking political power in the name of religion and were among the most potent forces for the partitioning of the subcontinent among the successor states of India and Pakistan. They have also been the background against which communal riots have repeatedly taken place in India in recent times (Bagchi, 1991).

The linking up of the financial markets of the major centres of the world, the collapse of the socialist bloc of Eastern Europe, and the imposition of financial liberalization and virtually unhindered mobility of capital imposed on the debt-encumbered countries by the IMF–World Bank as the watchdogs of transnational banks, have led to an unholy integration of the centres of political power in the world as a whole. The failure of economic growth strategies based on an unequalizing spiral, as in the case of Brazil (Morley, 1982), for example, has aided this integration at the top.

The upper layers of the bureaucratic, military, and technocratic élite found that since they could not defend a monopoly of power at their home base, it was in their interest to promote the mobility of both financial and human capital. One extremely ironical feature of the current neo-liberal campaign for free trade in goods and services is that it explicitly rules out free trade in the basic service provided by human beings—that is, their capacity for productive labour. Free mobility of finance and a somewhat less free mobility of highly trained womanpower and manpower, combined with almost totally restricted mobility of unskilled labour, ensure that labour remains fragmented in vast swamps of low-productivity employment, unemployment, and insecure living. Under this dispensation, capital also has access to a vast global reservoir of labour power ready to be deployed when, if ever, labour shortage is faced anywhere in the world.

Predatory commercialization and the prevalence of landlordism in alliance with speculative commerical capital has already exacerbated the fragmentation of society in multilingual, multi-ethnic, multi-religious countries, of which India is perhaps the leading example. The long caesura in the growth process because of the contradictions of retarded or dependent capitalism in most 'Third World' countries had led to pauperization and lumpenization of large sections of society. The impoverishment process was further accelerated through the IMF-style structural adjustment programmes and consequent criminalization of society in most Latin American countries (Branford and Kuchinski, 1988; Dornbusch and De Pablo, 1989). Similar aggravation of social fragmentation as a result of the neo-liberal policies pursued by the currently incumbent government of India is very much in the offing.

Many analysts in India have rightly emphasized the cultural diversity of the peoples living in India (*see*, for example, Pandey, 1991). Such an appreciation of the diversity of experience, ideology and memory is needed in order to be able to evolve a political culture that will allow all the culturally diverse groups to live in peace within a democratic polity and democratic society.[20] However, a purely epistemological analysis of cultural diversity is not enough to combat the politics of communalism or ethnic authoritarianism. An exposure of the numerous ways in which anti-democratic forces, of which communalism is a leading example, distort and invent history, and create false memories (Bhattacharya, 1991; Thapar, 1991) has to be combined with a discourse of democratic and egalitarian politics. That discourse has then to be translated into the daily practice of Indians as citizens (Samaddar, 1991).

As I have emphasized above, the discourse of socialist planning was caught up in a statist, authoritarian discourse, and was thereby discredited, for many vested interests could use that language to pursue private ends and subvert social goals by means of which the discourse of planning was legitimized. A very similar fate overtook many organizations of resistance and solidarity such as trade unions: the latter tended too often to be used in order to pursue short-term sectarian goals. However, the vastness of India and the richness of the heritage of the diverse groups of people living there can throw up alternative forms of struggle providing alternative models of political practice.

One such alternative form was provided, for example, by the Chhattisgarh Mines Shramik Sangh (CMSS), which later became part of the Chhattisgarh Mukti Morcha (CMM). This was an organization of mine workers in Madhya Pradesh which took up not only the cause of the permanent workers in the mines, but also that of temporary workers and contract labourers in mines and forests in a huge area covering several districts of the State. The CMSS was led by Shankar Guha Niyogi, a trade union leader, who realized that workers' struggle for higher earnings had to be combined with continual reconstruction and redefinition of their lives (Guha Niyogi, 1992). Against tremendous odds, he persuaded the workers to construct their own schools, their own hospital, and to reclaim their environ-

[20] For the classic analysis of the distinction between a democratic polity and a democratic society, *see* de Tocqueville 1848, Vol. 1, Chapters III–IV, XIII–XIV, XVII; and Vol. 2, Third Book ('Influence of democracy on manners properly so called').

ment in the teeth of opposition not only from private contractors, but also from managers of state-run mines, government officials, and, of course, policemen. Many workers were arrested and killed in police firings or that by gangsters. But they did not yield (Seabrook, 1992). Guha Niyogi and his comrades also undertook campaigns to help workers give up drinking, because they realized that getting them addicted to drink was one of the surest ways of rendering them docile and ineffective.

Women are among the worst victims of predatory capital and the repressive state apparatus. The CMSS mobilized them effectively to fight drunkenness which was most prevalent among men and to reconstruct their environment, the cumulative deterioration of which especially affected the health and survival of children. Women workers became equal comrades-in-arms of their struggling male counterparts. Guha Niyogi disturbed too many vested interests, for too long, and he was killed in the early hours of 28 September 1991 by gunmen identified as the agents of some major business tycoons in the area. His real assailants are yet to be tried, but his killing elicited nationwide protests, and it has led to the strengthening rather than the weakening of the resolve of his surviving comrades to carry on the struggle.

Just as there are strong forces in India working against the building up of a democratic society and the preservation and completion of a democratic polity, there are also other forces fighting them. In a country in which hitherto repressed groups are voicing their demand for a place in the sun, the language of struggle is often couched in purely statist terms—pitching new sets of authoritarian slogans against the authoritarian practices of a quasi-democratic state apparatus. In order to get out of this spiral of bloody struggles only to be caught in the clutches of another bunch of authoritarian state apparatuses, it is necessary to voice clearly the demand for demarcation of the spheres of public and private interests—in the areas of religion, regional cultural practices, property rights, and so on. It is also necessary to appreciate that the rejection of exclusive chauvinism need not preclude the adoption of inclusive, solidary nationalism.

As I have argued earlier, giving the focus of investment 'a habitation and a name' is necessary for it to act steadily to improve productivity, expand markets, and improve the infrastructure both for production and for day-to-day living. The post-Enlightenment discourse of the uniqueness of the atom may be a very convenient camouflage for a global power structure which does not want any nodes of resistance to be built up anywhere. But, in order to be able to conceive the possibility of a *human* existence for the majority of the people in a society buffetted by forces of predatory commercialization and premature financial development, it is necessary to retrieve, redefine and reconstruct the logic of a public interest which can integrate the myriad atoms into a social conglomerate which is not all the time threatened with extermination through internally-generated explosions.

REFERENCES

AKERLOF, G.A., 1970: 'The Market for "Lemons": Quality, Uncertainty and the Market Mechanism', *The Quarterly Journal of Economics* 84 (Aug.): 3, pp. 488–500.

ALAM, J., 1985: 'Dialectics of Capitalist Transformation and National Crystallisation: The Past and the Present of the National Question in India', in PANIGRAHI (Ed.), 1985, pp. 396–431.

———, 1985a: 'Class, Community and Nationality Formation: A Theoretical Exploration through Two Case Studies' (Occasional Paper No. 70) (Calcutta: Centre for Studies in Social Sciences).

ANON, 1989: 'Bastar: Development and Democracy', *Economic and Political Weekly* 24 (30 Sept. and 7 Oct.): 39 and 40, pp. 2188–90; and pp. 2237–41.

ASIAN DEVELOPMENT BANK, 1990: *Asian Development Outlook 1990* (Manila: Asian Development Bank).

BAGCHI, A.K., 1960: 'Aspects of Government Economic Policy in a Partially Planned Economy', *The Economic Weekly* 12 (Jan.): 4–6, pp. 237–40.

———, 1963: Private Investment and Partial Planning in India (unpublished Ph.D. thesis, University of Cambridge).

———, 1964: 'Protection and Rationalisation: A Problem of Economic Policy in Underdeveloped Countries', *Arthaniti* 7 (Jan.):1, pp. 1–14.

———, 1966: 'Tariff Protection, Controls and Shadow Prices', *Indian Economic Review* (New Series) 1 (April): 1, pp. 22–44.

———, 1971: 'Notes Towards a Theory of Underdevelopment', *Economic and Political Weekly* 6 (Jan.): 3–5, pp. 351–66.

———, 1982: *The Political Economy of Underdevelopment* (Cambridge: Cambridge University Press).

———, 1984: 'Towards a Political Economy of Planning in India', *Contributions to Political Economy* 3 (March), pp. 15–38.

———, 1985: 'Hijacking the Indian Economy', *The Herald Review* (Bangalore) (5–11 May), pp. 39–44.

———, 1987: 'Introduction' to Special Issue of *Political Economy: Studies in the Surplus Approach*,* 3 (2), devoted to East Asian capitalism, pp. 115–32.

* published in Torino

———, 1987a: 'Industrialisation' in EATWELL, MILGATE, AND NEWMAN (Eds.), 1987. Vol. II, pp. 797–803.

———, 1987b: *Public Intervention and Industrial Restructuring in China, India and Republic of Korea* (New Delhi: ILO–ARTEP).

——— (Ed.), 1988: *Economy, Society and Polity: Essays in the Political Economy of Indian Planning in honour of Professor Bhabatosh Datta* (Calcutta: Oxford University Press).

———, 1989: 'Capitalism, Uneven Development and the Foundations of National Liberation', in LIMQUECO (Ed.), 1989, pp. 87–97.

———, 1989a: 'State Intervention and Industrialisation' (Lal Bahadur Shastri Memorial Lecture) (Institute of Public Enterprise, Hyderabad) (mimeo).

———, 1990: 'Public Sector Industry and the Political Economy of Indian Development' (Occasional Paper No. 119) (Calcutta: Centre for Studies in Social Sciences).

———, 1990a: 'The IMF view of international economic policy and the relevance of Sraffa's critique of economic theory', in BHARADWAJ and SCHEFOLD (Eds.), 1990, pp. 494–515.

———, 1990b: 'An Economic Policy for the new government', *Economic and Political Weekly* 25 (10 Feb.): 6, pp. 317–20.

———, 1990c: 'Planning for employment', *Economic and Political Weekly* 25 (21 April): 16, p. 819.

———, 1991: 'Predatory Commercialisation and Communalism in India', in GOPAL (Ed.), 1991, pp. 193–218.

———, 1992: 'Transnational Banks, U.S. Power Game and Global Improverishment', *Economic and Political Weekly* 27 (30 May): 22, pp. 1133–6.

BANURI, T., (Ed.), 1991: *Economic Liberalisation No Panacea: The Experience of Latin America and Asia* (Oxford: Clarendon Press).

BAXI, U., 1986: 'Sociology of Law' in INDIAN COUNCIL OF SOCIAL SCIENCE RESEARCH, 1986, pp. 69–195.

BHADURI, A. and S. MARGLIN, 1987: *Profit Squeeze and Keynesian Theory* (Helsinki: World Institute for Development Economics).

BHARADWAJ, K. and B. SCHEFOLD, (Eds.), 1990: *Essays on Piero Sraffa: Critical Perspectives on the Revival of Economic Theory* (Delhi: Oxford University Press).

BROWN, B., 1987: *The Flight of International Capital: a Contemporary History* (London: Routledge).

CARR, E.H., 1970–72: *Socialism in One Country 1924–26, Vols. 1–3* (Harmondsworth: Penguin Books).

CHAKRAVARTY, S., 1987: *Development Planning: The Indian Experience* (Delhi: Oxford University Press).

CHANDRA, B., 1969: *The Rise and Growth of Economic Nationalism in India* (New Delhi: People's Publishing House).

———, 1979: 'Jawaharlal Nehru and the Indian Capitalist Class' in his *Nationalism and Colonialism in India* (New Delhi: Orient Longman), pp. 171–203.

CHANDRA, N.K., 1988: *The Retarded Economies: Foreign Domination and Class*

Relations in India and other Emerging Nations (Bombay: Oxford University Press).

CHATTOPADHYAYA, R., 1985: The Idea of Planning in India, 1930–1951 (unpublished Ph.D Thesis, Australian National University, Canberra).

————, 1988: 'Liaquat Ali Khan's Budget of 1947–48: the Tryst with Destiny', *Social Scientist* 16 (June–July): 6–7. Nos. 181–2 (June–July) pp. 77–89.

CORNIA, G.A., R. JOLLY, AND F. STEWART, (Eds.), 1987: *Adjustment with a Human Face, Vol. 1, Protecting the Vulnerable and Promoting Growth* (Oxford: Clarendon Press).

————, 1987a: *Adjustment with a Human Face, Vol. 2, Ten Country Studies* (Oxford: Clarendon Press).

D'COSTA, A.P., 1989: Capital Accumulation, Technology and the State: The Political Economy of Steel Industry Re-structuring (unpublished Ph.D Thesis, University of Pittsburgh).

DHAR, B. and C.N. RAO, 1992: 'Dunkel Draft on TRIPS: Complete denial of developing countries' interests', *Economic and Political Weekly* 27 (8 Feb.): 6, pp. 275–8.

DHAWAN, R., 1989: 'Introduction' to GALANTER, 1989, pp. xiii-c.

DIAMOND P and E. ROTHRCHILD (Eds.), 1978: *Uncertainty in Economics: Readings and Exercises* (New York: Academic Press).

DIXIT, A. and A.S. KYLE, 1985: 'The Use of Protection and Subsidies for Entry Promotion and Deterrence', *American Economic Review* 75 (March): 1, pp. 139–52.

DIXIT, A. and V. NORMAN, 1980: *Theory of International Trade* (Cambridge: Cambridge University Press).

DORNSBUCH, R., and J.C. Pablo, 1989: 'Debt and macro-economic instability in Argentina', in SACHS (Ed.), 1989, pp. 37–56.

DOSI, G., 1984: *Technical Change and Industrial Innovation: The Theory and an Application to the Semi-conductor Industry* (London: Macmillan).

DRÈZE, J. and A. SEN, 1989: *Hunger and Public Action* (Oxford: Clarendon Press).

EATON, B.C. and R.G. LIPSEY, 1978: 'Freedom of Entry and the Existence of Pure Profit', *Economic Journal* 88 (June): 3, pp. 455–69.

————, 1979: 'The Theory of Market Pre-emption: The Persistence of Excess Capacity and Monopoly in Growing Spatial Markets', *Economica* 46 (May): 182, pp. 149–58.

————, 1989: 'Product Differentiation' in SCHMALENSEE and WILLIG (Eds.), 1989, *Vol. 1*, pp. 723–68.

EATWELL, J., M. MILGATE, and P. NEWMAN, (Eds.), 1987: *The New Palgrave Dictionary of Economics*, 2 Vols. (London: Macmillan).

ETHIER, W.J., 1982: 'Decreasing Costs in International Trade and Frank Graham's Argument for Protection', *Econometrica* 50 (Sept.): 5, pp. 1243–68.

FOUCAULT, M., 1972: *The Archaeology of Knowledge* (London: Tavistock).

GALANTER, M., 1989: *Law and Society in Modern India* (Delhi: Oxford University Press).

GALBRAITH, J.K., 1952: *American Capitalism: The Concept of Countervailing Power* (Boston: Houghton–Mifflin).

GAMBETTA, D. (Ed.), 1988: *Trust: Making and Breaking Co-operative Relations* (Oxford: Blackwell).

GHOSH, D.N., 1992: 'Bank profitability and priority sector lending: From populism to impressionism', *Economic and Political Weekly* 27 (22 Feb.): 8, pp. 387–8.

GOPAL, S. (Ed.), 1991: *Anatomy of a Confrontation: The Babri Masjid–Ramjanambhoomi Issue* (New Delhi: Penguin Books (India)).

GOULD, S.J., 1977: *Ever since Darwin* (Harmondsworth: Penguin Books).

————, 1984: *Hen's Teeth and Horse's Toes* (New York: W.W. Norton).

GOVERNMENT OF INDIA, 1947: *Report of the Advisory Planning Board* (Delhi: Government of India).

————, 1990: *Approach to the Eighth Five Year Plan* (Delhi: Government of India).

GRAHAM, F.D., 1923: 'The Theory of International Values Re-examined', *Quarterly Journal of Economics* 38 (Nov.): 1, pp. 54–86.

GUHA NIYOGI, S., 1992: *Sangarsha O Nirmán** (Purnendu Basu and Shankar Sanyak, eds.) (Calcutta: Anustup).
* in Bengali.

HELPMAN, E., 1989: 'The Non-competitive Theory of International Trade and Trade Policy', *Proceedings of the World Bank Annual Conference on Development Economics* (Washington DC: International Bank of Reconstruction and Development), pp. 193–216.

INDIAN COUNCIL OF SOCIAL SCIENCE RESEARCH,* 1986: *Survey of Research in Sociology and Social Anthropology, 1969–1971, Vol. II* (New Delhi: Satvahan).
* ICSSR

JONES, R.W. and P.B. KENEN, 1984: *Handbook of International Economics*, Vol. 1 (Amsterdam: North Holland).

————, 1985: *Handbook of International Economics*, Vol. 2 (Amsterdam: North Holland).

JONES, S.R.H., 1987: 'Technology Transaction Costs and the Transition to Factory Production in the British Silk Industry, 1700–1870', *Journal of Economic History* 47 (March): 1, pp. 71–6.

KAHN, R., 1954: 'Some Notes on Liquidity Preference' [Manchester School] reprinted in KAHN, 1972, pp. 72–96.

————, 1972: *Selected Essays on Employment and Growth* (Cambridge: Cambridge University Press).

KALDOR, N., 1935: 'Market Imperfections and Excess Capacity', *Economica* 2 (Feb.): 5 (New series), pp. 35–50.

————, 1957: 'A Model of Economic Growth', *Economic Journal* 67 (Dec.): 4, pp. 591–624.

————, 1972: 'The Irrelevance of Equilibrium Economics', *Economic Journal* 82 (Dec.): 4, pp. 1237–55.

KANEDA, H., 1989: 'Rural Resource Mobility and Inter-sectoral Balance in Early Modern Growth' *in* WILLIAMSON and PANCHAMUKHI (Eds.), 1989, pp. 367–89.

KENNEDY, W.P., 1987: *Industrial Structure, Capital Markets and the Origins of British Industrial Decline* (Cambridge: Cambridge University Press).

KEYDER, C., 1990: 'Manufacturing in the Ottoman Empire and Turkey: The,

Course of Transition 1900–1950', a paper presented at the Conference on Manufacturing in the Ottoman Empire and Turkey (1500–1980), Fernand Braudel Center, Binghampton: State University of New York, (mimeo).

KEYNES, J.M., 1936 *The General Theory of Employment, Interest and Money* (London: Macmillan).

KOSAMBI, D.D., 1946: 'The Bourgeoisie Comes of Age in India' (reprinted) in KOSAMBI, 1977, pp. 10–18.

———, 1977: *Exasperating Essays (Exercises in the Dialectical Method)* (Calcutta: India Book Exchange).

KRISHNA, K., 1990: 'The Case of the Vanishing Revenues: Auction Quotas with Monopoly', *American Economic Review* 80 (Sept.): 4, pp. 828–36.

KRISHNA, R., 1963: 'Farm supply response in India–Pakistan: A Case Study of the Punjab Region', *Economic Journal* 73 (Sept.): 3, pp. 477–87.

KRUGMAN, P., 1979: 'A Model of Innovation, Technology Transfer, and the World Distribution of Income', *Journal of Political Economy* 87 (April): 2, pp. 253–66.

———, 1989: 'Industrial organisation and international trade', in SCHMALENSEE and WILLIG (Eds.), 1989, Vol. 2, pp. 1179–1223.

KRUGMAN, P. and E. HELPMAN, 1985: *Market Structure and Foreign Trade* (Cambridge, MA: MIT Press).

———, 1989: *Trade Policy and Market Structure* (Cambridge, MA: MIT Press).

LIMQUECO, P., (Ed.), 1989: *Partisan Scholarship: Essays in Honour of Renato Constantino* (Manila: Journal of Contemporary Asia Publishers).

LORENZ, E.H., 1988: 'Neither friends nor strangers: informal networks of subcontracting in French industry' *in* GAMBETTA (Ed.), 1988, pp. 194–210.

LYNN, L., 1984: 'Japan Adopts a New Technology: The Roles of Governments, Trading Firms and Suppliers', *Columbia Journal of World Business* 19 (Winter): 4, pp. 39–45.

MALAVIYA, H.D., 1954: *Land Reforms in India* (New Delhi: All India Congress Committee).

MCNEILL, W.H., 1983: *The Pursuit of Power* (Oxford: Blackwell).

MARKOVITS, C., 1985: *Indian Business and Nationalist Politics 1931–39: The Indian Capitalist Class and the Rise of the Congress Party* (Cambridge: Cambridge University Press).

MENON, U., 1992: 'A licence to plunder?', *Frontline* (Madras), 9 (15–26 Feb.): 4, pp. 26–9.

MITRA, A., 1967: 'Class Relations and Growth of Output' *in* SINGH and SINGH (Eds.), 1967, pp. 451–8.

MODY, A., 1990: 'Institutions and Dynamic Comparative Advantage: The Electronics Industry in South Korea and Taiwan', *Cambridge Journal of Economics* 14 (Sept.): 3, pp. 291–314.

MOHANTY, M., 1989: 'Changing Terms of Political Discourse: A Poser', *Economic and Political Weekly* 24 (16 Sept.): 37, pp. 2069–72.

MORITA, A., E.M. REINGOLD, and M. SHIMOMURA, 1987: *Made in Japan: Akio Morita and Sony* (London: Fontana/Collins).

Morley, S.A., 1982: *Labour Markets and Inequitable Growth: The Case of*

Authoritarian Capitalism in Brazil (Cambridge: Cambridge University Press).

MORRIS, D.J., P.J.N. SINCLAIR, M.D.E. SLATER, and J.S. VICKERS (Eds.), 1986: *Strategic Behaviour and Industrial Competition* (Oxford: Clarendon Press).

NARAYANA, D., 1992: 'Directed credit programme: a critique of [the] Narasimham Committee Report', *Economic and Political Weekly* 27 (8 Feb.): 6, pp. 255–8.

NURKSE, R., 1944: *International Currency Experience* (Geneva: League of Nations).

ODAKA, K., K. ONO, and F. ADACHI, 1988: *The Automobile Industry in Japan: A Study of Ancillary Firm Development* (Tokyo: Kinokuniya).

OHKAWA, K and H. ROSOVSKY, 1973: *Japanese Economic Growth: Trend Acceleration in the Twentieth Century* (Stanford: Stanford University Press).

OKIMOTO, D.I., T. SUGANO, and F.B. WEINSTEIN, 1984: *Competititve Edge: The Semi-conductor Industry in the US and Japan* (Stanford: Stanford University Press).

PANDEY, G. 'In Defence of the Fragment: Writing About Hindu–Muslim Riots in India Today', *Economic and Political Weekly* 26 (March): 11–12, pp. 559–72.

PANIGRAHI, D.N., (Ed.), 1985: *Economy, Society and Politics in Modern India* (New Delhi: Vikas).

PATNAIK, P., 1988: *Time, Inflation and Growth: Some Macroeconomic Themes in an Indian Perspective* (Calcutta: Orient Longman).

PERRAKIS, S. and G. WARSKETT, 1986: 'Uncertainty, Economies of Scale and Barrier to Entry', *in* MORRIS *et al.* (Eds.), 1986, pp. 58–74.

PIORE, M.J. and C.F. SABEL, 1984: *The Second Industrial Divide: Possibilities for Prosperity* (New York: Basic Books).

PRASAD, P.K., 1989: *Lopsided Growth: Political Economy of Indian Development* (Bombay: Oxford University Press).

RAKSHIT, M.K. (Ed.), 1989: *Studies in the Macro-economics of Developing Countries* (Calcutta: Oxford University Press).

ROMER, P., 1986: 'Increasing Returns and Long-Run Growth', in *Journal of Political Economy* 94 (Oct.): 5, pp. 1002–37.

RUBINSTEIN, W.D., 1983: 'The End of "Old Corruption" in Britain, 1780–1860', *Past and Present* (Nov.): 101, pp. 55–85.

SACHS, J. (Ed.), 1989: *Developing Country Debt and the World Economy* (Chicago: University of Chicago Press).

SAGHAFI, M.M., and C.S. DAVIDSON, 1989: 'The New Age of Global Competition in the Semi-conductor Industry: Enter the Dragon', *Columbia Journal of World Business* 24 (Winter): 4, pp. 60–70.

SAMADDAR, R., 1991: 'Sampradáyikata-ke Bojba Ar Rokha',* *Anustup (Shabingsha Varsha*, 26th year) (*Sharadiya*/Autumn): 1.
* in Bengali

SCHMALENSEE, R. and R. WILLIG, (Eds.), 1989: *Handbook of Industrial Organisation*, 2 Vols. (Amsterdam: North-Holland).

SEABROOK, H., 1992: 'The re-conquest of India: the Victory of International Monetary Fundamentalism', *Race & Class* 34 (July–Sept.): 1, pp. 1–16.

SEN, A., 1977: 'Rational Fools: A Critique of the Behavioural Foundations of

Economic Theory', *Philosophy and Public Affairs* 6: 4, pp. 317–44 (reprinted in Sen, 1982, pp. 84–106).

———, 1982: *Choice, Welfare and Measurement* (Oxford: Blackwell).

———, 1992: 'Dunkel's design', *Frontline* (Madras) 9 (15–26 Feb.): 4, pp. 23–5.

Sen, S. and R.U. Das, 1992: 'Import liberalisation as a tool of economic policy in India since [the] mid-eighties', *Economic and Political Weekly* 27 (21 March): 12, pp. 585–94.

Simon, H.A., 1982: *Models of Bounded Rationality*, Vols. 1 and 2 (Cambridge, Massachusetts: MIT Press).

Singh, B. and V.B. Singh, (Eds.), 1967: *Social and Economic Change: Essays in Honour of D.P. Mukerji* (Bombay: Allied).

Srinivasan, T.N., 1989: 'Comment' on Helpman 1989, *Proceedings of the World Bank Annual Conference on Development Economics* (Washington DC: International Bank of Reconstruction and Development), pp. 217–21.

Sudarshan, R., 1983: 'Stability of the Constitutional Order and President's Rule in India: A Political Perspective' (a paper presented at the seminar on South Asian Polity and Economy), University of East Anglia, Norwich (mimeo.).

Taylor, L., 1991: 'Economic openness: Problems to the century's end', in Banuri (Ed.), 1991, pp. 100–47.

Thapar, R., 1991: 'A Historical Perspective on the story of Rama', *in* Gopal (Ed.), 1991, pp. 141–63.

Thorner, A., 1981: 'Nehru, Albert Mayer and Origins of Community Projects', *Economic and Political Weekly* 16 (24 Jan.): 4, pp. 117–20.

Thorner, D., 1976:* *The Agrarian Prospect in India* (New Delhi: Allied).
 * Second edition.

de Tocqueville, A., 1964*: *Democracy in America*, 2 Vols. (12th edn.) (Transl. and Ed. Phillips Bradley) (Bombay: Popular Prakashan).
 * Originally published in 1848.

United Nations, 1947: *Economic Development in Selected Countries: Plans, Programmes and Agencies* (New York: Department of Economic and Social Affairs, United Nations).

Van Olferen, K., 1990: *The Enigma of Japanese Power* (New York: Vintage).

Visvesvarayya, M., 1920: *Reconstructing India* (London: P. S. King).

———, 1934: *A Planned Economy for India* (Bangalore: Bangalore Press).

Williamson, J.G., 1989: 'The constraints on industrialisation: some lessons from the first industrial revolution', in Williamson and Panchamukhi (Eds.), 1989, pp. 85–105.

Williamson, J.G. and V.R. Panchamukhi, (Eds.), 1989: *The Balance Between Industry and Agriculture in Economic Development* (New York: St. Martin's Press).

Economic Planning in India: • 2

PRAMIT CHAUDHURI

Writing about Christianity, Gibbon reminds us that:

> The theologian may indulge the pleasing task of describing Religion as she descended from Heaven, arrayed in her native purity. A more melancholy duty is imposed on the historian. He must discover the inevitable mixture of error and corruption which she contracted in a long residence upon earth, among a weak and degenerate race of beings. [Gibbon 1960: 143.]

An economist attempting an assessment of economic planning in India would do well to bear this injunction in mind. Much of the illusion of power and control over resources, and much of the disillusion amongst the critics of economic planning in India, spring from an idealized view of the planning process which is to a degree removed from the reality. Part of this chapter will emphasize some of the strategic and institutional limitations of economic planning in India which may account for its limited success. I do not, however, aim at a full and detailed evaluation of economic planning in India, a kind of 'Whither Indian Planning?' exercise, periodically so popular. What follows is a somewhat basic assessment of the nature of Indian planning, emphasizing some of its particularities and limitations. I wish to stress that an awareness of these is necessary in order to discuss what role planning has played or might continue to play in shaping economic activities in India.

The political context of economic planning in India

There are two elements underlying my approach. One is that I have defined the term 'planning' narrowly, in the sense of the economic strategy embodied in the Plans, which assigned a particular role to the public sector in productive activities, which in turn expressed itself in particular levels and configurations of development expenditure. This is in contrast to the use of the term that makes planning serve as a shorthand for economic policy generally and which consequently attributes all successes and failures of economic policy to planning and the planners. If we are to think about the experience of planning as something specific to economic policy in India and a small number of developing countries, then the distinction I have imposed is essential. I wish to emphasize that the domain of planning policy was circumscribed in a particular way in India for reasons that are partly historical and partly political, which in turn accounts for much of its failure to achieve some of the major objectives. The experience reflects the uniqueness of the Indian situation and I have not sought to compare the Indian planning experience with that of other socialist countries.

Economic planning as a process of allocation of resources interacts with some of the social and economic aspects of society, in this case Indian society, and may modify or be modified by them in an interactive process. Economists may take these aspects to be structural parameters within which planning has to operate. However, it would be unrealistic to assume in the present context that it was not the purpose of planning to shift at least some of these parameters to achieve some of the major objectives. That the poverty of India was in a large measure due to what would today be called the structural rigidities imposed by colonial rule was very much a part of political consciousness. These may have varied from the lack of an industrial base to the exploitative and stagnant nature of agrarian society. It is therefore important to discuss the origin of economic planning in India, as much as its objectives, its strategy, and tactics, and the institutional framework within which it operated (Hanson, 1966).

As is well-known, the origins of planning in India can be traced back to Visveswarayya, the National Planning Committee (NPC), and the Bombay Plan (Chattopadhyay, 1985; Chaudhuri, 1988; Thakurdas *et al.*, 1944). It is useful, in some ways, to begin with the NPC, partly because the basic strategy is very clearly enunciated there, and partly because Nehru was amongst those who provided the thrust for both pre- and post- Independence activities relating to planning. The main elements of the NPC approach can be briefly stated (Chaudhuri, 1988). Colonial rule embodied the spirit both of imperialism and capitalism, imposing simultaneously political and economic subjugation. India's endemic poverty was its legacy. It was not possible to get rid of poverty without getting rid of colonial rule. Economic and political independence has to go (or, rather, come) together or not at all:

> You seem to separate the three objectives—India's independence, the creation of a socialist state, and the solution of the problems of India's poverty and unemployment and you suggest that the last named should be tackled first. I am afraid that this whole conception of our struggle is wrong. If all of us in India devoted ourselves to fighting poverty under the present system—political and economic—we shall not get rid of it. If we could get rid of it the problem is a simple one and the need for *swaraj* is not very apparent. [Nehru, 1936.]

Hence, Swaraj would lead towards socialism and industrialization, industrialization towards 'self-reliance', weakening if not breaking a link in the chain of international capitalism. Moreover, industrialization had to take a particular form of concentration which came to be called 'heavy industry', which is more precisely termed 'core industries', or more romantically 'mother industries' by Subhas Chandra Bose and Jawaharlal Nehru at the time. The need to defend political freedom once gained also required such a strategy (Nayar, 1972).

This transformation of the economy in the industrial sector had to be pioneered by the state, through the creation of public sector industries of different degrees of control and authority. Indian capitalists, being capitalists, could not be expected to act in the national interest, as they would put private profit before social benefit. Within the agrarian economy, the allies of the colonial rulers were the *zamindar*s whose feudalistic exploitation drove the poor peasants and landless agricultural labourers into poverty and bondage. Land reform that would destroy the power of the landlords was essential to substantially reduce rural poverty and to ensure a tolerable quality of life. It was industrialization that, following the Western historical pattern, would ensure the growth of the economy and ease the pressure of population on land. We may note in passing that the latter required a high rate of labour absorption outside the agrarian economy. The views of the NPC were neither the dominant nor the consensual view of the Indian National Congress. However, they did substantially influence the formal approach to planning in India in many ways.

The Indian capitalists who were the authors of the Bombay Plan, were no doubt fearful of Nehru's socialism. However, many among them, like Tata and Thakurdas, would have been surprised to be told that their economic interests were identical to those of the British capitalists, or that they were 'compradors', or that they did not care about Indian poverty. They *shared* the view that industrialization of the Indian economy was essential to the alleviation of Indian poverty, and the Bombay Plan had things to say both about profit and poverty. The Plan discussed minimal nutritional norms that had to be achieved in order to reduce poverty in India, as well as the need to promote industrialization. It differed rather obviously from the NPC statements in not wanting a dominant public sector, but did not wish the Government to keep out of the industrialization process. The latter was to finance the provision of the necessary infrastructure, the expansion of industries such as steel which required massive investment, and implicitly if not explicitly offer them protection against foreign capital. The role of the state was to be the 'handmaiden' to Indian industry, bearing a large proportion of the costs and risks of development but taking much less of the surplus. Not so different some would say from the role that the Indian state came increasingly to play within the industrial sector, of promoting a form of state capitalism for the benefit of the industrial bourgeoisie. For these reasons, Indian business was quite happy to finance the Indian Congress (Markovits, 1985).

A few points need to be stressed about these antecedents to planning for development in India. Firstly, the newly independent Indian state was to have political as well as economic objectives, which were at least for some of the time seen to be interrelated. This is a point that I shall refer to below. Secondly, both the ruler-elects who distrusted Indian capitalists, and the latter, shared the common view that the principal centre of economic activity in an independent India was to be the process of industrialization and a common antipathy to foreign capital. They differed in their attitudes to the role that public sector industries were to play in the economy.

Thirdly, there was an asymmetry in the nature of the strategies for industrial growth and agricultural development. Indian capitalists required institutional reform of a radical kind, through land reform that was necessary to break the stranglehold of the landlords upon the peasants. These were pre-HYV days, and the problem of rural poverty, of below subsistence living, was seen very much as a problem of maldistribution of agricultural output rather than that of an impoverished productive base in agriculture that would require the investment of capital and skills to transform. The problem was one of exploitation rather than low productivity. It is difficult to see why colonial rule could not have been instrumental for both of these effects, a combination of ideological and fiscal imperatives helping to impoverish both the production base and the *ryots*. Leaving aside the development of irrigation in the Indo-Gangetic belt, there was little impetus in British India for investment in agriculture or to foster technological progress.

The tendency of nationalist leaders to ignore these aspects, while emphasizing the poverty of the peasants, was to have important effects for agricultural development policy in independent India.

Economic planning in India has endlessly engaged in discourse with the social scientists of our times. We must not infer from this, however, that it was equally a matter of obsession for the politicians who came to run the Indian state. Their chief concern was overwhelmingly political—the need to protect the 'nation' (i.e. the state, or by inference, the people) from real or perceived external and internal threats, to maintain internal stability and cohesion. This is a point made early on by Nayar, in the context of defence expenditure (Nayar, 1972). One is reminded of this in a recent book by Brass (1990). Economic matters absorb tiny proportions of even such significant economic pronouncements as Bose's Presidential address to the Haripura Congress, or Nehru's at Faizpur (Ghose, 1975).

In the statement on Fundamental Rights, 'Equality' is chiefly equality before the law, 'Freedom' relates to freedom of speech or religion—the emphasis being on abstract ethical concepts rather than more concrete rights in spheres of economic activity. There is, of course, the injunction against exploitation of labour, especially child labour below the age of 14 years, again hinting at the ethical rather than the economic significance of a repugnant practice. The emphasis on working through constitutional remedies, on an orderly progress towards social change as opposed to violent and radical transformation of society, is of some significance (Banerjee and Chatterji, 1957). I do not know whether the origins of these political attitudes are to be attributed to Gandhian influence within the Congress or to a legacy of the Partition and its violent associations; or whether they are a harbinger of socio-political alliances yet to be formed. What is very clear is that within a fairly brief period of time (between the mid '30s and the early '50s), policies such as the confiscation of *zamindari* land became politically quite out of the question. It is equally relevant that the first significant Constitutional amendment should relate to the implementation of a compensatory land redistribution programme.

The Directive Principles of the Constitution, an early statement of policy objectives, are more explicitly concerned with social and economic issues. They enunciate broad statements of intent such as a 'just social order for the promotion of welfare', but also quite specific commitments to particular goals. Some of these are of the 'Basic Needs' variety, such as a living wage and a decent standard of life, equal rights for men and women, equal pay for equal work, increased levels of nutrition and health, maternity leave, and the promotion of education and the economic interests of the Scheduled Castes (SCs) and Scheduled Tribes (STs). The one that was most noticeably honoured in the breach was the objective of achieving free and compulsory education for all children under the age of 14 by 1961.

The Directive Principles do not contain detailed, explicit statements relating to the formation of a socialist state. The two most significant principles in this context are that 'the ownership and control of material resources [should be] so distributed as to best serve the common good' and (to ensure) that 'the operation of the economic system does not result in concentration of wealth and the means of production to the common detriment'. In passing, we may note the dialectical tension between the traditional and the modern in the desire to organize agriculture and animal husbandry along modern scientific lines and to ban the slaughter of cows.

What I have suggested so far is that it is important for us to recognize that planning in India has its own history, which is a part of a larger process. When it was planning for development, the Government of India was also faced with major political objectives which put a premium on constitutional procedures and orderly processes of change. This meant that whatever was said, what was done—in actual fact—was to situate the strategy of planning within given structural and institutional parameters. This posed from the beginning a major problem for the agrarian sector.

In the absence of any major technological breakthrough, institutional reform of agrarian relations was central to the development of agriculture after a long period of stagnation. The problem of industrial progress was far more tractable and could be approached in a more technocratic fashion, through capital accumulation and modernization of techniques. Any conflict that there was lay between the Indian capitalists and the state in defining the role of the public sector *within* an industrialization strategy. Both were reluctant to become subservient to foreign capital. There was a recognition from the early days that reduction of mass poverty had to be a major objective of economic policy, although no very clear strategies were worked out beyond the need to let India proceed unhindered by colonial rule, along the historical Kuznets highway.

Two general comments can be made about the process of economic planning that was initiated in the '50s. Firstly, there was wide agreement within the country amongst different socio-economic classes that planned development, with the government playing a key role in productive activities, was necessary for India. Planning, like 'India's China War', was popular at the time, however much it came to be criticized later. This was emphasized early on by Lewis, in a sympathetic assessment of Indian economic policy (Lewis, 1962).

Two eminent economists, both with intimate experience of policy-making in India, have attested the need for planned development quite convincingly (Chakravarty, 1987; Patel, 1986). Mozoomdar (1989), in an article I will need to refer to again, also points to the presence of such a consensus, stating rightly that it no longer exists. Of course, there were dissident voices amongst the Gandhians in the Congress, or extreme free-marketeers (e.g. B. R. Shenoy). They were not influential, the ideological edge to their critical voices being sharper than the analytical, with the exception, perhaps, of the work of Brahmananda and Vakil (1956).

Internationally, strident and persistent criticism came largely from the United States, reaching a crescendo in the report of the Bell Mission preceding the devaluation of June 1966. Somewhat paradoxically, criticisms of Government policy by the World Bank and the International Monetary Fund (IMF) considerably quietened in later years, as dissatisfaction with planning within India grew harsher. In my view, the earlier widespread support for planning in India was entirely justified, taking two elements of such a strategy to be active intervention by the government in patterns of resource allocation and a substantial drive towards creating an industrial production base. However, problems remain, as we shall see.

The second point arises in the form of a qualification of the first. When we say that planning had popular support, whose support do we have in mind? Obviously, the Congress raj was highly popular at the time. But within the narrower context of strategy, we must mean the support voiced by a small part of the population, the educated, 'better off', articulate minority. In such a context, how do planners and politicians know what people want, what is the range of social choices that they desire, how much bread today is worth how much bread-and-butter tomorrow? How do they identify 'Basic Needs'? An unplanned market economy, given its constraints, often gives an unequivocal answer in terms of profits and losses, or queues, or butter mountains. It is not a question here of choice between a planned and a market economy; it is that a planned economy requires some mechanism for the transmission of certain kinds of information about needs and choices if it is not to go badly off course. Such a mechanism has to be provided by some politically initiated process of 'grass-roots' participation.

The main point to make about the political economy of planning in India is the failure of the planners to perceive such a need, let alone to make provisions to supply it. The reason for this lapse was absurdly simple, in a sense. It was assumed that a political party that had struggled to bring Independence to the masses was in some sense the embodiment of the general will, representing the welfare of the masses in actions it performed in their name. Even if honestly conceived, such an approach is not conducive to an identification and prioritization of the basic needs of a poor and illiterate population (Chatterjee, 1989). Since the process of resource allocation does not operate within a political vacuum, in the absence of any countervailing power it would to some degree reflect the needs of those groups better able to articulate their needs. Whatever the nature of such a bias might be—urban or *kulak* or *bania* bias—it became almost an institutional characteristic of the planning process from its very inception (Government of India, 1952: 7–8).

Objectives of economic planning

The formal objectives of Indian Development Plans have remained remarkably constant over the years, although the relative emphases and the means through which they were to be achieved have not. The three objectives that we need to identify in the present context are the poverty objective, the adoption of a 'socialist pattern' of development, and the aim of 'self-reliance'. Each of these need some elaboration.

The poverty objective is clearly stated in the First Five Year Plan:

> The central objective of planning in India at the present stage is to initiate a process of development which will raise living standards and open out to the people new opportunities for a richer and more varied life . . . The urge to economic and social change ... comes from the fact of poverty and of inequalities of income, wealth and opportunity [Government of India 1952: 7–8].

The document goes on immediately to say:

> The elimination of poverty cannot, obviously, be achieved merely by redistributing existing wealth. Nor can a programme aiming only at raising production remove existing inequalities. The two have to be considered together . . . [Government of India 1952: 8.]

Nehru expressed the same point of view in saying that in a poor country there was only poverty to redistribute. This is an early recognition of the 'growth *vs* poverty' trade-off. The underlying philosophy, that in the long run the elimination of poverty in less developed countries requires a sustained and satisfactory rate of economic growth as well as redistributive mechanisms to operate in society, is unexceptionable. What it leaves out of consideration are both the relative emphases to be put on growth and redistribution, and the policies that would attempt to reconcile the two. I shall turn later to a discussion of what these policies were.

We may note in passing that the growth–poverty objective was restated in both the Second and the Third Five Year Plans but with the emphasis shifting towards the need for economic growth first in order to reduce both unemployment and poverty. Moreover, the growth strategy was far more clearly outlined in the Plans than was the policy towards redistribution. Indeed, Chapter II of the Second Five Year Plan remains one of the best statements of the so-called Mahalanobis strategy (Government of India, 1956). On the other hand, the First Five Year plan lays much greater emphasis on the need for institutional reforms, especially in agriculture, than does the Second (Chatterjee and Sen, 1988).

An explicit strategy for a reduction in poverty was first worked out in the Fifth Five Year Plan in a way that was quite consistent with the language of planning that the Plans have always adopted, where alternate strategies are worked out in terms of alternative patterns of investment (Chaudhuri, 1988: 278). The chosen solution to poverty eradication was a technical one, that would operate through the manipulation of market forces rather than through institutional or political reform aiming at shifting the parameters within which such forces operate. Some of these ideas were contained in a seminal paper *on the poverty* implications of planning, which was written as early as 1962 (Pant, 1962).

The other two objectives, a 'socialist pattern' and 'self-reliance' are both highly ambiguous. Much has been written on the need for a scientific socialism; the need for a flexible rather than a dogmatic interpretation of the concept to suit the Indian social climate. This approach, which is no less legitimate than the later controversies on what 'feudalism' or 'capitalism' *meant* in the Indian context, suffers by possessing only negative attributes, in that it does not tell us what a socialist approach might mean in policy terms when applied to India. It is useful to do that.

A socialist approach to development, or even the allocation of resources in general, might be exemplified by a 'command economy', where the allocation of resources is directly determined by administrative fiat. This could not, obviously, be the interpretation that could be put upon the term in a 'mixed economy', although some ill-informed criticisms of Indian planning and its *dirigiste* tendencies have been made as though this were the case. In a mixed economy, socialist planners could operate according to Lange–Lerner rules, that is to say, simulate a set of socially optimal price–cost relationships to direct the allocation of resources. 'Assume a virtue, if you have it not', is the

Lange's back-handed compliment to the market economy. Although social cost–benefit approaches to investment allocation did become a part of the planners' tools in India, it certainly did not play a strategic part in Indian planning. We are, then, left with three possible interpretations of the term (Brus, 1972).

One is to say that socialism is about a concern for establishing a more equitable distribution of wealth and income than that existing at the time. However, such a concern must exist at the level of formulation and implementation of policies. We have already seen that at least in one sector of the economy, the all-important agricultural sector, land redistribution *without compensation* was ruled out at a very early stage. Insofar as land reform was implemented as policy, it fairly clearly did not have the effect of reducing rural inequality. It is clear that a concern for employment-creation as such, cannot itself constitute a form of socialism, less so if labour-intensive technologies are in some sense optimal in any case. Even if the Indian planners had gone out to promote employment-generating policies, which they have not, that would not have constituted a form of socialism. In this sense, the more recent employment-generating programmes in the rural sectors, the many variants of IRDP, are in essence 'socialist', although they are not today emphasized in that sense. Socialism has lost its glitter.

The other possible interpretation that could be put upon the term relates to the attempt to 'capture the commanding heights' of the economy, which is hinted at in the Directive Principles, and was later incorporated in the two Industrial Policy Resolutions in a modified form. The approach of the Industrial Policy Resolutions (IPR, 1948, 1956) was foreshadowed in the NPC, when it attempted to classify industries into three categories, according to the degree of government control over them. We should note, however, that even for the Schedule A industries, it was the future developments that were to be the exclusive responsibility of the state (Government of India, 1956: 45). In any case, this point will have to be taken up again in our discussion of the relative roles of the public and the private sectors in the economy.

This leaves us with the last, a highly instrumental interpretation of socialism in the Indian context, which relates to the fiscal stance of the government, in terms of its tax policy, and levels and patterns of welfare expenditure. Indian tax policy has never been notably progressive, and with the exception of famine and drought relief, and the IRDP programmes referred to earlier, public expenditure is not significantly welfare-oriented. Unlike Sri Lanka, India has not had its rice rations, although it has had its Fair Price Shops.

'Self-reliance' is also a somewhat ambiguous concept, and can be accorded either a High or a Low cultural interpretation. It is given the former interpretation in the pre-Independence discussions, when the desire to break away from imperialist–capitalist domination meant the loosening of ties with the international economy. As the following quotation shows, the capitalist international economy stood for exploitation of nation by nation:

> ... the basic principle of Planned Economy for India ... [is] to meet primarily its own demands and not planned deliberately with a view to competing in foreign markets ... this in order to avoid all economic welfare or imperialism ... [AICC Papers, 1939, Pt. 3, Ch. 4, quoted in Chaudhuri, 1988: 278].

The Gandhians had, with similar logic, hit upon the idea that machines exploited labour. Industrialization, based on the machine, was anaethema to them. However, this ideological fissure could be bridged, for a benevolent socialist state could be held to internalize the process of creation of surplus value through exploitation—the people, so to speak, exploiting themselves for their own good, with friendly machines through the instrumentality of the state.

The Low reading of this objective puts it in terms of a balance of payments constraint, which requires the planners to maintain some consistency between the growth of output that is planned for, the demand for foreign exchange that it generates, and the supply of it that can be made available through exports or foreign capital inflows. One might debate at length whether or not the import substitution policies pursued by India inevitably tightened the foreign exchange constraint.

I would like to stress here that what had begun as part of a social philosophy, ended up as a form of economic necessity. One area of activity where one might have expected a government bent on transforming the economy through its own efforts to put particular emphasis was that of developing an independent technological capacity. With the exception, perhaps, of some defence activities, the government has done little to encourage 'self-reliance' in technology; indeed one might say that in some instances it has actually hindered such a development.

I have so far argued that amongst the many objectives of Indian planning, the one that could be expressed most unambiguously was that of promoting economic growth in order to raise average per capita income. There was also a clearly delineated strategy for pursuing that objective, whether one thought it to be correct or erroneous. This was industrialization along the Mahalanobis path. The others, important though they were perceived to be, reducing poverty and inequality, promoting a 'socialist pattern of development' and 'self-reliance', were ambiguous in conception and ill-defined in terms of their means of achievement.

Institutional constraints on planning

Whatever the objectives, how were they to be pursued through the planning process? Some critics of Indian economic policy paint a picture of the Indian state and its planning activities as all-embracing and powerful (Bauer, 1971; Lal, 1983). This is highly misleading, indeed a slightly comic tilting at windmills. A more balanced approach requires us to maintain at least two sets of distinctions between planning and the planners, on the one hand, and alternate sets of decisions and decision-makers, on the other. For the first, we need to distinguish between economic planning proper, as embodied in the planning process in the narrow sense, and general policies of economic management pursued by the government. As things worked out for the Indian economy, the latter came to dominate the planning process, leading at times to its abandonment.

The operation of a balance of payments constraint, or of a wage goods constraint, or the conflicting demand from competing groups for limited resources, gave rise to different forms of instability from time to time, which had to be responded to. This need, in turn, often led to a shift of power over the allocation and utilization of resources away from the Planning Commission to other organs of government, say, the Ministry of Finance. It would be foolish to insist that these instabilities were wholly caused by planning, although particular forms and degrees of destabilizing pressures within the Indian economy obviously resulted from the particular strategies and policies followed. For example, the Mahalanobis strategy might have tightened the balance of payments constraint, especially in the early days, whilst an excessive reliance on direct, non-price controls might well have made some of the problems more difficult to handle.

Mozoomdar (1989) has recently drawn attention to the fact that economic policy was made through the participation of a number of governmental organizations in the process, including the Planning Commission. These were, for example, the Ministry of Finance, or Commerce and Industry, or the Reserve Bank of India, the first being of particular importance. In addition, heavy reliance on direct methods of control, through an elaborate licencing procedure, led to the setting up of an interlocking committee structure, involving the various ministries and the Directorate-General of Technology and Development (DGTD).

There are two points to be made here. One is that some problems of economic management have to be faced by all types of economic system, planned or unplanned, while some are specific to the former (e.g. the 'permit raj' was contingent upon the desire to use direct controls rather than the price mechanism). The other is that the relative autonomy and power over the allocation of economic resources was contingent partly upon the personalities involved—Nehru warmed towards planning, Mrs Gandhi was cool towards it—and partly on the relative strength of the forces for growth and for stability.

The persistence and the magnitude of foreign exchange scarcity in particular strengthened the hands of the Ministry of Finance over the allocation of a key resource and over budgetary policy. In recent years, power has been increasingly concentrated within the Prime Minister's secretariat, away from the formal ministerial structure of government. Of course, one should not carry these distinctions too far; the inclusion or exclusion of particular projects from the Plans was obviously a matter of great significance.

The other distinction that I think it is important to maintain is that between economic policies that are part of an economic plan and those that are of a more general nature, which are not necessarily of any lesser significance. I have earlier drawn attention to the fact that there was, from the beginning, an asymmetry between agricultural policy and industrial policy. In theory, if not in practice, institutional reform of agrarian tenurial relationships was a fundamental part of the strategy. This was to be, let us say, the keystone of the arch through which the rural poor were to be led out of poverty. It was particularly important before any technological breakthrough in agriculture, and I believe that it is still of fundamental significance.

However, policies towards land reform were never part of any Plan *strategy*, in spite of the pages devoted to it. Indeed, as the Constitution stood, it was not a sphere of activity for the central government, although it is difficult to believe that a government that took the matter seriously could not have made a better show of progress. Indeed, the Plans never provided for alternative sets of policies that would be contingent upon a failure to implement land reform. In a somewhat similar fashion, IRDP programmes, which are now seen as the core of the government's poverty-eradication policy, are called special programmes, running on a parallel axis. Gadgil once described the process as planning without a policy framework, where the set of policies that would ensure the fulfilment of the Plan objectives were not clearly specified (Bagchi, 1970).

Impact of agricultural policy on planning

A great deal has been written about the government's agricultural policy, which it is unnecessary to repeat here. I do not think that the broad outlines of the story are controversial. A line can be drawn roughly around the years 1965–67, marking the 'Green Revolution' watershed. In the earlier period, it can be argued that Indian agricultural production was bounded by a technological constraint, which made it difficult to benefit the rural agricultural poor, either as consumers or producers of foodgrains, by increasing total output. One possibility lay in extensive investment in irrigation but, given the way in which water for irrigation tends to be shared out in India, it is unlikely that it would have benefited the poor as a class, although it would no doubt have brought prosperity to many backward regions.

The principal avenue through which the poor could be led to a better existence lay in increasing their share of agricultural assets and output. The failure of land reform either as land redistribution or as improvement in tenurial condition, as recorded in the Ladejinsky Report, attests to the fact that the government policies were not aimed at benefiting the poor in this period, but the more substantial occupancy tenants turned landowners. The issues that still need research are the timing and the manner in which the pro-peasant attitudes of the Indian National Congress (INC) were abandoned in a relatively short period, say between 1949 and 1952.

The possibilities opened up by the High Yield Variety (HYV)-technology changed agricultural production into something better than a zero-sum game, even within the given institutional constraints. Few would now deny that the relative distribution of the gains of new technology has been biased, through a conscious design of policy towards the higher income groups within the agrarian economy. Policy in this period also has not been pro-poor, with perhaps fewer justifications. What has been argued about is the identity and class affiliations of these higher income groups. Whether they are the so-called *kulaks* or rich peasants, or bullock-capitalists, the middle peasants (Byres, 1981; Rudolph and Rudolph, 1987), it seems to me unlikely that government policy has from the beginning been consciously designed to benefit large sections of the rural population, and that the steady diffusion of the new technology, and its associated benefits, downwards into the middle income strata simply demonstrates the inherent dynamism of capitalist development in Indian agriculture, which is no respector of either government manipulations or the rigid class categories of the researcher. Such an interpretation is quite consistent both with the empirical evidence and the central thrust of classical theory, from Adam Smith to Karl Marx. The truth is that the poor have been incidental beneficiaries of agricultural growth in the latter period, as consumers of an increased foodgrains supply.

Both the necessity and the failure of land reform arose out of an existing pattern of distribution of ownership and user rights in existing forms of wealth, i.e. agricultural land. Hence, the conflict of interest expressed by that distribution had to be faced. In the industrial sector, the problem was less acute, for new sources of accumulation could be directed towards the creation of new forms of wealth, through a process of industrialization. In the industrial sector, it appeared to be much less of a zero sum game that had to be played out, the need for prior institutional reform seemed to be less, and the possibility of avoiding a clash of economic interest between different groups greater. The creation of an expanding public sector could be seen as an ongoing process of institutional change, any conflict of interest being resolved through the delineation of the relative spheres of activity of the public and the private sector. Once again, the problem of matching the objectives of the planners with those of different groups of society was not explicitly acknowledged.

Planning in practice

Here, the discrepancy between what was planned and what was achieved came to be of critical importance. Indian planning is neither of the command economy kind, with the state directly setting detailed output targets for industries that have to be delivered, nor of the kind termed indicative planning. The core of Indian planning strategy has been the planning of investment expenditure—of laying down planned investment targets, especially for the public sector. The investment in turn should add to and remould the stock of industrial capital in the economy. The creation and distribution of new forms of wealth in the industrial sector depends, on the one hand, on the rate of accumulation of capital (i.e. the rate of savings or investment in the economy) and, on the other, on the relative shares in it of the public and the private sector. However, the accumulation of any surplus depends on the actual, and not the planned rates of overall savings and investment, and its sectoral distribution between the public and private sector.

The particular problems that this posed for a mixed economy with inadequate command over the allocation and end use of resources were identified, among others, by Bagchi (1970). Where there is overall excess demand for resources, the inability of the government to control the amount of resources that are utilized by the private sector, leads to an equivalent shortfall in the public sector.[1] When that happens, two consequences follow. In so far as the resources are used for consumption, there is a presumption, in the Indian context, that the share in consumption of the higher income groups will rise in

[1] Equally, the public sector may sometimes 'squeeze out' the private sector.

relation to that of the poor. In so far as the resources are used for investment, the anticipated institutional change in the industrial sector, the longed for 'socialist pattern of society' will not be achieved. This is what the economists, who hold that the Indian Plans have simply served the interests of the industrial bourgeoisie, have in mind. There is clearly much to be said for such a theory.

However, the point needs some elaboration. Firstly, if some key investment projects are designated for the public sector because they are seen to be of strategic importance, then a failure to meet public sector investment targets carries with it some presumption of strategic failure. Secondly, it is possible that the public sector will remain a net borrower of savings from the private sector, while at the same time failing to meet its investment targets. This is a consequence of low productivity and inefficient resource use within the public sector, for which it is a little hard to hold the class struggle responsible.[2] Thirdly, where the two sectors buy and sell each other's outputs, one may end up by taxing or subsidizing the other by charging too high or too low prices. Examples of this are artificially low prices for power and other services, subsidized fertilizers, or rural credit which are appropriated by large farmers, etc.

Fourthly, we must not leave out of account the government's anti-monopoly stance, which obviously operates against the interests of the Indian capitalists. The effectiveness of the government's anti-monopoly policy is a matter of some debate. It is perhaps a fair comment to say that it discourages profit-making behaviour more successfully than it does 'rent-seeking' activities. However, one might say that this is to a large extent offset by the government's policies for managing the balance of payments, which offers a highly protected market to both public and private sector enterprises. Fifthly, there are vested interests within the public sector, whose interests are not identical to either the state's or the private capitalist's, the so-called intermediate classes (Bardhan, 1984).

Lastly, where the economy is not 'at full employment', and output and profits are in part influenced by demand, it is clear that a higher level of activity in one sector will stimulate the other. A thriving public sector will necessarily stimulate activity and profits in the private sector. This need not be at all undesirable, especially if the government is willing or able to have a sensible policy towards corporate taxation. Equally, expanding the public sector by absorbing loss-making private firms is a strange way of creating a 'socialist pattern'. Today, the size of the public sector, measured in terms of share of

[2] It may, of course, be the case that the public sector investment projects are particularly risky or have particularly high capital–output ratios, which may be the reason why they have to be located in the public sector.

output in total GDP or of the capital stock, is very much larger than it has ever been. However, its strategic importance is less, the proportion of non-development expenditure and salaries and wages having steadily increased over the years in relation to its development component.

Who has benefited from planning?

In conclusion, let me deal with a broader issue which may help to clarify some assumptions that are implicit in my approach. How was planning meant to benefit, not in terms of officially designated target groups, but through the likely outcome of the set of policies embodied or implicit in the Plans. We cannot attempt to deal with the question without taking account of the methods of control that were used by the government to manage the economy, especially the non-price controls over foreign trade and private investment, the major components of the 'permit raj'.

In the historiography of Indian planning, the answer has tended to be that it was meant to benefit the industrial capitalists and the rich farmers. So far as the latter are concerned, I have already argued that they were the main beneficiaries of agricultural policy, both before and after the 'Green Revolution', although there has been a spontaneous tendency for the benefits of technical progress to be diffused downwards to the rural middle income groups.

The issue appears to be less clear when we consider the interests of the industrial capitalists, for the government's policy stance towards them was ambiguous. Some aspects of it were clearly pro-capitalist. I have already drawn attention to the inability of the planners to maintain the real share of resources for fulfilling investment targets within the public sector. The same point is underlined at the micro-level by the diversion of controlled goods, such as steel or cement, in various periods to non-priority uses. The same can also be said of the import-substitution policy, which guaranteed a large protected market to many relatively inefficient producers.

Another point which has received much less attention than it deserves is the very radical change in the attitudes of the government towards the industrial labour force over the years, beginning early on in the post-Independence period, and followed through in the encouragement of the growth of a pro-Congress trade union (Indian National Trade Union Congress or INTUC) leadership. On the other hand, a very wide range of bureaucratic controls, anti-monopoloy legislation, a certain lack of enthusiasm for expansion of capacity, and especially the government's lack of concern for the development of a viable indigenous technological capacity does not appear to be an obvious sign of a pro-capitalist bias.

I do not have the space here to deal with the hypothesis that all classes in India have suffered relatively through the growth of an 'intermediate class'. While it can be empirically attested that there has been a very substantial growth of public sector bureaucracy and its power, it is arguably a consequence of abortive capitalist development in India, as much as the cause of the abortion itself. A curious footnote to the history of planning in India is that while it is common practice in West European economies to recruit high level management from the private sector into public sector enterprises, both under socialist and non-socialist governments, this has never been the practice in India, where lack of efficient management has been a major problem in the public sector.

Perhaps the major indictment of planning in India is that it has failed to benefit the poor. Here it is important to draw a distinction between low income groups and the poor. For our purpose, such a distinction is not best drawn solely on the basis of income or expenditure per head or some nutritional intake norm, although such characteristics will often overlap the one I would like to propose. The distinction needs to be based on a structural characteristic which relates to the extent of the marginalization of the poor in the production structure. Such marginalization may come about through a number of channels, such as health and/or demographic status. The endogenous causes are lack of access to land for small farmers and lack of adequate employment opportunities for agricultural labourers.

Over the years, the realization that the poor were not sharing the benefits of economic growth has led to the initiation of a number of special programmes, referred to collectively as the Integrated Rural Development Programme (IRDP). It does not seem to me to be helpful to register blanket condemnation of them. While the experience of these programmes has certainly been very uneven, there is considerable evidence that in some States relatively low income groups have derived substantial benefits from participation in them over their duration.

It is however equally true that the poor, in the sense of the lowest income strata in rural society, who are not integrated into the normal pattern of productive activity either because they are 'dependants' or have nothing but unskilled labour to sell, have not. It is unrealistic to expect that, under Indian conditions, a substantial improvement in their condition will not require an element of social subsidy, which may have to be sustained over some length of time. There is no convincing evidence that, if the category is narrowly defined so as not to include all low income groups (i.e. excluding those who have some stock of skills and/or assets which they can be encouraged to use more productively, but including a significant proportion of the landless), the fiscal burden imposed by such subsidies will be unacceptably high, especially as much of it can be a form of transfer of subsidies from groups with much higher standards of living in both the public and the private sectors. It is not a trade-off that any poverty-conscious society ought to reject offhand on the grounds that it will imply the acceptance of a lower rate of growth.

It is sometimes argued that there are political forces operating within society which tend to increase the leverage of the poor within a democracy. There is obviously some truth in this when it comes to dealing with acute subsistence crises, such as famines, than to what has come to be termed 'food insecurity'. In a broader context, the argument applies more strongly with respect to what I have called the low income groups, rather than to the poor. The experience of 'Operation Barga' in West Bengal, or the experience of the IRDP, or the promise of job reservation in the public sector seems to me to underline the argument rather than refute it.

Whether there is a future for planning in India cannot be answered without reference to a particular view of its past. In some ways, planning in India has not been a tidy, complete exercise in resource allocation and management. Its objectives have often been unclear, its policies partial in extent and in execution, its instruments inexact. There are major areas of policy which have lain outside the planning exercise. Both in agriculture and in industry, the benefits of economic growth have accrued to those with control over land and industrial capital, at the expense of the rural and the urban poor.

However, many of these consequences arose out of the existing structure of society, the nature of the problems it faced, and the new range of problems that its development efforts generated. Many of these would have had to be faced in an unplanned market economy, and there can be no presumption, either in facts or in theory, that a liberalized free enterprise economy will deliver faster or better results in India (Chakravarty, 1987). Nevertheless, it remains of key importance in India that high and productive investment must be directed towards particular sectors and activities, which imply the husbanding of scarce resources from other uses. Examples might be investment in irrigation and soil conservation, afforestation, power generation, health, literacy and water supply. It is difficult to argue that these will come about without active and intelligent government intervention, or that the poor will be better off in its absence. It is a matter of secondary consequence whether the range of policy initiatives fit into a theoretical construct called economic planning, or not.

REFERENCES

BAGCHI, A.K., 1970: 'Long-term Constraints to India's Industrial Growth 1951–1968' in ROBINSON and KIDRON (Eds.), pp. 170–98.

———— (Ed.), 1988: *Economy, Society and Polity: Essays in the Political Economy of Indian Planning* (Calcutta: Oxford University Press).

BARDHAN, P., 1984: *The Political Economy of Development in India* (Oxford: Basil Blackwell).

BANERJEE, A.K., and K.L. CHATTERJI, 1957: *A Survey of the Indian Constitution* (Calcutta: A. Mukherjee).

BASU, D.D., 1952: *Commentary on the Constitution of India* (Calcutta: Sarkar).

BAUER, P., 1971: *Dissent on Development* (London: Weidenfeld).

BRAHMANANDA, P.R. and C.N. VAKIL, 1956: *Planning for an Expanding Economy* (Bombay: Vora).

BRASS, P., 1990: *The Politics of India since Independence* (Cambridge: Cambridge University Press).

BRUS, W., 1972: *The Market in Socialist Society* (London: Routledge & Kegan Paul).

BYRES, T.J., 1981: 'The New Technology, Class Formation and Class Action in the Indian Countryside', *Journal of Peasant Studies* 8 (July): 4, pp. 405–54.

CHAKRAVARTY, S., 1987: *Development Planning: The Indian Experience* (Oxford: Clarendon Press).

CHATTERJEE, P., 1989:* 'Development Planning and the Indian State' (mimeo.).

*Paper presented at a Conference on the 'Evaluation of Indian Planning' at the School of Oriental and African Studies, University of London (April).

CHATTERJEE, P., and A. SEN, 1988: 'Planning and the Political Process in India: Duality and Differentiation', in BAGCHI (Ed.), pp. 196–226.

CHATTOPADHYAY, R., 1985: The Idea of Planning in India 1930–1951 (unpublished Ph.D. Thesis, Australian National University, Canberra).

CHAUDHURI, P., 1988: 'The Origins of Modern India's Development Strategy', *in* SHEPPERDSON and SIMMONS (Eds.), pp. 272–81.

GHOSE, S., (Ed.), 1975: *Congress Presidential Speeches* (New Delhi: All India Congress Committee).

GIBBON, E., 1960:* *The Decline and Fall of the Roman Empire* (London: Chatto and Windus).

*D.M. Low (Ed.)

GOVERNMENT OF INDIA, 1952: *The First Five Year Plan* (New Delhi: Planning Commission, Government of India).

————, 1956: *The Second Five Year Plan* (New Delhi: Planning Commission, Government of India).

————, 1961: *Perspective of Development: 1961–1976* (New Delhi: Planning Commission, Government of India).

——————, 1962: *Implications of Planning for a Minimum Level of Living* (New Delhi: Planning Commission, Government of India).

——————, 1973: *The Approach to the Fifth Plan 1974–79* (New Delhi: Planning Commission, Government of India).

——————, 1974: 'Perspective of Development 1961–76: Implications for Planning for a Minimum Level of Living' *in* SRINIVASAN and BARDHAN (Eds.), pp. 9–38.
*PITAMBAR PANT, the then Chief of the Perspective Planning Division was the author of this paper (written in 1962).

HANSON, A.H., 1966: *The Process of Planning* (Oxford: Oxford University Press).

LADJEJINSKY, W., 1965: *A Study of Tenurial Conditions in Package Districts* (New Delhi: Government of India).

LAL, D., 1983: *The Poverty of 'Development Economics'* (London: Institute of Economic Affairs).

LEWIS, J.P., 1962: *Quiet Crisis in India* (Washington: Brookings Institution).

MARKOVITS, C., 1985: *Indian Business and Nationalist Politics 1931–1939* (Cambridge: Cambridge University Press).

MOZOOMDAR, A., 1989:* 'The Rise and Decline of Development Planning in India' (mimeo.).
*Paper presented at a Conference on the 'Evaluation of Indian Planning' at the School of Oriental and African Studies, University of London (April).

NAYAR, B.R., 1972: *The Modernization Imperative and Indian Planning* (New Delhi: Vikas).

NEHRU, J., 1936: *Nehru Papers* (G5A–1936) (New Delhi: Nehru Museum and Memorial Library).

PANT,* P., 1962: 'Perspective of Development 1961–76: Implications for Planning for a Minimum Level of Living', *in* SRINIVASAN and BARDHAN (Eds.), pp. 9–38.
*The author's name does not figure in the original publication which was a paper of the Perspective Planning Division of the Planning Commission.

PATEL, I.G., 1986: *Essays in Economic Policy and Economic Growth* (London: Macmillan).

ROBINSON, E.A.G. and M. KIDRON, (Eds.), 1970: *Economic Development in South Asia* (London: Macmillan).

RUDOLF, L.I. and S.H., 1987: *In Pursuit of Lakshmi: The Political Economy of the Indian State* (Chicago: The University of Chicago Press).

SHEPPERDSON, M. and C. SIMMONS, (Eds.), 1988: *The Indian National Congress and the Political Economy of India 1885–1985* (Aldershot: Avebury).

SRINIVASAN, T.N. and P.K. BARDHAN, (Eds.), 1974: *Poverty and Income Distribution in India* (Calcutta: Statistical Publishing Society).

THAKURDAS, P., *et al.*, 1944: *A Plan of Economic Development for India* (Harmondsworth: Penguin).

Continuity and Change in Indian Industrial Policy · 3

SANJAYA BARU

The industrial policy pursued by the National Front and Janata Dal (Samajwadi) governments, provides further evidence in favour of the argument that, in the realm of economic policy, there has been a certain continuity over the last decade despite several changes in government. The policy of 'liberalization', entailing selective deregulation of industry, delicensing, replacement of physical controls with fiscal ones, a more liberal import and foreign investment policy, and so on, was slowly introduced during the Emergency (1975–7) and has been persisted with ever since, despite changes in pace, emphasis, sometimes the direction itself and, of course, the rhetoric. The policy package announced by Rajiv Gandhi in 1985 came under severe attack from the Left in India and soon even nationalist and liberal economists joined in the attack.[1] Many of them hoped that with the coming to power of the National Front government there would be a

[1] The earliest critiques of the new economic policy were offered at two conferences in Delhi and Calcutta. Some of the papers read here and the resolutions adopted by a cross-section of liberal and Marxist economists are available in *Social Scientist*, Nos. 146–7 (July/Aug.), 148 (Sept.), and 151 (Dec.) 1985.

reversal of this process, especially with respect to import and foreign investment policy, but have been disappointed by the continuity in policy.[2]

Indeed, if Rajiv Gandhi's slogan of 'change with continuity' could be interpreted to mean a change in economic policy along with continuity of the political party in power, namely the Congress(I) party, the National Front leadership ensured 'continuity with change', that is, continuity as far as economic policy was concerned despite a change in political leadership. This could either be interpreted as a sign of weakness in the Indian ruling class, that despite a mandate for change they have proved to be incapable of offering any; or as a sign of strength, that despite changes in the political leadership in government there is continuity in economic policy—as one would, in fact, expect in a mature capitalist economy!

It is in this context that I would like to situate my discussion of industrial policy in India, examining the 'political economy' of change and continuity in that policy. My purpose is to understand the nature of industrial policy in India since Independence, the direction of change in that policy, and how that change can be explained within the framework of the political economy of Indian economic policy. Is industrial policy in India a product of ideology or social necessity? Was it a policy imposed from 'above' by an ideologically-motivated élite or a policy thrown up from 'below' by vested interests and pressure groups?[3]

It is obvious that there are two levels at which policy can be viewed, namely, policy as stated and policy as practised. The discussion becomes complicated when practice does not conform to theory, and even as the shibboleths that defined the framework of policy are not officially abandoned, radical departures are made in practice. Indeed, this is what has happened in India where the Industrial Policy Resolution of 1956 (IPR, 1956) has so far not been officially replaced, while new policy statements which are in conflict with at least some of the stated policy objectives of the former are issued from time to time, and in practice actual policy diverges even further from the stated one.

[2] The Union Budget for 1990–1, the first major economic policy statement of the National Front government attracted strong criticism because it did not depart (in terms of the overall fiscal strategy) from budgets of previous years. Cf. *Economic and Political Weekly* 25 (21 April 1990): 16 (Special Issue on the 'Union Budget, 1990–91').

[3] The argument that the public sector, an important component of post-Independence industrial policy, was the product of socialist ideology rather than of business class interest or necessity arising out of the failure of private investment has been put forward repeatedly, especially by its critics. But it is an argument that does not stand up to empirical scrutiny. For a recent exposition of this view, *see* Nayar, 1989. I return to this point later in the chapter.

Analysts who argue that industrial policy was ideologically motivated by Nehru's brand of socialism, for instance, are often referring only to the stated objectives of policy—as, for example, the objective of IPR, 1956 of establishing a 'socialistic pattern of society'. On the other hand, those who view policy as the outcome of bargaining between vested interests, the national bourgeoisie demanding (from the state) protection for its market, are referring to policy in practice rather than in theory. This confusion arises because, often, policy statements made in the early post-Independence period did reflect Nehru's social-democratic values. This was so not only in the area of industrial policy but also on such issues as agrarian reform, social welfare, education, and so on. In practice, however, the actual policy pursued has been the outcome of bargaining, lobbying, and so on, both by the ruling classes in general and the concerned interest groups in particular, and by their representatives in the Indian National Congress (INC)—an amorphous political party embodying a wide range of ideological orientations ranging from Fabian socialism to feudal values.

This chapter discusses both policy in theory and policy in practice though, as will be obvious, my primary concern is with the latter. I believe that in the early post-Independence period there was a fair degree of consensus on economic policy across a wide band of the ideological spectrum, excluding the far right, though there were differences of emphasis and priorities. However, over the years, actual policy has moved increasingly to the right, resulting in a breakdown of that consensus. If the distance between the centrist coalition and the political right has narrowed, that between the former and the political left has widened.

Evolution of a consensus

In the current debate on industrial policy between pro-liberalizers and anti-liberalizers there is a tendency to portray the policy of import-substituting industrialization pursued after Independence, along with the policy in respect of the creation of a public sector, as if these were 'left-wing' or 'socialist' policies, opposition to them being seen as the natural response of private enterprise. The next step is to characterize the 'leftist' stance as 'ideological', deriving from faith in central planning, and to portray the liberal stance as 'pragmatic'. What is rarely, if ever, conceded is that there is as much dogma in the standard neoclassical free trade view of industrial policy as there is in a purely *etat*ist posture. In actual practice, no country has pursued the classical free trade policy out of commitment to a dogma and against the material interests of that country and its capitalist class in particular. So also, commitment to *dirigiste* dogma was more often than not a product of historical necessity rather than ideological obsession.

The fact is that even in the case of India, post-Independence policy with regard to industry and trade was the result of the articulation, over a decade, of what was seen as being in the 'national' interest, meaning thereby the interests of both Indian business and the consumer. The protection of the home market, development of infrastructure, and the creation of manufacturing capability in bulky and indivisible investment and industries requiring long gestation periods were among the demands articulated as much by domestic businessmen as by nationalist and working class leaders.[4]

To say this is not to imply that industrial policy in post-Independence India was the product of a conspiracy of interests. Indeed, I reject any simplistic conspiracy theory of policy articulation even where explicitly identifiable and selfconsciously organized group or class interests may exist. In actuality, during the '30s and '40s, groups of Indian businessmen, economists, political leaders across the ideological spectrum were all concerned about industrial development and the role of the state in it. Several factors contributed to this general interest in industrial policy and the specific concern with the role of the state, ranging from the then perceived success of Soviet industrialization to the lessons drawn from the Great Depression and the Keynesian anti-slump interventionist policies. Furthermore, the *dirigisme* implicit in German and Japanese economic recovery in the inter-war period also influenced the thinking of nationalist leaders and domestic business. Any enquiry about the role of the state in economic development in India must be firmly rooted in the specific historical context in which such interventionism occurred.

I emphasize this last point because, once again, in contemporary critiques of industrial policy there is a tendency to ignore the historical record and view pro-interventionist and anti-interventionist positions in ahistorical terms. Thus, for instance, it is often presumed that current hostility of the corporate sector to the public sector is a 'natural' hostility and that the interests of the two are always in opposition. The entire 'crowding-out' literature on public and private investment is based on such a conception of the two as mutually antagonistic. On the contrary, at the time of Independence, and through the '50s, public investment was seen in practice as comple-

[4] This has been adequately documented by economists and economic historians; *see*, for example, Bagchi, 1972; Mukherjee, 1976; Bettelheim, 1968; Baru, 1988; Qureshi 1938; Thakurdas *et al.*, 1944; and Chattopadhyay, 1987.

menting or even fostering private investment and not as restraining or 'crowding it out'.

Furthermore, it is methodologically wrong to refer to the existing relationship between the public and private sectors, and the perception of business groups today regarding the role of the former, in order to arrive at an understanding of the historically specific role assigned to the public sector in the '40s and '50s. For instance, Nayar (1989) rejects the hypothesis that the public sector was built in order to subsidize the private sector on the ground that imports are often cheaper than intermediate products or capital goods produced in the public sector. That the public sector has today become inefficient, in global terms, does not negate the fact that, when it was set up, it was meant to subsidize the private sector. What has obviously happened, between times, is that other intermediate classes (mainly public sector employees) have gained leverage sufficient to secure wage increases which are not always in line with productivity increases, and the latter has been at least partly constrained by inadequate investment in technology upgradation stemming from the 'fiscal crisis' facing the government (Baru, 1985). Both these factors have contributed to the increased inefficiency of the public sector and to altered perceptions regarding its role within the industrial economy. It must be added that wage increases in the public sector have also contributed to wage demands in the private sector, and the latter's resentment towards the former is partly due to this 'wage drift' effect of public sector wage settlements.

It is, therefore, surprising how little attention the so-called 'Bombay Plan' receives nowadays in discussions on industrial policy. Most textbooks that introduce students to the Industrial Policy Resolutions of 1948 and 1956 do not even mention the Bombay Plan.[5] Even when it is referred to, only Part I of the statement which was published in January 1944, setting out the quantitative dimensions of the plan, is mentioned. Few refer to the text of Part II (published in December 1944), which includes a fascinating chapter entitled 'The State and Economic Organization', a virtual manifesto of state capitalism in India at the time.[6] While rejecting state ownership or management on

[5] P. Thakurdas, G.D. Birla, J.R.D. Tata, Ardeshir Dalal, Shri Ram, K. Lalbhai, A.D. Shroff, and John Mathai, *A Brief Memorandum Outlining A Plan of Economic Development for India* (Part I: Jan. 1944; Part II: Dec. 1944). Sir Ardeshir Dalal did not affix his signature to Part II because he was appointed Member, Planning and Development in the Government of India, in August 1944.

[6] The notion of 'state capitalism' in the Indian, indeed the post-colonial, context is different from the Leninist concept of the same name used in post-revolutionary Russia. In India, Bettelheim (1968), Chattopadhyay (1969), Kalecki (1972), and Patnaik (1975) have used the term differently to mean a regime under which the state fosters capitalist development through public investment in the economy. Interestingly, some economists have characterized the same regime (with little empirical justification) as a 'state socialist' (*sic*!) or a 'bureaucratic socialist' regime: cf. Bauer (1961) and Sen (1982), among others.

a wide scale, the document does not rule out state ownership altogether. It concedes that in certain industries, which are perceived as important to public welfare or security, state ownership may be necessary for quite a long time. However, it is suggested that

> If later on private finance is prepared to take over these industries, state ownership may be replaced by private ownership, but it is essential in the public interest that the state should retain effective control over them [Thakurdas, *et al.*, 1944 (Pt. II), 28].[7]

On the subject of state control of industry and planning in general the Bombay Plan concedes a wide berth to governmental intervention:

> An enlargement of the positive as well as preventive functions of the state is essential to any large scale economic planning [Thakurdas *et al.*, 1944 (Pt. II), 23].

In striking contrast to current postures on the plan and market, the document says:

> In discussing different types of economic organizations from the point of view of planning, it is perhaps worthwhile pointing out that the distinction which is generally drawn between capitalism and socialism is somewhat overdone ... [this distinction] has lost much of its significance from a practical standpoint .... In our view, no economic organization can function effectively or possess lasting qualities unless it accepts as its basis a judicious combination of the principles associated with each school of thought. [Thakurdas *et al.*, 1944 (Pt. II), 25.]

Furthermore,

> These principles may be summed up as follows: first, that there should be sufficient scope for the play of individual initiative and enterprise; secondly, that the interests of the community should be safeguarded by the institution of adequate sanctions agianst abuse of individual freedom; and thirdly, that the state should play a positive role in the direction of economic policy and the development of economic resources. It is from this angle that we approach the problem of determining the place of the State in a planned economy in India. [Thakurdas *et al.*, 1944 (Pt. II), 26.]

To us, all this obviously suggests that the involvement of the state in the industrial sector, both directly, through public investment in the infrastructure, capital and intermediate goods industries, and indirectly, through the introduction of appropriate trade and fiscal policies, was aimed at aiding the development of indigenous capitalism. It is for this reason that we see post-Independence industrial policy as promoting a 'state capitalist' industrial economy (Baru,

[7] It is obvious that the seeds of the concept of 'privatization' of the public sector were sown on these lines.

1988). The historical conjuncture was also such as to favour *dirigisme* for, not only had 'socialist' planning demonstrated its success elsewhere in ensuring rapid industrialization, but even the capitalist world had conceded the primacy of state interventionism in overcoming what were seen as the inherent flaws of the free enterprise system. This much was explicitly stated in the Bombay Plan.

What is of equal significance is the fact that this perspective received widespread support. Nehru was conscious of this even earlier when he wrote *The Discovery of India* and described the composition of the National Planning Committee, whose reports were in many ways a logical precursor to the Bombay Plan and the Industrial Policy Resolutions.

On the composition of the National Planning Committee, Nehru wrote:

> Among the members are well-known industrialists, financiers, economists, professors and scientists, as well as representatives of the Trade Union Congress and of the Village Industries Association. The non-Congress Provincial Governments (Bengal, Punjab and Sind) as well as some of the major states (Hyderabad, Mysore, Travancore, Bhopal) cooperated with this Committee ... Hard-headed big business was there, as well as people who are called idealistic and doctrinaire, socialists and near-communists. [Nehru, 1951, 372.]

Interestingly enough, even the rival plans that were written in response to the Bombay Plan—namely, the Gandhian Plan and the People's Plan—did not differ from the Bombay Plan on the question of the role assigned to the state in industrial development. The differences were on the attitude to labour, the rural economy, the village and small-scale sector, and allocations for education and health, but with respect to the relationship between state and enterprise there were no fundamental differences at the time. Admittedly, such differences did arise later, especially between the perspectives of the Communist Party and the Government of India on industrial and trade policy. But, in the '40s and the '50s, differences on industrial policy were not as sharp as they are today.

However, a subject of considerable disagreement at the time, with respect to industrial policy, was the attitude to foreign capital. The radical position was crystallized in a celebrated study entitled *American Shadow Over India*, written pseudonymously by L. Natarajan (1952).[8] In his foreword to the book, J.C. Kumarappa, the eminent Gandhian, observed:

[8] I give below certain information about L. Natarajan that came to my attention a few years ago. L. Natarajan is in fact the pseudonym of a US citizen, a white American who was working in the US State Department at the time, and whose sympathies were

> For over a century and a half, the USA has been spreading her financial nets the world over. India was somewhat in a protected position because of British vested interests and jealousies. Of late, this protective fencing has been disbanded and India presents an open field of ruthless exploitation. It is sad that this should happen under the leadership of a patriot of the order of Nehru. The wiles of the Americans, who go about their work with the aid of modern psychology, are a little too much for the straightforward, simple, trusting statesman. [Natarajan, 1952: xii.]

D. R. Gadgil, another nationalist economist, by no means a Marxist, expressed early concern about US economic interests in India and its likely consequences in the following terms:

> ... even if the unexpected happens and large investments are intelligently made [by the USA] for the development of countries like India these appear likely, in existing circumstances, to lead immediately not to desirable social objectives but to strengthening, on the one hand of the forces of monopoly and to an increase on the other, of social discontent and conflict [cited in Natarajan, 1952: 256].

Natarajan's study ended with pious hopes in the nationalism of the Indian bourgeoisie:

> The Indian national bourgeoisie fought British imperialism for many decades and often gave leadership to popular forces, as in the boycott of British goods. In the economic sphere, many capitalists showed exceptional resourcefulness and courage in establishing national industries in the face of severe imperialist restrictions. It is, therefore, not too much to expect from them a forthright stand against recent American encroachments. [Natarajan, 1952: 258.]

Despite such faith in India's 'national bourgeoisie' on the part of both Gandhian and leftist economists and political leaders, this faith was not reciprocated in full measure in the concrete actions of the business leadership which, time and again, was willing to accede to a variety of demands made by business interests from developed countries. Given the technological dependence of Indian industry and the global influence of major US and European corporations, India, like most post-colonial societies, succumbed fairly quickly to the temptation of

---

clearly with India and the Indian people, especially those on the Left. The book utilizes confidential information to which he had direct access as a member of the State Department. Natarajan also established contact with some functionaries of the CPI at the time and it is they, along with fellow communists in the USA, who finally made the publication of the book possible. As McCarthyist witch-hunting gathered pace, the author and his associates had to go underground. They have never resurfaced since. The book clearly established the overt and covert methods utilized by the American government and American business to upgrade US business presence in India and influence Indian policy in a variety of spheres.

Editor's Note: The late Daniel Thorner, who was suspected by some of his American colleagues of having written the book, himself strongly believed that it must have been the work of a committee (perhaps including both Americans and Indians).

compromising with foreign capital. Here, policy in practice did not inhibit such trends. Under the guise of export-promotion and under the pretext of earning foreign exchange, the government has always pursued a relatively liberal policy with respect to foreign collaboration, if not direct foreign investment.

To be sure, there were episodes of nationalist posturing and, by and large, the domestic market was protected from an unrestricted inflow of foreign goods, but the attitude of the government was not as explicitly anti-imperialist as is suggested, for instance, in Kalecki's well-known formulation on intermediate regimes.[9] The actual experience is better summed up by Patnaik (1975: 78):

> Today, in India or other third world countries, the bourgeoisie, arriving too late on the scene, is forced to ally itself with remnants of feudalism; and, having no colonies and being threatened by the increasing political consciousness of the people, it is reintegrated into the imperial structure. Compromising with feudalism, it is forced to compromise with imperialism. The fact that the Indian bourgeoisie seeks repeatedly to assert its independence testifies to its relative strength: the fact that it fails testifies to the objective limits to bourgeois revolution today.

Given this historical weakness, the Indian business class was fortunate to be able to utilize the Indian state in its attempt to secure some control over the domestic market. Clearly, there is nothing unique in this experience. Several developed capitalist countries have also witnessed extensive state intervention aimed at countering external competition and securing the market for the domestic bourgeoisie. In the Indian case, as for instance in the case of Egypt, the domestic bourgeoisie discovered that it had to ally itself with the socialist world in order to achieve its objective of promoting independent capitalist development, and such dependence on the socialist world externally offered a wider margin for manœuvre for the socialists at home. India's need to go to the Soviet Union for assistance in setting up steel plants in the public sector when the USA refused help, meant that the Indian Left in general and the left wing of the ruling party in particular, gained greater political influence domestically. This in turn meant that those who were already favourably

[9] Kalecki 1972, saw post-colonial Third World regimes (as in India, Egypt, and Indonesia) as being both anti-imperialist and anti-feudal and, since these regimes were dominated by the middle or 'intermediate' classes, they were in conflict with the domestic big business class which was in league with imperialism, and the landlord class which was still feudal. The strength of the so-called 'intermediate classes' at the time of Independence is exaggerated, in my view, by Kalecki, though over time there is no denying that such classes have gained in terms of political and economic clout. Kalecki's model is, however, a static one since he does not explore the dynamics of an intermediate regime as capitalist development occurs.

predisposed to a *dirigiste* state policy were well-placed to gain the upper hand.

This peculiar set of historical circumstances only helped to consolidate the national consensus that had evolved in the case of the relationship between state and enterprise. Furthermore, despite differences in perception relating to foreign capital, the divergence in the approach of the USA and the USSR to domestic economic policy, especially the latter's willingness to support the state capitalist strategy, helped in forging the consensus on domestic economic policy within India. The only group that clearly, and from the beginning, remained outside this consensus was the extreme right comprising a motley group of Hindu chauvinists and anti-Nehru conservatives who were united in their hatred of socialism of even the palest pink, whose political representatives were to be found in the Jan Sangh and the Swatantra Party, and whose ideologues were activists on the Forum for Free Enterprise!

While the CPI remained a fervent critic of the Congress Party's strategy of development, it was only during the late '60s that the 'extreme' Left position, mainly articulated by the Maoists, describing the Indian bourgeoisie as comprador in character, gained currency within them. While the Left in general became increasingly critical of the economic policy pursued by the centrist coalition, the gathering evidence of the failure of this strategy contributed decisively to its rejection by them. This eventually led to a rupture in the consensus to which reference has already been made.

Thus, despite its internal inconsistencies, the industrial policy that emerged in the immediate post-Independence period, a policy that was carved out even during the inter-war period, sought to create the basis for an independent capitalist economy in India even though, in practice, it was forced to yield considerable space to foreign capital. By comparison with the experience of other post-colonial economies, the space that was thus yielded was not significant in an absolute sense, but by comparison with the stated goals of the nationalist leadership the compromise was by no means insignificant. However, the policy of a mixed industrial economy coupled with import-substitution was based on a fairly wide national consensus, excluding the fringe on the right of the ideological spectrum.

It was this very limited consensus that began to break down as the Indian economy entered a phase of crisis during the mid-'60s. While the creation of a public sector and the rhetoric of socialism had further alienated the Right, the subversion of state interventionism by the monopoly bourgeoisie and the emergence of sharp points of inter-class conflict had begun to alienate the Left. An important factor that contributed to this alienation was the mounting evidence to the effect that state interventionism in the industrial sector had not, in practice, helped achieve some of the crucial objectives of the stated policy.

The reports of the Mahalanobis Committee, Hazari Committee, Dutt Committee, and so on, and the research of individual, Marxist, and liberal economists convincingly illustrated the manner in which government regulations aimed at reducing the concentration of economic power, regional inequalities, monopolistic practices, and so on were systematically subverted, and how state intervention had often contributed to the attainment of the opposite objective of strengthening the monopolistic and inegalitarian tendencies embedded in the growth process. Such evidence had obviously mobilized left-wing opinion against state interventionism of the kind pursued by government and indeed helped drive a wedge in the consensus on economic policy.

Breakdown of the consensus

Undoubtedly, until the mid-'60s, the wider consensus on industrial policy prevailed despite the fairly early breach that had opened up with regard to the role of foreign capital. The first signs of a breakdown in this consensus emerged after the wars with China (1962) and Pakistan (1965), the drought of the mid-'60s (1965–7), and the foreign exchange crisis which forced India to devalue the rupee and liberalize imports (1966). These experiences facilitated, if that is the appropriate word, the liberalization of industrial policy, allowing easier entry for foreign capital and offering subsidies to Indian firms exporting to hard currency areas.

Political compulsions of the late-'60s only delayed what was to happen in the mid-'70s, namely, a further liberalization of economic policy both internally and externally. Such political compulsions imposed on the government a path that was dizzying in its oscillations between the Left and the Right. The underlying economic reality had left the government little option but to offer greater concessions to both domestic and foreign business. However, the nationalist consciousness of the electorate and the ability of the Left to make political use of such sudden changes in policy led to new initiatives such as, for instance, the introduction of the Foreign Exchange Regulation Act (FERA) in the early '70s.[10] These and other political constraints were

[10] The nationalization of banks is often seen as an anti-industry and a leftist policy. In fact, Indian business has never suffered from a lack of funds and, despite the nationalization of banks, the private sector (especially the corporate sector) had easier access to finance during the '70s and '80s. What the nationalization achieved was a greater flow of funds to the rural rich, especially in the Green Revolution areas, to

removed during the internal Emergency (1975–7). The Emergency witnessed a decisive swing towards the Right in economic policy leading to the final rupture of the national consensus on industrial policy (Baru, 1983).

The policy of economic liberalization, introduced in a partisan manner by Mrs Gandhi (and her son Sanjay) during the Emergency, received wider political support when the Janata Party government continued along the same course (1977–9). This helped widen the consensual support for a policy which might otherwise have been more fully identified with the reprehensible Emergency regime and delegitimized in the eyes of the mass of the people.

Indeed, the fact that such patently 'pro-rich' policies, especially when they were seen somehow to serve the interests of foreign business, continued to be a political liability became amply clear when the campaign against India's application to the IMF for a loan under the Extended Fund Facility gathered momentum during 1982 and 1983. So much so, that Mrs Gandhi decided to retrieve political ground by making the government's decision not to draw the last tranche of the loan a campaign issue. The real political fallout of all this could not be put to the test because of Mrs Gandhi's assassination and the specific circumstances surrounding the 1984 election. Left-wing economists and political leaders were convinced that but for the Punjab crisis the campaign against the IMF loan could easily have cost Mrs Gandhi her electoral office; a belief that was partly substantiated by trends in opinion polls at the time.

The first response of nationalist and Marxist economists to changes in industrial policy in the mid and late-'70s was to view this as the strategy of survival of a beleagured bourgeoisie incapable of ensuring growth on the old basis and therefore willing to pursue a strategy that would involve compromise with imperialism and a closer integration with the world capitalist system. Raj (1976), Patnaik (1985), and other economists saw the 'Brazilian and the South Korean model of development' as the obvious parallel to the process being set in motion in India, even though at the time it was not appreciated how *dirigiste* the South Korean model itself happened to be.[11]

---

small-scale enterprises, and the provision of jobs for the urban middle class. In that sense, it was a move that served the interests of the 'intermediate classes' without hurting the private corporate sector in any significant way.

[11] In the mid to late '70s, World Bank economists such as Bela Balassa, were selling the 'South Korean model' as an example of the wonders of 'free market capitalism'. More recent analyses of the South Korean experience suggest that greater credit must be given to state interventionism and *dirigiste* policies than the World Bank economists were willing to concede in explaining the 'South Korean miracle'. *See*, for example, Lee, 1989.

Some, like Mundle (1981), implicitly differed from this view. They suggested that a process of capital reorganization was under way that would result in an alteration of the industrial structure, ensuring a higher growth rate even if not an egalitarian growth process, and a restructuring of India's industrial economy along lines reminiscent of Meiji Japan where large-scale and small-scale sectors were not in contention against each other, but the former was well placed for expansion by utilizing the benefits of its close links with the latter. The limitation of this argument derives both from the fact that the large-scale–small-scale linkage of the Japanese sort has not really occurred in India to any significant degree and, where it has, the linkage was often between a public sector large-scale unit and private sector small-scale ones.

In other words, the kind of capital reorganization that Mundle alludes to has not been used as a strategic policy of survival by the private corporate sector; rather, it is the result of a conscious policy of assistance to small enterprise pursued by the public sector. Secondly, Mundle's analysis of capital reorganization does not refer to the emergence of new business groups but only refers to the restructuring of existing capitals. I shall argue, below, that a restructuring of sorts has in fact occurred, but this has been more due to the deepening and widening of the capital base in the country than to the reorganization of the existing capital structure.

As the state began to retreat from the economy, however slowly, and the protectionist import-substituting industrial policy regime was being replaced by a more liberal policy regime, economists began to seek explanations for this change. One school of thought suggested that liberalization was motivated by the search for efficiency and stimulating growth (e.g. Ahluwalia, 1985). According to another school of thought, Indian capitalism had come of age and the Indian bourgeoisie had gained the confidence needed to face global competition and was therefore seeking to remove the restrictions imposed by the state against the free play of market forces (such a position was taken mainly by Trotskyist and neo-Marxist critics of industrial policy).

A third school of thought suggested that liberalization in fact constituted a desperate attempt of a bourgeoisie that was incapable of ensuring sustained growth on the basis of an import-substituting strategy. Moreover, the new strategy pointed to the bourgeoisie's inherent weakness rather than strength as a class, since the process involved a compromise with imperialism and an abandonment of the political systems of liberal democracy on which the earlier growth process had been predicated (Patnaik, 1985).

Rajiv Gandhi's government continued to pursue the liberalization policy, notwithstanding the criticism that it signalled Indian capital's capitulation to the demands of monopoly and/or foreign capital. In fact, the liberalization approach was selfconsciously advertised as a 'new economic policy' aimed at taking India into the twenty-first century—implying that its objective was the modernization of the economy.[12] Fortunately for the protagonists of liberalization, economic growth in general and industrial growth in particular had begun to revive from the late '70s onwards, and this was cited in support of the liberalization process on the basis of an as yet unsubstantiated argument that the very revival of growth was itself the product of liberalization. Whilst there is a statistically and empirically observed correlation, there is as yet no evidence to substantiate any causal relationship between the two.

In the event, however, the new growth process has conformed neither to the classical Brazilian format nor to the classical Japanese format, nor even to the South Korean format of state-sponsored modernization with the aid of foreign capital and technology. That it would approximate to one of these models was of course not the contention of any of the economists we have referred to above who were engaged in attempts to situate the Indian experience within the contours defined by these models of industrialization. However, none of them drew adequate attention to the possibility that new processes were under way in India that would initiate a revival of industrial growth along a typically Indian path.

In short, India was not replicating the industrial growth experience of any other country, but was generating a process peculiar to its endowments, all of which were by-products of its strategy of state capitalist development. That the revival of industrial growth was neither the result of modernization supported by the state and funded by foreign capital, nor the result of an indigenous process of re-organization of capital, nor even the result of the coming of age of an indigenous capitalist class, but was, in fact, the manifestation of the spread of capitalism (as a mode of production) is an argument that has not been effectively made so far in the literature. In my view, however, what we have witnessed in India during the '80s is the by-product of a process of state capitalist development during the post-independence period. No doubt this process has been confined to

[12] That this process was not as fast as some of the more ardent advocates of liberalization, both in India and abroad, would have liked should not detract attention from the persistence of the process itself. Even if a step backward was taken, for political reasons, it was after several steps forward had been taken and so the net movement was in the intended direction of greater liberalization of economic policy both internally and with respect to the external world. Early reactions that wrote off the process were clearly hasty and off the mark. Cf. Harriss (1987), Manor (1987), and Kohli (1989).

certain regions in the country in which a transition to agrarian and industrial capitalism has taken place, even when semi-feudal landlordism persists in others. Let me elaborate.

The legacy of state capitalism

In the literature on state and industrialization in India, far too much emphasis is placed on the direct role of the state in fostering industrial development. There is surprisingly little appreciation of the impact of public investment in other sectors of the economy (viz. agriculture, infrastructure, education, finance, and banking) on industrial development and policy.

I believe that this impact has been felt in three distinct ways. Firstly, public investment in agriculture and the development of agrarian capitalism have initiated the process of locally based industrialization in several regions of the country. This stands in direct contrast to the colonial pattern of industrial growth which was characterized by its enclave nature, delinked as it was from the immediate hinterland in which such industrialization (in terms of the accumulation process and the home market for industrial produce) was taking place. The best examples of such agriculture-led industrial growth are to be found in the Punjab–Haryana, western Maharashtra–south Gujarat, Andhra, Karnataka, and Tamil Nadu regions.

Secondly, public investment in higher education (especially professional education, such as medicine, engineering, and management) has created a large pool of skilled individuals who have been able to emigrate to the West (especially the United States) and who (along with the more recent and less skilled migrants to the Middle East and the traditional trading communities in Africa) have become an important resource base of capital accumulation and/or entrepreneurship and a means of access to technology. Those among them who have returned home, namely non-resident Indians (NRIs), along with their counterparts in India, consisting of the professional middle class (including small-scale entrepreneurs), have also become an important source of demand for consumer durables and non-durables.

Thirdly, public investment in the infrastructural sector, both in urban and rural areas, has pumped an enormous amount of money into the hands of a new contractor class (for the construction of roads, dams, and so on) which has also become a major source of capital, enterprise, and demand.

The legacy of state capitalism is not simply the creation of an industrial and technological base, as some of the early commentators have claimed. It has also included the generation of the potentialities for capitalist development in the countryside and the consequent creation of the base for indigenous capitalist development. That such a new capitalist class would clamour for liberalization of a policy regime that has served the interests of existing big business should be obvious. That this would not be confined to internal liberalization alone, since the NRIs are pressurizing the government to adopt a policy of external liberalization as well, is also clear. However, it is by no means obvious why the traditional big business groups have also become vociferous supporters of internal as well as external liberalization.

The conventional formulation that the FICCI represents the interests of the 'national bourgeoisie' can no longer remain unchallenged because it has been at least as vocal in its support for an inflow of foreign direct investment into India and import liberalization as the Associated Chambers of Industry and Commerce (ASSOCHAM) (traditionally viewed as the representative organization of the 'comprador bourgeoisie') has been.

To suggest that this only proves the comprador nature of all Indian business is begging the question and also goes counter to the actual historical experience of the last forty or fifty years. A possible explanation would hinge on the interest of big business in utilizing the opportunity provided by a process of liberalization, especially external liberalization, in order to restructure, rationalize, and rehabilitate its industrial enterprises.

It is a well-known fact that Indian big business has not invested in the upgradation of technology in most of the major product lines. If Indian industry is to keep pace with global industry, then at least in the medium-term of ten to twenty years it will have to depend on imported technology. The government in India, unlike the government in Japan or South Korea, has not pursued a vigorous policy of compelling Indian firms to be technologically dynamic. Consequently, Indian industry can catch the new technological bus only by becoming even more dependent and by utilizing the opportunity afforded by economic liberalization in order to upgrade its technological base. If this entails a compromise with foreign capital, so be it.

In sum, the revival of industrial growth in the '80s was not merely quantitative in character, with material production in existing or new industries increasing, but also heralded a qualitative change in the social and economic base of the Indian business class, with new entrants emerging as major players on the industrial scene. These include not only the small-scale entrepreneurs who have been assisted by public sector undertakings through their systems of ancillarization and subcontracting, but also big business (e.g. the Ambanis, the K.V.K. Raju family, the Chhabrias, the Mallya family). Apart from these better-known names, there are scores of less well-known, regionally-based, businesses (e.g. Ramoji Rao, Subbirami Reddy, and Anji Reddy in Hyderabad), professionals (e.g. electronics engineers and doctors who have invested in electronics, pharmaceuticals, food processing, and a large variety of consumer durables and non-durables).

The 'regional' character of this business class is defined by the fact that these groups have not invested outside their State of domicile and, more importantly, their emergence (if not subsequent growth) is entirely due to the support extended by the local political leadership and the State government rather than the central government or its financial institutions.

It is in response to such new sources of entrepreneurship that the policy of liberalization has proceeded apace. Another interesting dimension to the new policy is reflected in the fact that the emphasis on export promotion has helped the small-scale industrialists (rather than the traditional big business groups), who have proved to be the main exporters in India, to secure access to imported technology. They have thereby been helped to overcome entry barriers erected by existing groups. The case of the Ambanis (who are, of course, not in the small-scale sector) is an extreme example, and probably an exception inasmuch as they have also received generous political support. But there are several other less well-known business groups in other product lines which have also benefited from this policy.

It is a well-known fact that Indian big business has not invested in the liberalization, of the continuation of the extant consumer goods-led upper-income-groups-oriented industrial growth strategy, and who would like to see in India the unleashing of uninhibited capitalist development, unencumbered by the social–democratic ideological baggage as in the past. The impact of this group on policy, and more importantly, the representation of that policy in the media, has been particularly noticeable in recent debates on the Eighth Five Year Plan. The Planning Commission has been portrayed in these media presentations as being inclined to revert to an earlier regime of social democratic policy in opposition to the unabashed free-market capitalism of the new business groups.

Conclusion

It is true, however, that during the '80s the Indian experience of liberalization would appear to be rather limited and halting by comparison with that of several Third World countries (and even some of the East European economies). The impatience of some of the more ardent liberalizers proves the point. Undoubtedly, this testifies to the strength of the longer-established segment of the national big bourgeoisie, on the one hand, and to that of a tenacious bureaucracy (the intermediate class), on the other. These constitute the countervailing forces which have checked the liberalization process. In recognizing the continuity of economic and political influence enjoyed by the former, one must not remain blind to the change in terms of the weakening of its power and of the need for it to compromise, on the one hand, with foreign capital, and on the other, with the newly emergent (albeit regional) bourgeoisie. I have chosen to use the term 'regional bourgeoisie' in order to distinguish this group of business houses, based mainly in specific regions, from the mainstream of the 'metropolitan' business class whose field of operation is not bounded by any regional limitations.

Marxist critics of the liberalization process have tended to overemphasize the role of 'external pressures', partly because the critique of the liberalization process began with the campaign against the IMF Loan in 1982, when the threat of a loss of economic sovereignty became a central campaign issue and, in the process, the external pressure in favour of liberalization was exaggerated. Admittedly, such external pressures are by no means negligible and continue to play an important role in influencing domestic economic policy, as is again demonstrated by the recent episode of India being named under the US law *Super 301*. The picture that emerges is not a black and white one of national sovereignty either triumphing over or succumbing to imperialist bossism. There are shades of grey which illustrate a complex mix of dependence and independence where both internal and external interests mesh together to produce complicated policy outcomes.

It is my contention, emphasized here only because this has not been adequately recognized in the existing literature on liberalization, that if imperialism, to use a catch-all phrase, or foreign capital, has acquired a greater access to the territory carved out by the Indian bourgeoisie, it is at least partly because a new business class has emerged which hopes to utilize this access to foreign technology in order to overcome entry barriers erected by Indian big business groups. Similarly, they view internal liberalization also as a means of barrier removal. To the extent that the old business class continues to be an important segment of the bourgeoisie, there are elements of continuity in industrial policy, but to the extent that new groups have emerged and the economic and social base of capitalism in India has expanded and deepened in new regions of the economy, there are also elements of change in policy. Economists today are faced with the task of analysing in concrete terms the change in the character and composition of the Indian capitalist class. Old formulations can no longer suffice in view of rapid changes that are underway in India's industrial economy.

ACKNOWLEDGEMENTS

For useful discussions and suggestions I am grateful to John Cameron, N. Krishnaji, D. Narasimha Reddy, Rama Vaidyanathan Baru; and to Amiya Bagchi, A. Vaidyanathan, and other participants at the Bhubaneshwar conference.

REFERENCES

AHLUWALIA, I.J., 1985: *Industrial Growth in India: Stagnation since the Mid-Sixties* (New Delhi: Oxford University Press).

BAGCHI, A.K., 1972: *Private Investment in India 1900–1939* (Cambridge: Cambridge University Press).

BARDHAN, P.K., 1984: *The Political Economy of Development in India* (Oxford: Basil Blackwell).

BARU, S., 1983: 'Self-Reliance to Dependence in Indian Economic Development', *Social Scientist* 11 (Nov.): 11, No. 126, pp. 34–46.

————, 1985: 'The State in Retreat?' *Economic and Political Weekly* 20 (20 April): 16, pp. 703–6.

————, 1988: 'State and Industrialization in a Post-Colonial Capitalist Economy: The Experience of India', *Economic and Political Weekly* 23 (23 Jan.): 4, pp. 143–50.

BAUER, P.T., 1961: *Indian Economic Policy and Development* (London: George Allen & Unwin).

BETTELHEIM, C., 1968: *India Independent* (trans. W.A. Casswell) (New York: Monthly Review Press).

BHAGWATI, J., and P. DESAI, 1970: *India: Planning for Industrialization* (London: Oxford University Press).

BLACKBURN, R., (Ed.), 1975: *Explosion in the Subcontinent* (Harmondsworth: Penguin).

CHATTOPADHYAY, P., 1969: 'Aspects of the Growth of State Capitalism in India, in Kerala State Planning Board' *in* GOVERNMENT OF KERALA, 1969, pp. 71–89.

CHATTOPADHYAY, R., 1987: 'An Early British Government Initiative in the Genesis of Indian Planning', *Economic and Political Weekly* 22 (31 Jan.): 5, pp. PE19–PE29.

GOVERNMENT OF KERALA, 1969: *Alternative Policies for the Fourth Five Year Plan* (Trivandrum: Government of Kerala).

HARRISS, J., 1987: 'The State in Retreat? Why Has India Experienced Such

Half-Hearted Liberalization in the '80s?', *IDS Bulletin* 18 (Oct.): 4, pp. 31–8.

KALECKI, M., 1972: 'Social and Economic Aspects of "Intermediate Regimes"' in his *Essays on the Economic Growth of the Socialist and the Mixed Economy* (Cambridge: Cambridge University Press), pp.162–9.

KOHLI, A., 1989: 'Politics of Economic Liberalization in India', *World Development* 17 (March): 3, pp. 305–28.

LEE, K.U., 1989: *The Changing Role of Government in the Economic Development of Korea* (Seoul: Korea Development Institute).

MANOR, J., 1987: 'Economic Liberalization in India: Tried and Then Abandoned', *IDS Bulletin* 18 (Oct.): 4, pp. 39–44.

MUKHERJEE, A., 1976: 'Indian Capitalist Class and the Public Sector, 1930–1947', *Economic and Political Weekly* 11 (17 Jan.): 3, pp. 67–73.

MUNDLE, S., 1981: 'Growth, Disparity and Capital Reorganization in Indian Economy: Some Speculations', *Economic and Political Weekly* 16 (March): 10–12, pp. 393–408.

NATARAJAN, L., 1952: *American Shadow Over India* (Bombay: People's Publishing House).

NAYAR, B.R., 1989: *India's Mixed Economy* (Bombay: Popular Prakashan).

NEHRU, J., 1951: *The Discovery of India* (London: Bodley Head).

PATNAIK, P., 1975: 'Imperialism and the Growth of Indian Capitalism' *in* R. BLACKBURN (Ed.), 1975, pp. 51–78.

————, 1979: 'Industrial Development Since Independence', *Social Scientist* 7 (June): 11, No. 83, pp. 3–19.

————, 1985: 'Political Economy of Economic Liberalization', *Social Scientist* 13 (July-Aug.): 7–8, Nos. 146–7, pp. 3–17.

QURESHI, A.I., 1938: *The State and Economic Life* (Bombay: New Book Company).

RAJ, K.N., 1976: 'Growth and Stagnation in Indian Industrial Development', *Economic and Political Weekly* 11 (Feb.): 5–7, pp. 223–36.

SEN, A., 1982: *The State, Industrialization and Class Formation in India* (London: Routledge & Kegan Paul).

THAKURDAS, P. *et al.*, 1944: *A Plan of Economic Development for India* (Parts I and II) (Harmondsworth: Penguin Books).

'Liberalization' and the Indian Economy: Myth and Reality • 4

SUMIT ROY

Introduction

From the mid-'60s onwards the Indian economy confronted a falling rate of industrial growth, leading to 'stagnation' in comparison with previous periods. In this context, 'liberalization' symbolized an attempt to restructure the domestic and external economy by shifting emphasis from state to market-based policies. This started emerging from the late '70s onwards. This chapter focuses on analysing the major explanations behind 'stagnation', the role of liberalization policies, and their socio-economic implications.

Market-based explanations have been somewhat divorced from socio-political realities which have moulded the nature and impact of past policies. Other explanations offer a fuller understanding of the limits of state intervention to raise resources and invest, coupled with the constraints imposed by a skewed distribution of income. These have been underpinned by access to, and use of, the state to further the interests of the major propertied classes. Liberalization has primarily benefited sections of industrialists, bureaucrats, and urban professionals, while the rural and urban poor have been largely bypassed. Against a backdrop of worsening external debts and domestic political instability, the Indian state faces increasing domestic and international pressures to intensify the new policy thrust. This poses critical questions for development.

Background

Liberalization has to be set against the backdrop of the state's attempt to reshape the Indian economy in the period following Independence. The weight to be placed on planning *versus* the market, the role of the industrial and the agricultural sectors, and the choice of goals to pursue, were of major concern. Policies have unfolded in the context of 'federal' India's Centre–State relationship which has been undergoing change (Sathyamurthy, 1989). Intense competition between different States and regions for resources has persisted (Vaidyanathan, 1983: 952).

Apart from ideological conflicts between 'radicals' and 'conservatives', policies and their execution have been influenced markedly by divergent group and regional interests. The class composition of the ruling party has been heterogeneous and the 'conflicting interests of different classes, especially between larger peasants, organized labour, and the entrepreneurial classes, were often as important as ideological factors' (Vaidyanathan, 1983: 952). Landless labourers, tenant farmers, and small cultivators, though significant numerically, have not been effectively organized to influence policies in their favour. The rejection of 'extremist', which suggests radical, solutions by the Congress leadership, was a major factor in weaving into the Constitution provisions which limited the scope for any drastic changes in distribution of land and other property.

Central to the analysis of stagnation of the Indian economy, and the rise of liberalization, has been the falling rate of industrial growth since the mid-'60s. The growth-rate of industrial production fell from 7 per cent between 1951 and 1965 to 3.3 per cent between 1965–70; over 1970–7 it increased to 4.8 per cent, but again declined by 1.4 per cent between 1979–80 (Patnaik, 1979: 13; Harriss, 1983).[1] Since the mid-'60s, however, the rate of growth of some industries has been negative. Some economists have questioned the basis of the increase in the rate of growth in the '80s.[2]

[1] Conditions in agriculture have a major effect on the overall rate of growth of the economy. For instance, in 1987–8 drought adversely affected the growth-rate. In contrast, the 1988–9 harvest was excellent and the GDP was expected to have a favourable effect on growth-rate.

[2] E.g. Chapter 3.

The sharp deceleration in industrial production was accompanied by unutilized capacity in a number of industries, stagnation of investment, shortage of foreign exchange, and import of foodgrains (Government of India, 1969: 3).

Factors underlying stagnation: economic and political arguments

It is essential to analyse the validity of the key explanations for stagnation. The first explanation places blame squarely on state intervention in the economy and calls upon market forces to govern economic activity. The 'neo-liberal' school of thought has been a major influence in moulding this approach. Comparisons between India and the 'market paradise' within which East and South-East Asian economies are said to function show India in an unfavourable light. The public sector in India has been attacked for being 'flabby', with 'labyrinthine' controls on private industry (Roy, 1984: 56).

The second explanation centres on the nature of public investment and the structure of demand. A decline in public investment, a skewed distribution of demand and insufficiency of agricultural growth have been major contributory factors. In the first case, the fall in state investment, brought about by a shortfall in external resources, coupled with inability to tap domestic resources, has been of key importance. There have been backlogs in power generation, constraining transportation and fuel (Rubin, 1985: 945). Secondly, the pattern of income-distribution, and its consequences for consumption and saving have slowed down and limited the possibility of stimulating production of industrial goods. Lack of adequate agricultural growth has also limited the role of this sector in stimulating demand for industrial goods. These problems require an exploration of the relationship between economic policies and the underlying political structures (Roy, 1990: 88–91).

These explanations of economic stagnation need to be placed in the context of state, class, and government policies in India. The extent to which the state has acted for the dominant classes needs to be scrutinized. Unlike the many 'open' developing economies, India has been relatively closed, and hence shielded from changes in the world economy. Self-sufficiency has been pursued for both economic and political motives.

The diverse views on the nature of the Indian state cannot be explored here. One major analysis suggests that this comprises a coalition consisting of three proprietary classes: the capitalist class, the agricultural rich, and the professionals (including white collar workers). The third class, which is not usually included among the proprietary classes, is also a part of this coalition. The argument runs that 'activization of conflicts' in this dominant coalition, given India's democratic politics, has had adverse implications for economic growth. Intra-coalition conflicts are put forward as the explanation (in spite of a trend towards a higher savings rate) for the consumption, by current expenditures, of the bulk of resources, and for the inadequacy of public investment even in sectors as crucial as railways, power, and irrigation. The diverse interests of the dominant coalition of the proprietary classes pull in different directions, and when none of them can dominate, the allocation of resources leads to a proliferation of subsidies and grants to placate them all. This reduces available surplus for public capital formation (Bardhan, 1984; Varshney, 1984: 1516). Indian political economy emerges as an elaborate network of patronage and subsidies, contributing not only to the decline in public investment, but also bringing about chaos within the management structure of the Indian public sector. The nature of the relationship between Indian democracy and the factors which have given rise to such a coalition demand further investigation. Liberalization symbolizes an attempt to satisfy the diverse and competing demands of the propertied classes.

A leading exponent of the first explanation has used India to illustrate the bad effects of 'the dirigistic dogma' based on the belief that the price mechanism should be supplanted (and not just supplemented). 'Socialism' has been stigmatized as an excuse for a vast extension of the 'essentially feudal and imperial revenue economy' (Lal, 1988: 24).

Liberalization, moreover, has been advocated as a means of not only restoring efficiency and technological modernity but also removing the discretionary powers of bureaucrats. It is claimed that the power enjoyed by bureaucrats has given rise to bribery, corruption, and blackmarketing (Rao, 1988: 4–5). Reform of the bureaucracy may help to make it more efficient, but the role of bureaucracy cannot be viewed in isolation from the political structures.

The attack on the public sector has been misplaced. There has been a failure to grasp the fact that a large number of private units have had to be taken over by the state because of endemic sickness, exemplified by jute mills, mini steel plants, pharmaceutical and engineering units, and a large part of the textile industry. Yet, the private sector has been regarded as efficient (Bagchi, Ghosh, and Dasgupta, 1985: 12).[3]

The second explanation uncovers the fundamental inability of the

[3] It is important to guard against a purely *etat*ist posture which can lead to dogmatism. The Indian state has also played a major role in agriculture, infrastructure, finance, and services. *See* Chapter 3.

state to raise adequate resources and to act in the interest of the masses. First, the state was unable, in contrast to earlier periods, to tap appropriate sources of public investment. As has already been pointed out, this stemmed from a slowing down of industrial production. This in turn led to disincentives for private investment, given a direct relationship between public investment and private corporate investment (Rangarajan, 1982; Harriss, 1983). The state had relied primarily on a combination of foreign aid and deficit financing in order to implement its public investment policy. Foreign aid accounted for 43 per cent and deficit financing for 42 per cent of the total public sector plan outlays in the Second and Third Five Year Plans (Harriss, 1983). Flows of foreign aid as a percentage of national investment had increased from 5.9 per cent between 1950–1 and 1955–6 to 17.2 per cent between 1960–1 and 1965–6, and to 23.6 per cent between 1966–7 and 1969–70. But growth was accompanied by indebtedness (Harriss, 1983). The state could not afford to finance investment through foreign aid. This fell (as a percentage of GDP) from its peak of 3 per cent in 1966–7 to a low of 0.5 per cent in 1972–3, and 0.6 per cent in the late '70s (Ahluwalia, 1985: 104).

Increasing direct taxation of the propertied urban and rural classes could have compensated for the shortfall in resources. On this rests one of the central problems of the Indian political economy: namely, the quality of state intervention. A combination of state inertia in the matter of raising taxes and the mechanisms for preventing the execution of such policies, even when they are half-heartedly pursued, has seriously handicapped the use of fiscal instruments in encouraging growth and social justice.

In this context, agriculture, which has contributed about a third of the national income, has been virtually free from direct taxation. Direct tax on agriculture formed less than 1 per cent of total agricultural income, while land revenue was derived from a flat rate levy and was regressive. This led to considerable pessimism about the fact that 'in the field of direct taxes the possibilities of raising additional resources through income taxes, corporation taxes and wealth taxes are somewhat limited' (Government of India, 1980: 27).

This pessimism has to be located in the relationship between the Centre and the State. Agriculture is a State subject under India's federal constitution. Politics at the State level are dominated by rural interests. The Centre's attempts to persuade the State governments to break through such inhibitions have been feeble (Joshi and Little, 1987: 375). Increased productivity and incomes emanating from the 'new' agricultural strategy of the mid-'60s led to a considerable increase in the income of rich peasants. The state, however, has signally failed to use fiscal policy to extract resources from the latter.

Moreover, the 'new' agricultural strategy has reinforced the power of the rural rich by enabling them to further resist such state policies (Harriss, 1983; Shetty, 1978; Byres, 1981). The middle and rich peasants started to emerge, under the leadership of Charan Singh, as a major force in Indian politics (Byres, 1988).

Secondly, the nature of the demand structure has also been a critical factor behind deceleration of industrialization. The growth of capital goods industries (up to 1965) was helped by a strategy of import substitution which did not require fresh increases in demand. Subsequently, however, the growth of these industries was hampered by an inability to generate additional demand. The exhaustion of import substitution meant that further growth in capital goods industries could only come from an increase in demand. This was not forthcoming (Rangarajan, 1982: 590). The limited demand base for industrial goods is evident from the fact that the richest 10 per cent of the rural population accounted for 32.2 per cent of total industrial goods consumption. In sharp contrast, the poorest 50 per cent accounted for only 22 per cent of the total. In the urban sector the top 10 per cent purchased 39.3 per cent of the industrial goods, while the bottom five deciles absorbed about 20 per cent of the total. Markets for industrial consumer goods have been concentrated in the upper income segments (Rangarajan, 1982: 596). The state used indirect taxation, coupled with deficit financing, to complement its investment resources. This was regressive and contributed to a lack of growth of consumption goods, while reinforcing dependence on the public sector for protection and public investments (Patnaik, 1979: 6). The share of indirect taxes in total tax revenue increased from 61.9 per cent in 1955–6 to 70.7 per cent in 1965–6 (Harriss, 1983).

Constraints upon agricultural growth have blocked expansion of the home market in industrial goods. The inegalitarian nature of the agrarian structure has meant that any increase in agricultural growth has given rise to a smaller profile of aggregate demand, in comparison to the same growth within a more egalitarian structure. This is because the rich save a larger proportion of their income; it is applicable to the demand for industrial goods, since after the basic food requirements have been met, the bulk of the incremental income of the poor goes towards purchasing industrial goods, while the rich, leaving aside their higher savings, have a preference for imported luxuries. The demand originating from within the agricultural sector for domestically produced industrial goods grows more slowly in a situation where the distribution of land and other assets in the rural sector is highly skewed (Patnaik, 1985: 8).

Fluctuations in agricultural output have compounded the slow expansion of demand for consumer goods. These were closely associated with demand for industrial consumer goods. For instance, a sharp decline in agricultural output between 1965–6 and 1966–7 was mirrored in the growth-rate of consumer goods industries in 1966 and 1967. Similarly, the decline in agricultural output in 1972–3 and 1974–5 was closely related to the output of consumer goods in 1973–4 and 1975 (Rangarajan, 1982: 595). The rate of agricultural growth has also contributed to the slow expansion of demand for consumer goods (Ahluwalia, 1985: 51).

Liberalization policies: creation, support, opposition, and impact

This section looks at the ways in which liberalization policies were shaped, the support and opposition they aroused, and their socio-economic implications. Support for, and opposition to, liberalization have varied among different groups and classes.

The importance of internal (domestic) or external liberalization should not be underestimated.[1] But this is undergoing change, as the Indian state is attempting to resolve mounting economic problems, including debt. Decontrols and loosening of the state's hold over the economy have influenced the industrial and service sectors directly and the agricultural sector indirectly.

Among the factors motivating liberalization, there have been two of particular importance: (1) limits on the mobilization of investment resources for sectors such as steel and power; and (2) exposing the economy to internal liberalization, in particular industries, in terms of freedom from state controls (but the latter has been less prominent in easing entry of foreign capital which could compete with domestic capital). State control over the activities of national firms, including production, and decisions on production and expanding size, have been combined with limited devaluation and the lowering of import barriers (Kohli, 1989).

Liberalization started to surface in a big way during the late '70s under the Janata administration and was intensified under Indira Gandhi and Rajiv Gandhi in the '80s. A number of policy initiatives steered the economy towards liberalization. On completion of the loan agreement for SDR to the amount of US $ 5 billion with the IMF, the Indira Gandhi (Mark II) government undertook economic measures during 1981–2, which included the decontrol of steel and cement prices. Controls on both entry and expansion by national firms were

[1] Baru argues that liberalization in India has been a rather limited and halting process. *See* Chapter 3.

relaxed. There was greater emphasis on efficiency of investment, accompanied by a move from administrative to financial interventions (Kohli, 1989). In 1984, Rajiv Gandhi, the new Prime Minister, pledged to further the new approach. This was exemplified by the 1985–6 budget which aimed to reduce controls and regulations, lower corporate taxes, introduce a freer policy over technology imports, and confine public investment to essential infrastructure, such as energy (Rubin, 1985: 947).

The response to liberalization policies has to be put in the context of competition for state resources among the propertied classes.[5] Both internal and external factors have shaped liberalization. Developments following liberalization can be viewed as a product of the contradictions arising out of pursuing a path of uneven and dependent development. The use of deficit financing as a means of implementing public investment has led to a redistribution of income in favour of the propertied classes through inflation and transferring to these sections an incremental surplus in excess of the deficit. This also has implications for consumption demand and, in turn, investment. The lack of a radical transformation of agriculture has aggravated the problems of falling growth. Limits of the domestic market and the disproportionately high investible resources in the hands of the big bourgeoisie have made the latter demand greater freedom within and outside the country. The expansion of monopoly capital within the country is expected to proceed at the expense of other segments of the capitalist class and the state sector. But in terms of the external sector it is likely to function through collaboration with metropolitan capital in third countries (Patnaik, 1984, 1985, 1986; especially 1985: 8–13).

Rubin (1985) has argued that liberalization could relieve the state from its reliance on domestic classes for raising investment resources. This is questionable because it overlooks the costs and benefits of the policy among different groups of domestic capitalists.

Established monopoly capitalists are likely to favour selective liberalization while opposing blanket proposals, and especially import liberalization. The discourse between metropolitan capital and the domestic monopolists reveals the latter's enthusiasm for joint export ventures; the former has demanded a larger entry into India's market itself. Import liberalization or local production for the domestic market by multinational-dominated firms may be acceptable to entrenched manufacturers provided that it is at the expense of public sector units or small capitalists. However, monopoly capital would be

[5] The economic and political implications of liberalization have been debated in many seminars in India. These have expressed doubts about the policy (*Economic and Political Weekly*, 1989a).

against any blanket liberalization which could threaten its own entrenched position (Patnaik, 1985).

New houses may seek to break the existing monopolies in the domestic market with the aid of metropolitan capital.[6] They are likely to be more enthusiastic about 'opening up' the economy, in contrast to the traditional monopoly bourgeoisie with a strong vested interest in protecting their domestic markets. The class base of the state may be narrowed as a result of such changes. The likelihood of imposing industrial discipline may lead to 'new opportunities for Indian democracy' (Patnaik, 1985: 15). But it is uncertain whether this will take the form of a radical and sustained political opposition to liberalization.

The response of different segments of the industrial and rural classes to liberalization needs to be documented. Moreover, this has to be related to changing political conditions. As has already been pointed out, the liberalization policies have centred on the industrial rather than the agricultural sector. In the long run, their impact on the latter is expected to emerge more clearly.

Business groups have been major supporters of liberalization, particularly under the Rajiv Gandhi government. Their general response, for example, to the latter's 1985–6 budget was favourable. However, there has been uncertainty about the extent to which the economy should be opened up to external goods and capital. The interests of various groups within business have diverged. Those in import substitution industries are likely to oppose liberalizing imports while those producing for external markets may not. Many tend to combine production for the domestic and external markets. Those from more traditional business houses have indeed feared competition from the newly established industries with new technology which may be prepared to compete internationally (Kohli, 1989).

Members of the 'liberal' and other professions as well as the civil service gave positive support to Rajiv Gandhi (Kohli, 1989). This group formed the section of the domestic market which created the plank on which domestic demand was to be built. Thus, the urban proprietary classes have been treated favourably. They have been allowed to keep a proportionately greater share of income and to operate their businesses more freely. This would cushion them against the effects of the withdrawal of subsidies and protection.[7]

[6] New entrants to the business class have emerged as major players in the industrial scene. *See* Chapter 3.

[7] *See* Kohli (1989: 318) and Rubin (1985: 955). According to Paranjape 1988, liberalization has already been implemented in small doses and has led to a minor distribution of power and income between the politicians/bureaucrats, on the one side, and the traders/industrialists/big farmers, on the other.

A number of sections of urban and rural-based classes, including the intelligentsia, industrial labour, and the rural and urban poor, have opposed liberalization policies. The rank and file of the Congress have expressed their regret. The Congress leadership recognized that aspects of liberalization, such as the import of technology and encouraging multinationals, were unpopular among the party's grass-roots workers who bear the responsibility of confronting people deprived of basic needs. The policies of liberalization adversely affected support for the Congress, mainly among the middle peasants and the rural poor belonging to the Scheduled Castes. The 1985–6 budget, for instance, was seen as being biased in favour of the rich in general, and the urban rich in particular. Liberalization was undertaken against a background of increasing alienation of the middle peasants from the Congress-dominated Centre (Kohli, 1989).

Liberalization has critical implications for growth, external debt, and poverty. In this connection, the distinction between 'self-reliance' and 'self-sufficiency' is thought-provoking, but the call for domestic decentralization and exposure of India to the world economy to make it more efficient is questionable.[8] The focus on export-led growth, particularly under worsening global market conditions, has benefited a limited number of products of a few individual manufacturers, but the overall rate of growth is unlikely to be significantly affected (Chakravarty, 1987: 73).[9] The import intensity of domestic production, induced by industrial and import liberalization, has also reinforced dependence on international commercial borrowings. This carries obvious debt servicing obligations (*Economic and Political Weekly*, 1989).

India's changing relationship with the IMF is likely to exercise a major influence on the shaping of 'free-market' policies in developing countries. India drew Rs 8.15 billion in 1980–1 from the IMF Trust Fund and the Compensatory Financing Facility, and its acceptance in November 1981 of a large Extended Fund Facility amounting to SDR US $ 5 billion over 3 years for about Rs 50 billion provoked considerable debate. However, only US $ 3.9 billion of this was used before India terminated the agreement in May 1984 (Joshi and Little, 1987: 374).

India's recent loan of US $ 1.8 billion from the IMF has aroused new uncertainty about the implications of such a transaction. A second loan is also being sought. This has exposed the ways in which a number of stringent conditions can be imposed on domestic policies. These are expected to include the deregulation of industry,

[8] *See* Chapter 5 .

[9] *See also* Patnaik, 1986; Bagchi, Ghosh, and Dasgupta, 1985; Sau, 1982.

greater encouragement to foreign investment, the liberalization of foreign trade regimes, major reform of public sector enterprises (including privatization of a few), and some sweeping reforms of the financial sector. Some of these measures have already been set in motion.

The ratio of India's external debts to GDP at the close of 1984–5 was 17 per cent. This was low compared to most developing countries. The debt–service ratio, which measures the ratio of interest and principal payments due in a year to export receipts for that year, was 15.5 per cent (Joshi and Little, 1987). But India's external debts rose more than sixfold in the decade beginning March 1980. First, India's external debts at the close of December 1987 were US$ 50.4 billion, making it the fifth largest target debtor among the developing countries. Secondly, outstanding external debts grew by US$ 7 billion or 16 per cent between (the end of) 1986 and (the end of) 1987. Thirdly, borrowings in international financial markets rose approximately twice as fast as the total debt in 1987: i.e. by U.S. $ 2.7 billion or 31 per cent (*Economic and Political Weekly*, 1989). Fourthly, debt servicing, including interest payments, amortization and IMF repayments and interest, but excluding interest on NRI accounts, was 27 per cent of current exchange receipts (World Bank, 1989: 34).

According to Pranab Mukerjee, a former Finance Minister under Rajiv Gandhi and currently India's Commerce Minister, the debt–service ratio, excluding deposits by Non-Resident Indians (NRIs), was 23 per cent of current receipts in 1988–9. India's foreign debt amounted to 20 per cent of its GDP.

Social consequences of liberalization

Adverse external charges are transmitted to the domestic economy. These have accentuated poverty in many developing countries. Concern has also been voiced about the socio-economic impact of structural adjustment programmes (insisted upon by the IMF and World Bank) which have set out to restore the balance of payments equilibrium of highly indebted countries. (Roy, 1990a; Singer and Roy, 1993).

There are grave doubts about whether liberalization can alleviate India's poverty. Growth models have failed to tackle this problem. Indian planners have acknowledged that growth has failed to 'trickle down'. A number of programmes have been launched in attempts to reduce inequality and poverty. These set out to meet minimum needs, raise employment and reduce poverty. The Integrated Rural Development Programme (IRDP) has been a major example of such programmes.

From the Third Plan (1961/1962–1965/1966) onwards, attempts have been made to grapple with 'minimum needs' programmes. The Plan stated that the objective of India's development must necessarily be to provide the masses with the opportunity of leading a good life (Government of India, 1961: 1). Alleviation of poverty and the attainment of economic self-reliance were major aims under the Fifth Plan (Government of India, 1974: 1). A strategy was sought that could make a perceptible impact on poverty within a specified time frame (Chakravarty, 1987: 5). The Sixth, Seventh, and Eighth Plans have continued to emphasize this theme.

Attempts to reduce inequality in agriculture, the key sector in terms of employment, through land reform made only a limited impact. '*Garibi hatao*', a slogan coined in the early '70s by Indira Gandhi, provided the plank for political mobilization and public outlays of poverty-based programmes.

Programmes to reduce poverty have aimed at 'target' groups, including small farmers, marginal farmers, agricultural labourers, and Scheduled Castes and Scheduled Tribes. They have been hampered by basic weaknesses in design and execution (Vaidyanathan, 1983: 993). The failure of these programmes has been marked by a lack of structural change and political commitment (Paul and Subramanian, 1983: 350).

The number of people estimated as poor by minimum nutrition standards was approximately 273 million in 1983–4 (Rao, 1988). Casual wage earners, who lack assets and are dependent on their families for basic needs, have been the most vulnerable (World Bank, 1989). Sharp inequalities in inter-regional poverty have also prevailed, with the eastern and central States facing severe developmental and financial constraints (Rao, 1988).

Against this background, the vocabulary of 'poverty alleviation', though hollow sounding, has been absent from the rhetoric of liberalization. Indeed, the focus of liberalization has been openly anti-egalitarian, encouraging the production and import of durable consumer and other goods aimed at the rich and upper middle classes. Stimulation of demand through abolition of taxes has been a major instrument to boost incomes and hence demand of the upper and lower middle classes. The rural sector has been virtually ignored with an estimated mere 10 per cent of the population forming the target market (Rao, 1988).

Scheduled Castes and Scheduled Tribes are likely to be adversely affected by liberalization. Increased reliance on the private sector suggests that employment could grow in organizations where no places are reserved for them and where caste discrimination is pervasive, according to case studies. The problems of implementing policies to help such groups is illustrated by the abortive attempt of the National Front government, led by Viswanath Pratap Singh, in 1990 to allocate a significant proportion of jobs to the backward castes. This led to violent protests from the forward castes. Pressures to increase the efficiency of the public sector by eliminating redundant employees or limiting new hiring are also likely to intensify the existing violent communal conflicts.

Concern has also been felt over a deficit in the central budget as a result of government welfare programmes, such as the welfare of Scheduled Castes and Scheduled Tribes, and the supply of rural water and rural sanitation. Increased allocation to the States has been advocated in order to cover such programmes, but no influential person in government has suggested that these programmes should be given priority in the same way as power generation (Rubin, 1985: 955). Programmes that directly benefit the poorest communities in India may be sacrificed at the altar of growth.

To conclude, it is important to distinguish between the myth and the reality of liberalization, which has emerged as a panacea to resolve India's deep-seated problems. A more searching analysis is required of the diverse economic, social, and political factors which have stunted India's development. This is essential given India's worsening domestic political stability, fuelled by secessionist, inter-caste, and inter-communal friction, which has aroused uncertainty about the future. In essence, liberalization cannot be separated from the discourse on development and politics in India.

REFERENCES

AHLUWALIA, I.J., 1985: *Industrial Growth in India: Stagnation Since the Mid-Sixties* (Delhi: Oxford University Press).

BAGCHI, A., A. GHOSH, and S. DASGUPTA, 1985: 'Industrial policy and the economy', *Social Scientist* 13 (Sept.): 9, No. 148, pp. 3–15.

BARDHAN, P., 1984: *The Political Economy of Development in India* (Oxford: Blackwell).

BYRES, T., 1981: 'New technology, class formation and class action in the Indian countryside', *The Journal of Peasant Studies* 8 (July): 4, pp. 405–49.

———, 1988: 'Charan Singh, 1902–87: An Assessment', *The Journal of Peasant Studies* 15 (Jan.): 2, pp. 139–89.

CHAKRAVARTY, S., 1987: *Development Planning: the Indian Experience* (Oxford: Clarendon Press).

ECONOMIC AND POLITICAL WEEKLY, 1989:* 'External Debt: Growing reliance on commercial borrowing' 24 (4 Feb.): 5, pp. 216–7.

*Editorial.

————, 1989a** 'Liberalization road to economic ruination' 24 (19 Aug.): 33: pp. 1873–4.

**Report of a seminar by BM.

GOVERNMENT OF INDIA, *c.* 1961: *Third Five Year Plan: 1961/1962–1965/ 1966 1966* (New Delhi: Planning Commission).

————, *c.* 1969: *Fourth Five Year Plan: 1969–74* (New Delhi: Planning Commission).

————, *c.* 1974: *Draft Fifth Five Year Plan: 1974–1979* Vol. 1 (New Delhi: Planning Commission).

————, *c.* 1980: *Sixth Five Year Plan: 1980–85* (New Delhi: Planning Commission).

HARRISS, J., 1983: 'Indian industrialization and the state', paper presented at Workshop on State in South Asia, *Occasional Paper* No. 22 (University of East Anglia; Norwich) (mimeo.).

JOSHI, V. and I.M.D. LITTLE, 1987: 'Indian macro economic policies', *Economic and Political Weekly* 22 (28 Feb.): 9, pp. 371–8.

KOHLI, A., 1989: 'Politics of economic liberalization in India', *World Development* 17 (March), 3: pp. 305–28.

KUMAR, D. and M. DESAI, (Eds.), 1983: *The Cambridge Economic History of India, 1757–1970, Vol. 2* (Cambridge: Cambridge University Press).

LAL, D., 1988: 'Economic Growth in India', Discussion Paper (University College, London) (mimeo.).

PARANJAPE, H.K., 1988: 'Indian "liberalization": Perestroika or Salaami Tactics', *Economic and Political Weekly* 23 (Nov.): 45–7, pp. 2343–58.

PATNAIK, P., 1979: 'Industrial development in India since independence', *Social Scientist* 7 (June): 11, No. 83, pp. 3–19.

————, 1984: 'Market question and capitalist development in India', *Economic and Political Weekly* 21 (Aug.), 31–3: pp. 1251–60.

————, 1985: 'On the political economy of economic liberalization', *Social Scientist* 13 (July–Aug.): 7–8, Nos. 146–7, pp. 3–17.

————, 1986: 'New turn in economic policy: context and prospects', *Economic and Political Weekly* 21 (7 June): 23, pp. 1014–19.

PAUL, S. and A. SUBRAMANIAN, 1983: 'Development prospects for the poor: Do strategies make a difference?', *Economic and Political Weekly* 18 (5 March): 10, pp. 349–57.

RANGARAJAN, C., 1982: 'Industrial growth: another look', *Economic and Political Weekly* 17 (April): 14–6, pp. 589–604.

RAO, V.K.R.V., 1988: 'India since independence: retrospect and prospects'* (Indian Council of Social Science Research; New Delhi) (mimeo.).

*Keynote address to Seminar on 'India Since Independence'.

ROY, S., 1984: 'Pricing, Planning and Politics: A Study of Economic Distortions in India', Occasional Paper 69 (London: The Institute of Economic Affairs) (mimeo.).

————, 1990: *Agriculture and Technology in Developing Countries: India and Nigeria* (New Delhi: Sage).

————, 1990a: 'Structural adjustment programmes, the economy and the rural sector', *Africa Quarterly*, 30: 1–2, pp. 1–12.

RUBIN, B.R., 1985: 'Economic liberalization and the Indian state', *Third World Quarterly* 7 (Oct.): 4, pp. 942–57.

SATHYAMURTHY, T.V., 1989: 'Impact of Centre–State Relations on Indian Politics: An interpretative reckoning, 1947–87', *Economic and Political Weekly* 26 (23 Sept.): 38, pp. 2133–47.

SAU, R., 1982: 'Economic crisis and economic recovery' *Economic and Political Weekly* 17 (April): 14–6, pp. 573–88.

SHETTY, S., 1978: 'Structural retrogression in the Indian economy since the sixties', *Economic and Political Weekly* 13 (Feb.): 6–7, pp. 185–244.

SINGER, H.W. and S. ROY, 1993: *Economic Progress and Prospects in the Third World: Lessons of Development Experience since 1945* (Aldershot: Edward Elgar).

VAIDYANATHAN, A., 1983: 'The Indian Economy since Independence (1947–70)', in Kumar and Desai (Eds), 1983, pp. 947–94.

VARSHNEY, A., 1984: 'Political Economy of Slow Industrial Growth in India', Economic and Political Weekly 21 (1 Sept.): 25, pp. 1511–17.

WORLD BANK, 1989: *India: Poverty, Employment and Social Services: A World Bank Country Study* (Washington D.C.: The World Bank).

Liberalization and Technology Policy: Redefining Self-Reliance • 5

E SRIDHARAN

This chapter examines the shifts in India's technology policies during the '80s, particularly since 1985, with a view to determining whether a fundamental shift in the terms of discourse on technology is indeed taking place. I focus on certain central concepts, themes, objectives, and policy positions in the discourse on science and technology in India, and whether and in what ways they are being rejected, reshaped, or redefined by the recent changes in technology policy.

The chapter is divided into four sections. In section 1 a very brief historical overview of Indian technology policy since Independence is provided. This is divided into three phases, the last being the liberalizing phase of the '80s. Section 2 summarizes the conclusions and recommendations of the critiques that emerged during the '80s attacking the existing paradigm and policy regime, and providing an intellectual foundation for liberalizing policies. Section 3 deals with an analysis of the political economy of the shift in technology policies, identifying two sets of compulsions—economic and strategic—and how they dovetail with the recommendations of the revisionist critique. In Section 4, an attempt is made to delineate the shift in the terms of discourse taking place, focusing on the ongoing redefinition of the central concept of self-reliance.

Technology policy in India: A historical overview

India's technology policy can be roughly divided into three phases. The first, one of relatively liberal policies, lasted from Independence to the end of the '60s. The second phase, one of intensified import substitution, lasted from the end of the '60s to the early '80s. The third, one of liberalization of technology import, began in 1980 and gained momentum in the course of the decade, especially after 1985. I shall outline very briefly the main characteristics of the policy regime during these three periods, placing technology policy in the context of overall economic and industrial policies.[1] When referring to technology import policy, I refer to disembodied technology rather than technology embodied in capital goods.

The first phase of Indian technology policy, the Nehru phase, lasting until the late '60s was characterized by relatively liberal broad front import of technology without restrictions on the terms and conditions of contracts. It was the period of building up a diversified, heavy and capital goods industrial base by public sector-led import substitution. India had almost no technological base to begin with and no option but to import technology for industrialization on liberal terms. Science and technology policy concentrated on building a network of Research and Development (R & D) institutions, today the base for indigenous technological development. R & D spending concentrated heavily on vital areas where there was a threat of cut-off of technology and equipment, e.g. atomic energy and defence, and in other vital areas such as agriculture. Specific mission-oriented institutions were created for these areas. There was also a vast expansion of higher technical education, including specialized institutions such as the Indian Institutes of Technology (IITs), in order to generate the skilled workforce.

The R & D institutional infrastructure thus created consisted of two broad types of institutions: (a) general purpose broad front institutions and, (b) special-purpose mission-oriented institutions. The

[1] For a comprehensive history of science and technology policy in India up to the early '80s, *see* Nayar (1983). Also Government of India, Departments of Science and Technology, Atomic Energy, Electronics, Space, Telecommunication and Industrial Development, and Ministry of Defence; *Annual Reports* (various years); Planning Commission, *Annual Plan* (various years); Department of Science and Technology *R & D Statistics 1988–89;* Govindarajulu (1990); and Reserve Bank of India (1985).

Council for Scientific and Industrial Research (CSIR) is the principal representative of the first, a category in which the university system also falls. The second category of institutions comprise the Department of Atomic Energy (DAE), the Defence Research and Development Organization (DRDO), the Indian Council for Agricultural Research (ICAR), and the Indian Council for Medical Research (ICMR). After 1972, the Departments of Space (DoS) and Electronics (DoE) also joined the second category, subsequently followed by those of Environment, Ocean Development, and Biotechnology. All these are apex agencies, with networks of laboratories, institutions, and public sector enterprises under them.

The CSIR was the largest of these institutions and consisted of four types of laboratories:

1. discipline-oriented, such as the National Physical and National Chemical Laboratories (NPL and NCL);
2. engineering laboratories;
3. commodity research laboratories;
4. multipurpose regional laboratories.

The CSIR, whose objective was to lay the basis for broad front industrial technology development, was weakened by having no links with either production or teaching.

Almost all (i.e. 90–5 per cent) R & D in the '50s and '60s was carried out in the central public sector. State-level and private sector R & D were only marginal. Over 75 per cent of central R & D spending came under four apex bodies: the CSIR; and the three main mission-oriented bodies, namely the DAE, DRDO, and ICAR. During this period, the DAE was consistently in the lead, ahead of the CSIR, accounting for about a third of central R & D spending. Interestingly, despite Nehru's stress on industrialization, agricultural R & D was not neglected. The ICAR was ahead of the DRDO and just behind the CSIR in R & D spending until after the India–China war. Thus, the base was laid for the application and extension of Green Revolution technologies in the late '60s.

The second phase in Indian technology policy began in the late '60s and lasted until the early '80s. Towards the end of the '60s a number of developments came together to impel a change in technology policy towards greater indigenization, a thrust towards self-reliance, and strict controls on the import of technology and foreign private investment. This resulted in a strictly import-substitution approach to technology development across the board that lasted until the early to mid-'80s.

First, the experience of the 1962, 1965, and 1971 wars had

underlined, by the late '60s and early '70s, the importance of technological self-reliance for long-term security.

Second, the food and foreign exchange crises of the mid-'60s, and the persistently tight foreign exchange position, reinforced the emphasis on self-reliance defined as import-independence or self-sufficiency.

Third, by the late '60s, a diversified industrial base had been built. In the light of this, the terms of technology licensing and foreign investment agreements that had been struck during the previous two decades began to look excessively favourable to foreign capital. After all, they had been concluded at a time when India followed more liberal licensing and foreign investment policies due to its weak technological base. A need was now felt to renegotiate the terms of foreign collaboration in order to strengthen indigenous capability, assure technology transfer, remove restrictive clauses in collaborations, and save foreign exchange.

Fourth, the greater intellectual influence of the Left and the small-scale industry lobby since the end of the '60s had an impact on technology policy thinking. The Left argued essentially that excessively liberal import of technology and foreign investment had had adverse consequences in that they were injurious to:

1. the balance of payments;
2. indigenous technological development; and, therefore,
3. sovereignty; and
4. they were often a way of dumping obsolete and inappropriate technologies.

The above influences, along with a perspective hostile to rapid economic growth, brought about a technology policy in the 1969–74 period that strictly controlled the import of technology. The controls included restricting royalty rates to a ceiling of 5 per cent of sales, a 40 per cent tax on royalties, and a five year contract duration limit; they also discouraged compensatory lump-sum payments, repetitive import of technology, export restrictions and other restrictive clauses that were so common in the earlier period. The new regime generally insisted on greater association of Indian technical personnel to absorb the technology and the Indian party's right to sub-license the technology on mutually agreed terms. The *Patents Act* was amended in 1970, reducing the duration of patents from 16 to 14 years (7 years for food and drug patents), making licensing compulsory after three years if a patent remained unexploited and generally reducing protection for foreign patent holders. It also favoured outright purchase rather than licensing or foreign equity.

All these changes were made within the overall shift against business and foreign firms represented by the *Monopolies and Restrictive Trade Practices Act* (MRTP) of 1969 and the *Foreign Exchange Regulations Act* (FERA) of 1973. The newly nationalized banks (1969) were directed to favour small-scale industry in sanctioning credit. In 1973 the list of items reserved for small-scale industry was expanded from 51 to 147, and in 1977 the list was further expanded to 180.

The Janata government hugely expanded the list from 180 to 504 items. It also came forward with a new policy on technology import that was seemingly more restrictive but left loopholes for the government to be more liberal if it so desired, though, in practice, it turned out not to be the case. A 1969 policy statement had drawn up three lists of industries according to whether

1. there was no foreign collaboration,
2. there was only technical collaboration, or
3. both financial and technical collaboration were permitted.

In 1978, this was replaced by a 200-product list covering 22 industries in which no foreign collaboration was permitted. This included pharmaceuticals and electronics. However, the government allowed exemptions for cases in which domestic technology was too closely held and for export promotion, but did not use these loopholes. It allowed, however, slightly more liberal import of capital goods.

In 1980, the Congress government, although it expanded the list of items reserved for small industry to 834, did begin to liberalize, albeit in fits and starts, the country's trade, industrial licensing, and technology import policies under economic, strategic, and intellectual influences that I shall spell out below. The electronic components industry was allowed to import technology and capital goods in 1981 and the telecommunications end-user equipment industry was opened to private enterprise in 1984. Beginning in 1980, a major R & D and industrial thrust was launched in defence and related high-tech areas such as nuclear, space, aeronautics, electronics, computers, and telecommunications.

In January 1983, the government issued a 'Technology Policy Statement' which contained two substantive points. It noted that technological self-reliance did not mean technological self-sufficiency, thereby justifying import of technology where appropriate. It also proposed the creation of a National Register on Foreign Collaboration (Government of India, 1983). However, this gradually liberalizing third phase of the '80s is not marked by a formal or substantive policy change or legislation.

The spate of liberalization measures announced by the minority Congress government led by P. V. Narasimha Rao (since taking power in July 1991) as a part of a directional change in economic strategy, takes liberalization of technology import further as part of a general invitation to foreign direct investment in India. Automatic permission is now to be given for foreign technological collaboration in 34 high priority industries up to a lump-sum payment of 10 million rupees, 5 per cent royalty for domestic sales and 8 per cent for exports (net of taxes) over a 10-year period from the date of agreement, or 7 years from commencement of production. These 34 industries are also open to 51 per cent foreign equity if the latter covers the foreign exchange requirement for imported capital goods. In all other industries, automatic permission for foreign technology will be given if no free foreign exchange is required for any payments. Thus, the latest measures take the liberalization of technology import in the third phase of technology policy further and represent a step backward from the earlier effort to detach technology from foreign equity. I shall elaborate on technology policy in this latter phase in the following two sections.

Looking back at the three phases of technology policy in terms of R & D spending as a percentage of GNP and the number of foreign collaboration approvals per annum as indicators, it is clear that there are roughly three phases and that the period since 1980 (and especially since 1985) has been one of accelerated R & D and technology import.

Table 5.1.

Trends in the Growth of Science and Technology (S & T) Expenditure in India

Five Year Plan Period	Total S & T Expenditure (Rs million)
1	200
2	670
3	1440
4	3730
5	13810
6	37160
7	75350
8 (projected)	20,0000

NOTE: S & T expenditure covers R & D expenditure plus related scientific and technological activities. R & D expenditures are internationally more comparable (Govindarajulu, 1990: M35).

SOURCE: Government of India, 1988: 37ff.

Table 5.2

Trends in National Expenditure on Research and Development (R & D) Activities

Year	R & D Expenditure (Rs million)	As Percentage of GNP
1950–1		0.02*
1955–6		0.12*
1958–9	229.3	0.18
1965–6	683.9	0.31
1970–1	1396.4	0.38
1975–6	3566.9	0.53
1980–1	7605.2	0.62
1985–6	20687.8	0.89
1987–8	29369.4	1.00†
1988–9	34718.1	1.00†

NOTE: *Estimated figures for S & T, *not* R & D expenditures using Power Law Model †, ie. based on Quick Estimates of GNP.

SOURCE: Govindarajulu 1990; Government of India, 1990: 48

Table 5.3

Changing Pattern of R & D Expenditure by Government and Industry

Year	Percentage share by		
	Central Government	State Government	Private Sector
1950–1	100	Nil	Nil
1965–6	91	5	4
1970–1	81	9	10
1975–6	81	7	12
1980–1	76	8	16
1984–5	80	7	13
1985–6	80	8	12
1988–9	83	7	10

SOURCE: Computed from Government of India, 1990: 47.

Table 5.4

Changing Pattern of Distribution of R & D Expenditure by Major S & T Agencies (percentage)

Year	*Agencies*							
	DAE	CSIR	DRDO	ICAR	DST	DOS	Department of Environment	Department of Biotechnology
1958–9	41	27	8	20				
1965–6	30	21	14	10				
1970–1	32	24	20	21				
1975–6	24	17	23	15	2	16		
1980–1	17	16	18	22	9	13		
1986–7	14	11	27	10	7	19	6	1
1988–9	14	10	17	9	7	20	7	2

SOURCE: Computed from Nayar, (1983 (Vol. I)), 248–9; Government of India, 1990, 49.

Table 5.5

Foreign Collaboration Approvals by Government

	1	2	3	4	5	6
Calendar Year	Total No. of Approvals		Total No. of Approvals with Fin.-cum-Tech. collaboration		Total No. of Approvals with Tech. collab-oration only	
	DSIR	IIC	DSIR	IIC	DSIR	IIC
1948–55	284					
1956–	82					
1957–	81					
1958–	103					
1959–	150					
1960–	380					
1961–	403		165		238	
1962–	298		124		174	
1963–	298		115		183	
1964–	403		123		280	
1965–	241		71		170	
1966–	202		49		153	
1967–	182		62		120	
1968–	131		30		101	
1969–	134		29		105	
1970–	183	183	32	32	151	151
1971–	245	245	46	46	199	199
1972–	257	257	36	37	221	220
1973–	265	265	34	34	231	231
1974–	359	359	55	55	304	304
1975–	271	271	40	25	231	246
1976–	277	275	39	40	238	235
1977–	267	267	27	27	240	240
1978–	307	307	44	44	263	263
1979–	267	267	32	32	235	235
1980–	526	526	65	74	461	452
1981–	389	389	56	57	333	332
1982–	588	590	113	113	475	477
1983–	673	682	129	129	544	553
1984–	740	752	148	151	592	601
1985–	1041	1026	256	238	785	788
1986–	960	957	256	240	704	717
1987–	903	853	259	242	644	611
1988–	957	926	289	282	668	644
1989–		605		193		412
1990–		666		194		472
Total	14118	10678		2285		8383

SOURCES: Government of India (1989) for columns 1,3,5; 'Indian Investment Centre' (unpublished) for columns 2,4,6.

Table 5.5

Foreign Collaboration Approvals by Government
(continuation of Table 5.5)

7a	7b	8a	8b	9a	9b	10
Approvals		Rejections		Other Disposals		Out of cases Disposed of
No.	%	No.	%	No.	%	No.
284						
82						
81						
103						
150						
380						
403						
298						
298						
403						
242						
202						
182						
131						
135	63					
183	48					
245	60					
257	56					
265	—					
352	50	233	34	107	16	692
244	52	139	30	88	18	471
232	54	101	23	99	23	432
242	59	67	16	105	25	414
273	64	57	13	99	23	429
230	50	73	16	156	34	459
449	63	98	14	164	23	711
243	76	78	24	-	0	321
346	78	72	16	27	6	445
417	69	150	25	34	6	601
465	66	208	30	31	4	704
621	80	131	17	21	3	773
637	70	206	22	73	8	916
603	75	155	19	48	6	806
650	86	73	10	35	4	758
364	74	125	25	6	1	495

Government of India (Ministry of Industry), various years; Government of India (Department of Industrial Development), various years* (for columns 7a, 7b, 8a, 8b, 9a, 9b and 10).

*NOTE (Table 5.5)

(a) Ministry and Department Annual Report figures give only the purely foreign collaboration cases disposed of by the Foreign Investment Board, Secretariat for Industrial Approvals, Department of Industrial Development, Ministry of Industry.

(b) IIC figures are the comprehensive figures, including both the purely foreign collaboration cases as well as composite and other cases with FC component disposed of by the Export Oriented Undertaking Board, Kandla Free Trade Zone Committee, Santa Cruz Electronics Export Processing Zone Committee, Special Approval Committee for Non-Resident Indians (NRIs), and other administrative ministries under delegated powers.

(c) The Government of India (1989) figures exceed the IIC's approval figures due to some Technical Development Fund cases being included. The discrepancy is not material to the argument.

(d) All cases of financial collaboration can be considered financial-cum-technical collaboration except for a few cases since 1987 which do not materially affect the argument, namely that there was a rising trend in foreign technical collaboration.

(e) The total of 14,050 approvals for 1948–90 in Column (1) includes the (IIC) figures for 1989 and 1990 in Column (2).

Table 5.6

Statement showing remittances made to Foreign Companies & Industrial R & D

(Rs crores)

Year	Profits		Dividends	Royalty		Technical Know-How
	(1)	(2)	(1+2)	(3)	(4)	(3+4)
1965–66	13.50	19.40	29.90	2.95	6.98	9.93
1966–67	14.47	28.77	43.24	5.13	10.43	15.56
1967–68	15.95	32.70	48.65	4.32	14.68	19.00
1968–69	12.96	30.25	43.21	4.78	17.97	22.75
1969–70	12.72	31.14	43.86	5.80	13.05	18.85
1970–71	13.12	43.48	56.60	5.23	20.63	25.86
1971–72	9.94	38.87	48.81	5.86	13.90	19.76
1972–73	15.94	39.08	54.62	7.33	11.33	18.66
1973–74	21.91	37.51	59.42	6.21	14.08	20.29
1974–75	7.19	18.46	25.65	8.46	12.56	21.02
1975–76	20.36	24.84	45.20	10.49	25.66	36.15
1976–77	19.39	48.47	67.86	15.88	37.80	53.68
1977–78	10.13	68.01	78.14	19.50	28.14	47.64
1978–79	10.24	54.35	64.59	12.65	55.52	68.17
1979–80	14.37	50.92	65.29	9.53	43.97	53.50
1980–81	12.10	55.92	68.02	8.88	104.93	113.81
1981–82	12.16	58.92	71.08	15.99	270.70	286.69
1982–83	19.12	70.31	89.43	39.72	258.58	298.30
1983–84	20.00	62.11	82.11	27.60	314.89	342.49
1984–85	16.68	74.58	91.26	28.49	300.60	329.09
1985–86	11.80	75.20	87.00	23.50	367.90	391.40
1986–87	10.60	85.50	96.10	40.10	358.40	398.50
1987–88	9.30	92.00	101.30	60.00	459.00	519.00
1988–89	42.00	126.00	168.00	190.00	317.00	507.00

Table 5.6

Statement showing remittances made to Foreign Companies & Industrial R & D

(continuation of Table 5.6)

	Interest by private sector in foreign currency	Total	R & D Expenditure by public & private sector industry
	(5)	(6)	(7)
1965–66	—	42.83	
1966–67	—	58.80	
1967–68	—	67.65	
1968–69	—	65.96	
1969–70	9.28	72.26	
1970–71	12.80	95.26	
1971–72	12.13	80.70	
1972–73	15.60	88.88	
1973–74	16.27	95.98	
1974–75	36.70	83.87	
1975–76	24.65	106.00	
1976–77	25.11	146.65	80.18
1977–78	22.70	148.48	95.70
1978–79	31.44	164.20	131.07
1979–80	25.22	144.01	165.56
1980–81	22.32	204.15	193.76
1981–82	41.08	398.85	254.55
1982–83	80.23	467.96	289.41
1983–84	81.51	506.11	337.98
1984–85	123.91	544.26	436.82
1985–86	218.70	697.10	425.99
1986–87	318.90	813.50	505.50
1987–88	320.00	941.30	575.51
1988–89	389.00	1054.00	697.87

NOTE: 1987–8 and 1988–9 figures are provisional and mostly rounded off to the nearest whole number.

SOURCE: Compiled from (unpublished) Reserve Bank of India 'Statistics' (various years); Government of India (Department of Science & Technology), *Research & Development Statistics* (various years).

Table 5.7

Industry-wise Foreign Collaborations Approved

Serial No.	Industry	1982	1983	1984	1985	1986	1987	1988	1989	1990
1	Metallurgical industries	49	20	26	53	45	29	27	30	26
2	Electrical equipment (including electronics)	107	129	157	205	175	183	183	99	88
3	Telecommunications	7	7	3	36	37	16	23	37	69
4	Transportation	28	39	63	101	53	39	38	30	22
5	Industrial machinery	107	115	138	152	108	132	141	59	75
6	Machine tools	29	44	34	32	13	10	21	9	24
7	Industrial instruments	29	37	56	52	20	47	43	35	38
8	Chemicals (other than fertilizer)	53	62	69	69	107	84	96	66	66
9	All others	479	229	206	326	419	313	354	240	258
Total		888	682	752	1026	977	853	926	605	666

SOURCE: Indian Investment Centre Data Bank (unpublished).

We see from Table 5.5 that the number of foreign collaboration approvals fell sharply from 403 in 1964 to 242 in 1965 and remained under 400 for all the years through 1979, exceeding 300 only in two years during this period. Thereafter, the number of approvals rose sharply in the '80s, peaking at 1041 in 1985, Rajiv Gandhi's first year in office, and falling below 400 only in 1981. The average number of approvals was 356 for the 1960–4 period, falling to 178, 262, and 278 respectively for the 1965–9, 1970–4, and 1975–9 periods, reflecting the restrictive second phase in policy. Thereafter, the average for 1980–4 and 1985–8 rose to 583 and 965 respectively (by DSIR figures). Going by IIC figures, which for the years since 1985 are slightly though not significantly lower (less inclusive) than DSIR figures, we obtain averages of 588 and 873 for 1980–4 and 1985–9 respectively. These figures reflect a rise in foreign technical collaborations during the '80s, and especially since 1985, the peak year for approvals. As is evident from Table 5.7, the most important increases in import of technology occurred in electronics, telecommunication, transportation, and chemicals, with 1985 being the most common peak year.

This pattern of liberalization of technology import in the '80s, especially after 1985, is corroborated by the figures of remittances by Indian companies to their foreign collaborators (Table 5.6). While profits, dividends, and royalties are a function of existing collaborations at any given time, the column on technical know-how payments more accurately reflects current infusions of technology. It shows relative stagnation from the late '60s to the mid-'70s, a jump from 1975–6 during the Emergency, a sharper jump in 1980–1 from Rs 44 crores in the previous year to Rs 105 crores rising to Rs 368 crores by 1985–6. Royalty outflows also show a progressive increase during the '80s.

If one looks at the number and percentage of approvals in the less inclusive approvals list of clearances for foreign collaboration issued by the Ministry of Industry's Secretariat for Industrial Approvals' Foreign Investment Board, one gains an idea of the prevailing attitude towards approval of foreign collaboration. The number of approvals fell from 403 in 1964 to less than 300 every year through 1979, with the average number of approvals for the five year periods 1960–4, 1965–9, 1970–4, and 1975–9 at 364, 178, 262, and 245 respectively. In the '80s, this rises to 364 and 655 for 1980–4 and 1985–9 respectively, being at over 600 every year since 1985 (except for 1989). More significantly indicative of liberalization as a policy is that the percentage of approvals granted by the Foreign Investment Board was in the range of 48 per cent to 64 per cent of cases disposed of every year from 1969 to 1980, never falling below 66 per cent in any year since 1980 (being within the 70–90 per cent bracket in most years).

These figures confirm the periodization of foreign technological collaboration policy given in this chapter. It needs to be noted that while the increase in the '80s was no doubt partly due to the economy and industry moving into a higher growth path, the figures for numbers and payments of collaborations are more than proportionate in their increase and can be taken to reflect a liberalizing policy. It also needs to be noted (namely, Notes to Table 5.5) that almost all foreign collaborations contain technology agreements due to long-standing policy on foreign collaboration, except for a few cases which do not materially affect the argument.

However, as a result of the restrictive post-'60s policy's preference for outright purchase or licensing of technology rather than having it bundled together with foreign investment, the number of purely technical, as opposed to financial-*cum*-technical, collaborations show relatively faster growth in the number of approvals until 1981. It is also noteworthy that (Table 5.6) R & D expenditure by public and private industry was seldom more than twice, usually much less than twice, the royalty-*cum*-technical know-how payments remitted by industry to foreign companies. This contrasts with the Japanese history of spending several times as much on domestic R & D as on technology import, and tends to support the argument that lack of effort at absorption of technology was the key weakness of India's technological development. Lastly, the total number of items of technology import and payments for their import are insignificant compared to several other industrializing Third World countries.

It can be seen from Table 5.2 that while there has been a steady rise in R & D expenditure as a percentage of GNP in the post-Independence period, the 1980–1 to 1988–9 period registers a steep rise from 0.62 per cent to 1.00 per cent. Furthermore, it is projected to rise to 2 per cent of GNP by 1995. In terms of the distribution of central R & D expenditure among major agencies and major activities, it can be seen (Table 5.3) that, while the DAE and CSIR have steadily lost their share over the decades, the DRDO and Space have sharply increased theirs throughout the '80s. The DRDO share jumped after the mid-'60s, and that of Space after the establishment of the DOS in the early '70s. In the '80s, the share of the DRDO increased from 18 per cent to 27 per cent by 1988–9, and that of Space from 13 per cent to 20 per cent by 1988–9. The bulk of the Central R & D spending increase in the '80s was concentrated on defence and space (including defence and aerospace electronics and the missile programme). If one takes into account that Atomic Energy still had a 14 per cent share as late as 1988–9, the defence-related complex of activities accounted for over half of central R & D expenditure. Also noteworthy is the emergence of the high-tech Department of Biotechnology and Department of Environment with shares of 2 per cent and 7 per cent respectively by 1988–9. We shall further analyse these shifts in the following sections.

Intellectual foundations of the shift in technology policy in the '80s

The liberalization of technology import and of foreign direct investment regulations in the '80s, especially marked after 1985, and the accompanying shift in the terms of discourse on science and technology, foreign direct investment, and the concept of self-reliance have their intellectual foundation in a number of studies on technology import policy carried out in the '70s and early '80s.[2] These studies

[2] *See* the series of articles on: 'Foreign Technology Suppliers' View of Indian Industry' in *Economic and Political Weekly* 20 (Nov. 1985: 45–7), especially the introductory and concluding articles by Ashok V. Desai, the two articles by Don Scott-Kemmis and Martin Bell, and the one by Ghayur Alam. *See also* Desai (1988) *in* Lucas and Papanek (1988); and Fransman and King (1984), especially the papers by Lall and Desai; also *World Development* 12 (May/June, 1984), 5/6 (Special Issue on Exports of Technology by Newly-Industrializing Countries), Sanjaya Lall (Ed.), especially 'Over-

were motivated by the manifestly poor performance of Indian technology which had been falling further and further behind world levels since the mid-'60s, and its reflection in the poor performance of Indian manufactured exports compared to that of several other LDCs during the '60s and '70s. The fact that this was the case despite the nationalist technological policies since the '60s, mainly aimed at achieving autarkic self-reliance in technology development, made the prevailing intellectual paradigm and policy regime increasingly suspect. At the same time, too, the fact that India embarked on several high-tech and defence-related indigenization programmes in the '80s made the technological gap a very serious issue and increased considerably the receptivity of technology policy circles to the revisionist critique.

The principal analytical project in India was that of the Indian Council for Research on International Economic Relations (ICRIER), National Council for Applied Economic Research (NCAER) studies concluded by a group of scholars led by Desai, which also included a number of European scholars. Also influential were the comparative studies of less developed countries (LDCs) with special reference to multinational corporations, technology transfer, technology development, and exports carried out by scholars based in Europe, especially the group coordinated by Sanjaya Lall. Many of these scholars were associated with the Institute of Development Studies (IDS), Sussex, and the Research Policy Institute, Lund. By the mid-'80s, these studies had contributed to undermining substantially the paradigm and policy regime that had taken root since the late '60s, of intensive import-substitution and self-reliance in technology which went hand in hand with hostility to foreign technology and direct investment. Supporters of the existing paradigm and policy regime were put on the defensive by India's relative technological and export stagnation compared not only to that of the developed world but even to several advanced developing countries, many of which had begun to overtake India in the technological level (depth) of their industry despite this having been a strong point claimed for Indian policy.

Specifically, the work done by the ICRIER–NCAER scholars and the Europe-based comparativists challenged the existing import-substituting paradigm and policy regime on several points of fact and interpretation.

---

view' and 'India', both by Lall; *World Development* 13 (March 1985): 3 (Special Issue on Microelectronics, International Competition and Development Strategies); Kurt Hoffman (Ed.), for an analysis of technology issues in the electronics industry. The above literature is not exhaustive but representative of the best work that is directly critical of (or gives evidence critical of) India's import-substituting technology policy.

First, far from having been swamped by foreign technology, India had received only a tiny trickle in comparison to the flow of technology into most other LDCs as measured by the number of technical collaboration agreements. It had also received only a tiny fraction of the flow of Foreign Direct Investment (FDI) into the third world. Thus, in comparison to LDCs in general it was a very small importer of technology and recipient of FDI.

Second, in terms of the outflow of payments on account of technology import, and also on account of profits and dividends, India lagged far behind most other major LDCs. Taken together with the first point, this implied that over the '70s and '80s, India was marginal in the flow of technology and investment to the third world.

Third, the inflow of technology and the liberality of its import should be judged not only by the number of collaboration agreements but also by their content (or depth). In these terms, technology transfer agreements can be graded into four levels, i.e. transfer of, (1) operating skills at the simplest level; (2) manufacturing technology; (3) designs and design methodology; and (4) R & D facilities. Measured qualitatively in this manner, the technology inflow into India is still less impressive (or alarming, depending on one's viewpoint), since Indian collaborations have tended to be shallow. They were very often limited to putting the technology imported into production rather than transferring the design methodology needed to understand the process of technological change in order to be able to fully absorb, adapt, and innovate, let alone spearhead research and development in the technology. Thus, both quantitatively and qualitatively, technology import had been very limited.

Fourth, this was because of the restrictive conditions imposed since the late '60s and early '70s on technology import and foreign direct investment. Restrictions on royalties, duration, and equity participation, and the discouragement of larger one time payments have led to the transfer of smaller and shallower technology packages, even if of current vintage. This is so even though, in a great many cases, the Indian firm has been willing and able to pay more if necessary and the supplier has been willing in principle to transfer design methodology and R & D capabilities. In cases where the supplier has been willing to transfer technology, subject to greater equity participation, the restrictions on the latter have acted as a dampener. Thus, whilst restrictive policies may have lessened the outflow of payments and helped conserve foreign exchange, they have also impoverished the content and vintage of technology packages transferred. A price has been paid for such saving in terms of enhancement of the competitiveness of Indian industry. An additional point to note is that the restrictions on royalties and duration, and the 40 per cent tax on the former, have checked the outflow of payments to a much smaller extent than expected because supplier firms compensate themselves by raising one time fees.

Fifth, the import of technology has not dampened innovative activities among importing firms. Nor has non-import led to a spurt in innovation. On the contrary, and in line with expectations of the importance of critical threshold sizes and scale economies in R & D, it was found that R & D performance was closely associated with the number of collaborations; and, conversely, technology-importing firms tended to engage in R & D to a greater extent than non-importers—thus strongly suggesting that the import of technology stimulated rather than retarded R & D. It was not, of course, a necessary condition: non-importers of technology were responsible for considerable minor innovation too. Thus one of the key contentions, or rather assumptions, of the opponents of technology import, that such import retarded indigenous technological effort, does not appear to be supported by the studies. This point was buttressed by the fact that several East Asian LDCs and Brazil had created domestic firms with greater technological depth and export capability despite far more liberal import of technology.

Sixth, multinational corporations (MNCs) are not the major source of technology, including imported technology. Rather, most of the collaborations have been with specialist oligopolies which are not necessarily either large or multinational in operations, but whose strength lies in the mastery of technology in specialized product and process niches.

Seventh, technology imports have not conferred monopolies on the Indian partner, contrary to the prevailing widespread view. Rather, there has been considerable successful imitation and indigenous minor innovation spurred by scale adaptation and market adaptation (to Indian conditions), especially in differentiated products. There has also been repetitive import of technology despite the control regime. The reason for the lack of monopolies is that market structures have not been typically concentrated in most industries to the point where monopolies emerge and also that most product designs and know-how are perfectly appropriable and can be copied.

Eighth, the poor quality of collaborations was ultimately due to the fact that India's highly sheltered import-substituting industrialization and anti-MNC regime resulted in a lack of sufficient competitive pressures to innovate and upgrade the imported technology other than from relatively minor adaptive (scale and market) innovations. It was not competitive enough to motivate firms to attempt to match international standards of price and quality, since foreign competition was effectively barred and the macroeconomic policy regime had an anti-export bias. As a consequence, firms are not motivated to negotiate for, and in most cases not willing to pay more for, deeper technology packages, since they can operate profitably in the domestic market with basic technology packages. Thus, whilst the restrictive regime may have forced firms to settle for poorer technology packages and older vintages, even where the suppliers were willing to part with more (and sometimes buyers were willing to pay more), this is only a proximate cause. The ultimate cause of the shallower packages imported lay in the regime-induced lack of competitive, including foreign competitive, pressure to constantly upgrade technology and become internationally competitive.

Ninth, the policy slant against big firms, and in favour of small-scale industry, hampered the growth of precisely the firms that could effectively search and negotiate for better technological packages and which could effectively absorb technology. Also, foreign firms are more willing to collaborate and transfer technology to large firms which command large sales since royalties are fixed as a percentage of sales. So, Indian industrial policy inhibited the kind of firm that could effectively negotiate for and assimilate technology.

Tenth, the criticism was made that while India's policy of self-sufficiency had created a broader and deeper technological base than most LDCs in the '70s, this deepening of technology in the direction of basic research and advanced products constituted a misdirection of technological effort that could not be commercialized and in fact was not, at the same time, totally unable to keep up with the level of technology in advanced fields in developed countries. A premature effort at deepening thus falls between two stools and in fact resulted in the '80s in India being overtaken by other LDCs such as South Korea and Brazil in technological depth in high-tech areas such as electronic components, computers, and telecommunications. The revenues and stimulus of successful international competition were absent in the case of Indian firms, whereas in the countries cited above, such revenues, based on more appropriate technological efforts that paid off commercially, could be used to advance rapidly to higher technological levels under competitive pressures and assisted by more liberal foreign collaboration.

The result was that, in the '80s, even the technological depth and diversification, that was the strong point of the Indian strategy of self-reliance was increasingly overtaken by other newly industrializing countries (NICs), while the gap between India and the developed world increased. Lall (1984b: 563) has made the point that protection of technological learning was indiscriminate and excessive in the Indian case and led to a misdirection of effort that did not, indeed could not, yield the commercial pay-offs that were necessary for long-term self-reliance in technological capabilities (Lall, 1984b: 562–3).

Eleventh, and closely related to the above critique of over-protection of technological learning, was a devastating critique of government expenditure on industrial R & D, i.e. of the CSIR national laboratories and the National Research and Development Corporation (NRDC). The CSIR and NRDC were basically unable to produce commercially feasible technologies that would interest industry, hence there was very little purchase of technologies by the latter from the former. Part of the reason for this was the deliberate fostering of a 'research institute' culture without linking research itself either to teaching and generating R & D personnel or to industrial production. This Fabian and Soviet-inspired statist set-up was unable to engage in commercially relevant R & D. Their income from sale of technology is thus a tiny proportion of their total budget.

The pro-liberalization revisionist critique calls for the reorientation of the national laboratories away from technological import substitution, and indigenization of imported technologies towards supporting large-scale, long-range, cooperative R & D by industries in technologies where the interests of competing firms would not clash, i.e. in the technologies of the future that are not immediately exploitable *à la* R & D consortia in Japan and the EU. CSIR laboratories have largely official boards and are not even located near industrial centres. Moreover, they are not allowed to sell innovations directly to industry.

In sum, the revisionist critique of the technological import-substitution policy regime and dependency paradigms derived from Marxist reasoning that it is based on, recommended the following, by no means exhaustive policy measures. It recommended a considerable liberalization of restrictions on the import of technology and drastic simplification of procedures. Limits on royalties and duration should be removed as also on one time payments. Foreign equity participation should be allowed far more freely, FERA being suitably amended for the purpose. R & D should be encouraged through tax concessions, investment subsidies, and other financial support to firms, and by reorienting the CSIR network of laboratories towards long-range cooperative R & D in close collaboration with industry.

All this should be done in the context of liberalization of controls on private enterprise and freeing the forces of competition in the domestic market. For this a series of measures, such as lowering trade barriers, delicensing and broad-banding, removal of MRTP and FERA controls on large and foreign enterprises, de-reservation of products for the small-scale sector and industries for the public sector, exposing the public sector to competitive pressures, removal of restrictions on mergers and rationalization, easing exit and so forth, would be necessary.

It was argued that the liberalization of technology import would not be damaging to indigenous technological effort, or to the terms of foreign collaboration in the medium- to long-term, if Indian industry is subjected to competitive pressures, first domestic then foreign. Under such pressures, Indian firms themselves will strive to import the latest technology in deeper packages, to unpackage technology import, do follow-up R & D in appropriate directions that would pay off commercially, minimize payments, and generally improve the terms of collaboration in their own interest, which would then increasingly be defined as developing international competitive capability. Under such competitive pressures, freer technology import would go towards stimulating local innovation and pay off in exports, narrowing the gap with world technology.

In the process it would enhance, not undermine, technological and economic self-reliance, unlike the restrictive regime which had led to a widening gap and growing uncompetitiveness of Indian industry. The revisionist critique of the protectionist technology policy since the '60s and its recommendation do not thus reject but, rather, radically redefine the concept of technological self-reliance which has been a staple of the discourse on science and technology policy in India since Nehru. I elaborate on this below in the final section.

Parallel to this revisionist critique arising from transfer of technology studies, there also developed a critique of the existing technology policy regime and actual technological development in India. It found fault with the import substitution paradigm, not so much from the point of view of the benefits of foreign collaboration, but from that of a failure to assimilate/absorb, further build upon and promote diffusion of technology that may have been originally imported. This critique differs with the transfer-of-technology literature, in that the former concerns itself too much with

> the terms of the transfer, neglecting the more important consideration that the subsequent functioning of the technology which determines its value to the recipient, depends largely on the latter's own technological effort [Lall, 1987: 232].

As Bagchi (1987: 143) puts it:

> Larger imports of capital goods [i.e., embodied technology] *per se* do not go into the heart of the problem, viz., that of inadequate absorption of imported or domestically produced technology.

Here, it is not primarily restricted access to foreign technology that hampers technological advance but the failure of assimilation and diffusion. This critique attacks the failure to assimilate and build upon imported technology as the Japanese, for example, did, spending several times as much on domestic R & D as on technology import at the firm level. Again, as Bagchi (1987: 146) puts it:

> What, however, differentiates India from such a country as Japan is that the foreign collaboration agreement is not backed up by adequate expenditure on local effort for absorbing the technology.

This perspective emphasizes the critical importance of technological learning at the firm level and the need for appropriate protection of such learning processes for absorption and mastery of technology. It lays the blame for India's poor technological and export performance on such failure to assimilate and further develop technology.[3]

In the case of the public sector, which dominates large parts of India's basic and capital goods industries (the latter being the carriers of technological progress), failure of technological assimilation and innovation is blamed on the poor quality of state entrepreneurship and management deriving from inadequate autonomy and accountability and poorly spelt out objectives. For example, in a study of the public sector steel industry, Sengupta (1984) describes the lack of technological dynamism leading to growing relative inefficiency by world and even other LDC, standards, despite India's head start among LDCs. He holds the following factors to be responsible for this:

1. political factors, such as dependence on project aid foreclosing technical and procurement options, and location and labour policies adding to costs; and
2. organizational factors such as lack of coordination between levels of decision-making on issues such as plant size and technology, capacity utilization and pricing policy, and the lack of professional management inherent in the lack of autonomy and of a system of accountability.

These organizational factors were responsible for the poor quality of state entrepreneurship and resulted in a lack of balancing investments in plant and equipment, poor quality of raw materials, faulty design, and general lack of cost-consciousness and innovation (*see also* Desai (1972) for Bokaro Steel; Dhar (1984) for fertilizer; and Khanna

[3] For a non-exhaustive but representative selection of work that emphasizes the importance of technological learning, assimilation/absorption, indigenization, and diffusion of technology, and the general failure of Indian industry in these respects, *see* Lall (1987) for cement, steel, textiles, and power equipment; Sengupta (1984) for steel; Khanna (1984) for petrochemicals; Chaudhuri (1984) for pharmaceuticals; and Siddhartan (1984) for a 'neo-Schumpeterian' perspective. *See also* Bagchi (1987, especially 143–8) for a discussion of Indian technology together with a discussion of South Korean and Chinese policies on the absorption of and building upon imported technology.

(1984) for petrochemicals in the context of the two factors mentioned above). At the same time, also at work were macroeconomic factors such as the size and character of the market and lack of competitive pressure resulting in inadequate R & D investment to absorb and indigenously develop imported process and product technologies. Sengupta's critique of the steel industry's experience is from an assimilationist/diffusionist perspective, focusing on the critical importance of follow-up R & D. But he does not follow through with the logic of his argument, relating the lack of R & D investment and cost consciousness to lack of competitive pressure.

In his studies of cement, steel, textiles, and special cases, including Bharat Heavy Electricals Ltd. (BHEL), the public sector power equipment maker (Lall, 1987) also stresses the key role played by technological learning and the need to protect and foster it. But, he places it in the context of the incentives and pressures of a competitive environment which are seen as the driving forces of technological assimilation and diffusion at the firm and industry levels. While *appropriate* protection of technological learning for functions such as project execution and process and product engineering is vital, the driving force is the pressure/incentive system of competition deriving from limiting the *degree and duration* of protection, exposing firms to competition *when they can respond to it without being swamped by it*. A fine balance needs to be maintained between protection of technological learning and exposure to competition, domestic and foreign. India's blanket import-substitution, including restriction on domestic competition, was conducive to neither R & D efforts nor assimilation or diffusion of technology. Lall cites the case of BHEL whose technological and overall performance improved after *de facto* exposure to competition since 1978 due to global tendering for power equipment for foreign-financed projects.

To return to the larger theme of the shift in the terms of debate and the terms of discourse in technology policy during the '80s, the movement has been in the direction of easier access to foreign technology, notwithstanding the powerful critique of the failure of the restrictive Indian Substitution Industrialization (ISI) regime from the assimilation/diffusion perspective as distinct from the revisionist perspective derived from technology transfer studies outlined earlier. In my view, this has been so for the following reasons.

First, there is no necessary contradiction between the two perspectives except for their caricatural extremes. Indeed, they share much ground in common. Both perspectives differ from the crude neoclassical free market and the crude dependency theory. They can very roughly be categorized as 'neo-Schumpeterian' and emerge from the decline of neoclassical and dependency theories. They are concerned with the dynamics of the process of technological change, central to all economic change within the firm, and conditioned by the pressures and incentives of the kind of competition they face; in turn, both are dependent on market structures as well as influencing them (Fransman and King, 1985). The notion that restrictions on technology import are the major cause of technological sluggishness and their liberalization the major source of technological progress is not upheld by the revisionist critique. Rather, the stress is on the stimulating effect of domestic and foreign competitive pressure, including freer access to foreign technology, that is believed to promote active search, acquisition, assimilation, and diffusion of technology as firms specialize for actual and expected competition in their product lines over the long run. That is, competitive environments together with freer access to complementary (foreign) technology reinforces and pushes learning processes towards technological mastery and constant upgradation for competitiveness through R & D efforts.

Likewise it is *not* the assimilation/diffusion perspective's case that the terms of technology import be necessarily tightened and that indigenous technological effort unconditionally protected. In fact, as Sengupta (1984: 214) has put it:

> To remove this gap the entrepreneurship of the government has to play a role in developing R & D activities and confidence in our own technological capabilities. By this we do not mean, however, that India need not import any technology from outside ... there will remain the necessity for the basic know-how type of licensing agreement with foreign collaborators which can help us understand the know-why of a process or the complete design philosophy of an equipment system.

Or as Lall (1987: 161) puts it, in a broad overview of India's technological development, while referring to power equipment maker BHEL, one of the most successful public sector enterprises:

> It shows the vital role of internal effort in assimilating, adapting and rationalizing complex technologies ... It also shows the limitations of R & D in such a firm... the mastery of a certain level of know-why does not provide the know-why of the next generation.... continuous injection of advanced technology from the developed world seems necessary if the firm is to stay competitive.

In short, the assimilation/diffusion perspective, while stressing the appropriate protection of technological learning processes through indigenous R & D, is not at all incompatible with freer access to foreign technology and a more competitive environment, provided indigenous technological capabilities are at a level at which they can respond to competitive stimuli without being wiped out. As Sengupta (1984: 214) has stated:

> Foreign collaboration is productive in the long run only when it is considered as an input for the R & D activity of the importing country.

Therefore, in terms of policy influence, the assimilation/diffusion perspective stressing indigenous R & D is not seen to be necessarily incompatible with, and not surprisingly, has not arrested the actual trend in policy towards liberalization of technology import, influenced *inter alia* by the revisionist transfer of technology literature.

A second and more important reason for the continuing shift in the terms of debate and actual policy towards easier access to foreign technology has been the fact that technology policy is necessarily a part of the overall industrial and economic policy regime. If the latter shifts towards de-regulation and opening up towards the outside world, liberalization of technology import will inevitably be part of the new package. In addition, external liberalization and gradual integration with the world economy imposes its own priorities on policy, in that competitiveness will be top priority, especially in a worsening balance of payments and external debt situation, not to speak of the conditionalities accompanying IMF loans. Hence, as in technology policy, so too in overall industrial strategy, it is inevitable that the terms of debate and terms of discourse will shift from import substitution to competitive capability. This is what has been happening in Indian economic policy over the past decade, gradually and in fits and starts, with its fallout on the terms of debate in technology policy and on the concept of self-reliance, as I shall discuss below.

Economic and national security compulsions behind the shifts in technology policy during the '80s

The shift towards liberalization of technology import and associated foreign direct investment in the '80s (especially after 1985), combined with the stepping-up of R & D expenditure and the launching of an array of high-tech development programmes and industries, especially those related to long-term defence needs, was in retrospect motivated by two sets of compulsions *in tandem* with the rethinking process described above.

Economic compulsions were one source of the gradual shift. The foremost among these was the deteriorating balance of payments due to poor export performance since the late '70s (Nayyar, 1988). In addition, there was also the problem of slow industrial growth since the mid-'60s, compared to the earlier Nehru period. As has been noted by a host of observers, the rise in the savings rate was cancelled out by the rising incremental capital–output ratio since the mid-'60s, reflecting the inefficiency of Indian industry. This, in turn, was due to its growing technological obsolescence and was reflected in its uncompetitiveness (compared to several other developing countries) in manufactured exports. These problems were particularly pronounced in the public sector and in the sphere of heavy and capital goods industries (Ahluwalia, 1985; Varshney, 1984).

The awareness of serious technological backwardness in Indian industry, the slow growth of exports and the growing balance of payments difficulties—reflected in the IMF loan application of 1981—led to the government taking a series of steps with a view to liberalizing the economy. In the process, liberalization of technology import and FDI was also recommended and pursued. All four major commissions'/committees reports, namely the Alexander, Tandon, Dagli, and Hussain Committees of the late '70s onwards, all recommended liberalization of trade and industrial policies. The Sondhi Committee Report on Electronics (1979) also recommended liberalizing foreign collaborations.

Defenders of the restrictive regime on technology import were put on the defensive by the facts, finding it impossible to argue away any longer the failure of the regime to deliver in terms of technological catch-up and export competitiveness of Indian industry. Rather, the technological gap had inexorably widened. The resulting poor export performance and Balance of Payments (BOP) deficits necessitated greater foreign borrowing and debt. Thus, the restrictive, import-substituting technology policy regime seemed to be ultimately resulting in worsening debt and dependence, not to speak of technological lags, i.e. the very opposite of the self-reliance that it was designed to achieve.

By the mid-'80s, however, the cumulative effect of the above BOP and technological upgradation compulsions along with the technology policy and the largely pro-liberalization conclusions to which the slow growth debates led had generated a powerful momentum for freer import of technology and freer foreign equity collaboration. It is important to note that the policy impulse came from within the state apparatus rather than from organized business.[4]

However, liberalization of foreign technological collaboration would have been in the interest of Indian industry, given that the import-substituting policy framework, combined with licensing and MRTP/FERA barriers to entry and expansion, encouraged horizontal diversification into new products through import of technology rather than specialization, technology assimilation, and innovation. Hence, it would also lead to repetitive import of shallow technology packages.

[4] *See* Kohli (1989) for an elaboration of this argument which is strongly supported for the electronics industry in Sridharan (1989).

If one accepts this logic, one would expect Indian industry to lobby for easier import of technology. So, when we say that policy impulses originated within the state apparatus we merely mean that no outside lobby, to our knowledge, can be pinpointed as constituting *the* source of policy changes. However, this too should not be surprising since liberalization of technology import is a general, across-the-board, rather than specific sector-or target-oriented policy, whose effects on firms'/business houses' interests cannot be known *a priori* even to them. Technology import policy conditions, such as royalty, one time fee, duration, and other clauses belong to this category and hence do not become a focal point for lobbying or provoke a sharp reaction, leaving policy-makers with considerable discretion.

Focused lobbying would concentrate on the details of case-by-case approvals, analogous to industrial and import licensing that take place within the terms of the announced policy framework on foreign technological collaboration, rather than on the latter itself in principle. It should not be surprising if policies with such characteristics tend to be left to the discretion of policy-makers in the state apparatus.

These economic compulsions to modernize Indian industry's technological capabilities dovetailed neatly with strategic compulsions to develop high technology and defence industries indigenously for long-term security in the '80s. Together they gave a boost to the liberalization of technology imports.[5] It is also important to note, as has already been pointed out, that this trend towards freer, and increasing foreign collaboration (especially after 1985), went hand in hand with a step up of R & D expenditure, particularly in high-tech areas. Once again, both impulses came from within the state apparatus and not from organized business.

India's strategic environment deteriorated sharply from the beginning of the '80s due to a number of regional, international, and military–technological developments. The Soviet invasion of Afghanistan (December 1979) triggered off American military aid to Pakistan on a huge scale. This included extremely sophisticated weaponry to which India did not have access—such as F-16 aircraft, Sidewinder and Harpoon missiles, naval craft, etc. Throughout the decade of the '80s, there was also the threat of transfer of the force multiplier (Airborne Warning and Control System (AWACS)). This was in addition to the steady progress made by Pakistan in acquiring nuclear capability, thus matching, by early 1987 or so, India's earlier proven capabilities.

[5] The argument concerning the strategic compulsions propelling both technology import and indigenous high-tech programmes is made in detail for the electronics industry *in* Sridharan (1989: pp. 342–435).

In addition to these developments on the Pakistan front, China also steadily modernized its forces, including those in Tibet, throughout the decade with the aid of Western military technology. The United States chose to ignore Pakistan's nuclear programme while lavishing military and economic aid on that country. It also remained closely aligned with China against the Soviet Union for most of the decade, allowing it some access to advanced military technology. These developments threatened India's hegemonic role in South Asia, believed to have been secured after the 1971 war and the 'peaceful' nuclear explosion in 1974. Great pressure was exerted on the military élite to counter these apparent threats to India's security/hegemony with a stepped-up nuclear and missile programme and with an overall defence build-up. But such a programme would require extremely advanced technological capabilities in the new high-tech electronic warfare environment of the '80s and would also require the capability to keep up technologically in the long run.

Furthermore, military developments further afield in the Indian Ocean and West Asia also began to exert significant influence on Indian security for the first time, with implications for defence production and technology policy. During the '80s, the United States upgraded its base at Diego Garcia in the Indian Ocean and also established the Central Command with a Rapid Deployment Force (RDF) with responsibility for military operations throughout the area, ranging from Pakistan to Somalia. Pakistan was to play a role in this new development in return for massive aid and arms transfers. It soon began to play host to American aircraft and naval vessels at its ports. Further west, several West Asian countries (including Saudi Arabia, Iraq, Iran, Syria, and Israel) began to develop, acquire, or augment their existing missile capabilities, potentially bringing India within their range. Though not posing a direct threat to India, these developments gave rise to enormous pressure to step up high-tech and defence technology programmes, especially in the nuclear, missile, aeronautic, naval, and space areas.

But, a stepped-up defence and high-tech thrust, such as that started in 1980 and greatly expanded in 1985, would require for its long-term sustainability, a broad-based capability in the microelectronics-based complex of industries such as semi-conductors, computers, communications, and related areas. These are also increasingly vital not only to warfare, but to manufacturing and managerial efficiency across industry and the economy in general. India was epochs behind the world leaders in these areas; the restrictive import-substituting regime had led to steadily and visibly widening gaps in these fast-moving technologies. Technological upgradation in the defence-related sectors would need major infusions of advanced technology from abroad in the short run.

Furthermore, a nuclear or missile programme would run the risk of cut-offs of advanced technology by the West, led by the United States. The Reagan Administration had, from the onset of the second Cold War in 1981, drastically tightened controls on exports of strategic and other advanced technologies and equipment, and given the Pentagon a virtual veto over an expanded list of these. In 1984, for the first time, it was not only the Soviet bloc that was targeted for the squeeze but also a 'grey list' of 12 to 15 non-communist states suspected of being conduits—consciously or unconsciously—for the transfer of advanced technologies to the Soviet Union. India figured prominently in this list and suffered, in the '80s, several long delays in or rejections of clearance for the export of 'sensitive' items. This had begun even earlier with nuclear fuel for Tarapur Atomic Power Station under the Carter Administration, following the 1974 nuclear test and the 1976 Glenn and Symington *Amendments*. America also pressurized its allies and neutral states in Europe to follow suit. As a consequence, India was affected in some cases, the most prominent being the American refusal to allow neutral Sweden to export the Viggen fighter to India as it contained American avionic components.

These national security compulsions dovetailed neatly, during the early to mid-'80s, with economic compulsions to upgrade India's technological base, especially in high-tech areas such as the electronics complex, and also with the pro-liberalized import conclusions of the technology policy debate and slow industrial growth debate. The 'Technology Policy Statement' (Government of India, 1983: 3) reflects these concerns in its aims, which are

> to (a) attain technological competence and self-reliance, to reduce vulnerability, particularly in strategic and critical areas ...

There was no alternative to massive import of technology in addition to stepping up R & D efforts. The technology debate's revisionist conclusion supported this and the belief that market forces were necessary to sustain innovation reinforced the economic and defence compulsions. The need for access to high technology also dictated accommodation to American interests. While India did not change its pro-Soviet policies on Afghanistan and Kampuchea much to America's annoyance, it did sign in 1984 and 1985, a two-part *Indo-US Memorandum of Understanding on Technology Transfer* agreeing to refuse access to the Soviet Union to any American high technologies that it might acquire. For its part, America was willing to accommodate India's requests for high technology, thus avoiding unnecessarily alienating the principal South Asian power, balancing its commitment to Pakistan, and building bridges with the Indian defence establishment (Thomas, 1990).

In my view, the collapse of the Soviet Union and erstwhile socialist states as a global counterweight to the West and the Gulf War of 1991 only serve to emphasize these strategic compulsions of the '80s in three ways:

1. the critical importance of high-tech for warfare;
2. the growing impact of West Asian developments on South Asian security; and
3. the pressure to appease the West, in particular the USA in a unipolar or, at least, Western-dominated world.

By 1985, after Rajiv Gandhi became Prime Minister, the political and intellectual stage had been set for a massive acceleration of foreign technical collaboration for industrial modernization, for global competitiveness, as well as for a broad-based defence production and technological base. Import of technology was to be backed by an increase in R & D expenditure, which took place. as has been already noted, especially in defence R & D, space, electronics, and communications. R & D expenditure as a percentage of GNP and the number of foreign collaborations climbed significantly during the '80s, and especially after 1985.

I have in this section pulled together the threads and shown how, starting from the late '70s, economic and strategic compulsions dovetailed with the trajectory of the technology policy debate to set the stage, by the mid-'80s, for a policy shift towards liberal import of technology in conjunction with liberalization of the economy. This policy shift apparently contradicted the concept of self-reliance as it was defined by the earlier intellectual paradigm and policy regime, and hence signified a change in the terms of discourse on technology in India. This aspect is elaborated in the following section.

I wish to make two concluding observations at this stage. First, the suggestion sometimes put forward that the liberalizing trend in foreign technical collaboration originated with the Janata government (1977–9), dominated by right-wing elements drawn from Congress (O) and the Jana Sangh, cannot be taken seriously.[6] The Janata phase did not set any definite trend on liberalizing the economy for domestic and foreign capital or in import of technology. The terms of discourse on technology at that time were rather different as I shall briefly explore in the next section. Second, the major policy initiatives

[6] Nayar (1983, (Vol. I): 394–6) argues that the Janata government was more liberal than its predecessor with respect to technology import. But, from the vantage point of 1990, there was hardly any difference between the two and, indeed, the real rise in the number of foreign collaboration approvals and technical know-how payments came later.

originated within the state apparatus, which is where the economic and strategic compulsions were felt, and not from organized business or intellectuals.

I ought to qualify my acceptance of the view that the autonomy of the state from the dominant class coalition, and its ability to define and impose its objectives on the dominant coalition and its fractions and indeed on society as whole, has been eroded over time. The decision-makers in the state apparatus (with their truncated autonomy) still continue to exercise considerable discretion in those policy arenas that are characterized by non-specific or across-the-board, industry-wide or economy-wide measures, whose effects on the particular interests of industrial firms or business groups can be known only after such policies have been in operation for a period of time. They do not therefore provoke lobbying for policy changes and leave policy-makers with considerable autonomy. Sections of organized business had been demanding various forms of liberalization for a very long time, but did not get what they wanted. The post-1980 liberalization of technology import was the product of compulsions felt by the political leadership and constitutes an exception to the trend towards the reduction of the relative autonomy of the state.

Self-reliance, self-sufficiency, and international competitiveness: A shift in the terms of discourse

Does the shift in technology policy in the '80s (especially since 1985), consisting of

1. a considerable liberalization in the terms of technology import; and

2. an increase in R & D expenditures (especially in respect of the advanced technologies in the electronics–aerospace defence complex)

signify a shift in the terms of discourse on technology in India? I shall argue in this section that it does so to the extent that it involves an ongoing process of redefinition of the key concept of self-reliance that has underpinned science and technology policy discourse in India since Independence. In addition and secondarily, it also involves a redefinition or rethinking of the concept of appropriateness or of appropriate technology and its associated intellectual paradigm and policy regime.

The concept of self-reliance has been, from the outset, central to the discourse on development in India. Self-reliance in science and technology has been a central part of self-reliant development. Its antecedents lie in the freedom movement and it has been frequently reiterated in Indian discourse on science, technology, and development. Nehru proclaimed self-reliance as one of the goals of science and technology policy as of development policy in general and, with Bhabha and Bhatnagar, built the infrastructure for self-reliance in science and technology during the '50s and '60s.

Self-reliance in the dominant discourse came to have a specific connotation. The Third Five Year Plan (1961–6) first mentioned self-reliance as a specific goal and equated it with import substitution, i.e. self-sufficiency, without reference to relative costs. It declared that

> ... a basic objective in the strategy of development is to create the conditions in which dependence on external assistance will disappear as early as possible [and] replacement of imports is essentially a question of developing the necessary capacity for production within the country [Government of India, 1961, 26–7].

The Fourth Plan (1969–74) extended this logic further, declaring self-reliance to be a major objective of planned development, and emphasized the import substitutionist (i.e. self-sufficiency) interpretation of self-reliance in development, extrapolating it with the policy specifically relating to technology. The following quotations from the Fourth Five Year Plan bring out the flavour of the thinking on self-reliance:

> In addition to the fight against poverty and economic inequalities, the Plan seeks to enlarge the area of self-reliance in terms of financial resources and technological *inputs*. Here too the public sector has an important role to play. Besides ... the public sector should increasingly base itself on domestic know-how. The public and private sectors have both been too ready to look for foreign collaboration not only for financial but for technological resources. [Government of India, 1972, vi; emphasis supplied.]
>
> We should rely more and more on our own machinery and technical know-how even though it may entail some initial risks and difficulties [Government of India, 1972, vi-vii].

More specifically:

> Self-reliance in the technological sense implies the existence and effective functioning of indigenous organizations for design, construction and engineering of projects as well as capability for design and development of machinery, equipment and instruments indigenously manufactured. At present there is unwholesome dependence on foreign agencies for these services. [Government of India, 1972: 48.]

And,

> ....care has to be taken that foreign collaboration has to be resorted to only for meeting a critical gap and does not inhibit the maximum

> utilization of domestic know-how and services... It is necessary to subject every proposal for foreign collaboration to fairly rigid tests. [Government of India, 1972: 312.]

From the late '60s onwards, technology policy thus entered a second phase, restricting foreign collaboration and equating self-reliance with import-substitution, in tune with the general turn towards intensified import-substitution in overall policy. As Nayar has shown in a detailed study, technology policy in this phase was *in actual practice* strongly motivated by self-reliance in the sense of self-sufficiency (1983 (Vol. 2): 239–305). Furthermore, this local production and technology generation was to be by domestic capital, public or private, not foreign-owned firms or joint ventures. In other words, *self-reliance came to mean self-sufficiency* in production and technology generation, import substitution as opposed to imports, licensing of technology, or foreign collaboration. This self-sufficiency interpretation of self-reliance led to the widespread prevalence, since the late '60s, of knee-jerk opposition to foreign technical collaboration and joint ventures. Such attitudes were greatly strengthened by the policy-makers' obsession with saving chronically scarce foreign exchange.

It is this self-sufficiency interpretation of self-reliance that has come under attack and is in the process of being redefined in the '80s and '90s by the critiques of the earlier restrictive policy regime. It is argued that self-sufficiency is neither necessary nor sufficient for self-reliance. Indeed, the equation of self-reliance with self-sufficiency and the consequent indiscriminate import substituting practice of technological and industrial self-sufficiency until the mid-'80s, under the restrictive policy regime, have actually greatly harmed the cause of self-reliance. Thus, the Sixth Five Year Plan (1980–5) was the first plan document to refer to 'efficient' import-substitution.

The following quotations capture the shift in thinking about self-reliance:

> Self reliance, as should be obvious, but often is not, does not necessarily mean, self-sufficiency in all sections of the economy. So long as the country is able to pay its way, it cannot be said to be dependent on others. [Government of India, 1980, xxi.]
>
> However, self-reliance can no longer take the form of indiscriminate import-substitution ... In the eighties, export promotion is as much a part of the drive for self-reliance as efficient import-substitution. [Government of India, 1980, 10.]

The policy announcements of July 1991 make more explicit the shift in the meaning of self-reliance. Self-reliance would henceforward be linked to competitiveness. Thus, for example, the Ministry of Industry declared that whilst the

> government would continue to follow the policy of self-reliance, there would be greater emphasis placed on building up India's ability to pay for imports through its own foreign exchange earnings. At the same time, foreign collaboration would be welcomed in investment and technology in order to increase exports and expand the production base requiring higher technology. [Government of India, 1991.]

The import-substituting technology policy regime has created a state of growing relative technological backwardness over the past two decades and made Indian industry so uncompetitive that India was threatened with a very serious balance of payments and external debt situation at the start of the '90s. Ironically, self-sufficiency may actually lead to a debt crisis compromising India's sovereignty, an outcome of which will be the very reverse of the intended goal of self-reliance.

The revisionist critique in the '80s of the technology policy regime of the '70s redefines self-reliance to mean the possession of the capability to compete in the world market and earn enough to pay for one's growing imports in the long run, keeping one's foreign debt under control. Self-reliance is increasingly being defined as international competitiveness and self-sufficiency is neither necessary nor sufficient, indeed harmful, to the development of a capability to compete in the world market. Competitive capacity in turn requires the capacity to absorb technologies continuously and build upon them to develop a domestic capability in technology generation. As the technology transfer studies of '80s and comparative studies worldwide argue, the development of such capability is helped by market-oriented policies at home and pragmatic policies on foreign collaboration with only limited and appropriate state support, and hurt by indiscriminate import substitution, state controls, and undue restrictions on foreign collaboration.

Such an implicit redefinition of self-reliance in science and technology also involves a distinct shift in the terms of policy discourse as regards a subsidiary strand of technological discourse in India, viz. the discourse of appropriate technology. This strand of discourse has its roots in Gandhi's anti-technological, pro-small-scale, agriculturally oriented positions and has been a subsidiary strand of Indian discourse on technology for a long time, gaining particular prominence since the end of the '60s, and especially during the Janata phase of 1977–9.

Given prominence by Schumacher's 'small is beautiful' thesis and some Maoist Chinese policies, this line of thinking was appropriated by those concerned with equity, environment, labour-intensity of technologies, and employment potential. It dovetailed neatly (since the end of the '60s) with the interests of the emerging new small-scale entrepreneurs and some of the positions of the Left, especially its hostility to big business houses and multinationals, and concern for equity and employment. Modern capital-intensive advanced technologies originating in the West and employed by large firms in India were said to be inappropriate to India's labour-surplus economy. They did not generate employment, were environmentally destructive, and were dis-equalizing in their effects. The answer was said to be 'intermediate' or 'appropriate' technologies that were labour-intensive, used local materials, did not require much capital, and could be operated on a small scale.

India's industrial policy regime since the late '60s reflects to some extent the accommodation of small industry's interests and the influence of such thinking. This has taken the form of enactment of legislation (MRTP and FERA) ostensibly aimed against big industrial houses and foreign firms. At the same time, small-scale industry has been encouraged by the reservation of an ever-increasing number of products (until the early '80s) as well as a host of other concessions.

However, all this changed in 1985 when Rajiv Gandhi embarked on a policy of deregulating industry. A number of restrictions on MRTP and FERA companies were lifted and a large number of products reserved for the small sector were de-reserved, and other concessions withdrawn. It is worth emphasizing that the process of implicitly redefining self-reliance as international competitive capability (relevant to the '80s, especially since 1985) inevitably implies an attack on the 'small-is-beautiful', appropriate technology discourse and on the policy framework and interest groups associated with it. For, in an integrating world economy where economies of scale at the firm level, in marketing and finance, and innovation involving the application of high technology are requirements for international competitiveness on a significant scale, the terms of discourse will inevitably shift against small-scale interests and low-tech ideology.

There is thus an ongoing shift in the terms of discourse in India's technology policy which started in the early '80s. The process originated in economic and strategic compulsions and dovetailed with the conclusions of technology transfer studies in India and on LDCs in general during the course of the decade. It is hard to identify a turning point, but Rajiv Gandhi's 1985 initiatives would appear to be the nearest to one. The key element of the shift in the discourse is the ongoing process of redefinition of the central objective of self-reliance as not self-sufficiency but international competitiveness. The new policies since July 1991 only carry forward this logic. This implies an ongoing regime transition towards high technology, large-scale industry, the private sector, greater foreign technical collaboration, freer domestic competition, greater R & D spending, and more outward-oriented policies. It remains to be seen whether and to what extent these tendencies will continue in the latter half of the '90s.

REFERENCES

AHLUWALIA, I.J., 1985: *Industrial growth in India: Stagnation since the Mid-Sixties* (New Delhi: Oxford University Press).

ALAM, G., 1985: 'India's Technology Policy and its Influence on Technology Imports and Technology Development', *Economic and Political Weekly* 20 (Nov.): 45–7, pp. 2073–80.

BAGCHI, A.K., 1987: *Public Intervention and Industrial Restructuring in China, India and the Republic of Korea* (New Delhi: ILO-ARTEP).

BELL, M. and D. SCOTT-KEMMIS, 1985: 'Technology Import Policy: Have the Problems Changed?', *Economic and Political Weekly* 20 (Nov.): 45–7, pp. 1975–90.

BHATTACHARJEE, A., 1989: 'Mega Business Worldwide', *Telematics India* 3 (Nov.): 2, pp. 42–51.

CHAUDHURI, S., 1984: 'Manufacturing Drugs without TNCs: Status of Indigenous Sector in India', *Economic & Political Weekly* 19 (Aug.): 31–3, pp. 1341–83.

DESAI, A.V., 1980: 'The Origin and Direction of Industrial R & D in India', *Research Policy* 9 (Sept.): 5, pp. 74–94.

——————, (Ed.), 1985: 'Foreign Technology Suppliers' View of Indian Industry', *Economic and Political Weekly* 20 (Nov.): 45–7, pp. 1973–2094.

——————, 1985a: 'Indigenous and Foreign Determinants of Technological Change in Indian Industry', *Economic and Political Weekly* 20 (Nov.): 45–7, pp. 2081–94.

——————, 1988: 'Technology Acquisition and Application: Interpretations of the Indian Experience' *in* LUCAS and PAPANEK (Eds.), pp. 163–84.

——————, 1990: 'Recent Technology Imports into India: Results of a Survey', *Development and Change* 21 (Oct.–Dec.): 4, pp. 723–49.

DESAI, P., 1972: *The Bokaro Steel Plant: A Study of Soviet Assistance* (Amsterdam: North-Holland).

DHAR, B., 1984: 'Factors influencing Technology Selection: Case Study of Thal–Vaishet and Hazira Fertiliser Projects', *Economic and Political Weekly* 19 (Aug.): 31–3, pp. 1385–95.

FRANSMAN, M. and K. KING, 1985: 'Conceptualising Technical Change in the Third World in the 1980s: An Interpretive Survey', *Journal of Development Studies* 21 (July): 4, pp. 572–652.

—————— (Eds.), 1984: *Technological Capability in the Third World* (London: Pinter).

GOVERNMENT OF INDIA [Department of Atomic Energy], (various years): *Annual Report* (New Delhi: Department of Atomic Energy, Government of India).

——————, [Department of Electronics], (various years): *Annual Report* (New Delhi: Department of Electronics, Government of India).

——, [Department of Industrial Development], (various years): *Annual Report* (New Delhi: Department of Industrial Development, Government of India).

——, [Department of Science and Technology], (various years): *Annual Report* (New Delhi: Department of Science and Technology, Government of India).

——, (various years): *Annual Report* (New Delhi: Department of Science and Technology, Government of India).

——, (various years): *Research and Development Statistics* (New Delhi: Department of Science and Technology, Government of India).

——, [Department of Space], (various years): *Annual Report* (New Delhi: Department of Space, Government of India).

——, [Department of Telecommunications], (various years): *Annual Report* (New Delhi: Department of Telecommunications, Government of India).

——, Ministry of Defence, (various years): *Annual Report* (New Delhi: Ministry of Defence, Government of India).

——, Ministry of Industry, (various years): *Annual Report* (New Delhi: Ministry of Industry, Government of India).

——, [Planning Commission], (various years): *Annual Plan* (New Delhi: Planning Commission, Government of India).

——, [Planning Commission], 1961: *Third Five Year Plan* (New Delhi: Planning Commission, Government of India).

——, 1972: *Fourth Five Year Plan 1969–74* (New Delhi: Planning Commission, Government of India).

——, 1980: *Sixth Five Year Plan 1980–5* (New Delhi: Planning Commission, Government of India).

——, [Department of Science and Technology], 1983: *Technology Policy Statement* (New Delhi: Department of Science and Technology, Government of India).

——, 1988: *Research and Development Statistics 1986–7* (New Delhi: Department of Science and Technology, Government of India).

——, 1990: *Research and Development Statistics 1988–9* (New Delhi: Department of Science and Technology, Government of India).

——, [Department of Scientific and Industrial Research], 1989: *Foreign Collaboration [A Compilation: National Register of Foreign Collaboration] 1988* (New Delhi: Department of Science and Technology, Government of India).

——, Ministry of Industry, 1991: *Statement on Industrial Policy* (New Delhi: Ministry of Industry, Government of India).

GOVINDARAJULU, V., 1990: 'India's S & T Capability: SWOT Analysis', *Economic and Political Weekly* 25 (17–24 Feb.): 7–8, pp. M35–M40.

HOFFMAN, K., (Ed.), 1985: Special Issue on 'Microelectronics, International Competition and Development Strategies', *World Development* 13 (March): 3, pp. 263–463.

KHANNA, S., 1984: 'Transnational Corporations and Technology Transfer: Contours of Dependence in the Indian Petrochemical Industry', *Economic and Political Weekly* 19 (Aug.): 31–3, pp. 1319–40.

KOHLI, A., 1989: 'The Politics of Economic Liberalisation in India', *World Development* 17 (March): 3, pp. 305–28.

LALL, S., (Ed.), 1984: Special Issue on: 'Exports of Technology by Newly Industrialising Countries', *World Development* 12 (May/June): 5/6, pp. 471–659.

————, 1984a: 'Exports of Technology by Newly Industrialising Countries: Overview', *World Development* 12 (May/June): 5/6, pp. 471–80.

————, 1984b: 'India' *in* Lall (Ed.), 1984, pp. 535–65.

————, 1987: *Learning to Industrialize* (London: Macmillan).

LUCAS, R.E.B. and G.F. PAPANEK, (Eds.), 1988: *The Indian Economy: Recent Developments and Future Prospects* (New Delhi: Oxford University Press).

NAYAR, B.R., 1983: *India's Quest for Technological Independence* (2 Vols.) (New Delhi: Lancers).

NAYYAR, D., 1988: 'India's Export Performance 1970–85: Underlying Factors and Constraints', in LUCAS and PAPANEK (Eds.), pp. 217–52.

RESERVE BANK OF INDIA, 1985: *Fourth Survey Report on Foreign Collaboration in Indian Industry* (Bombay: Reserve Bank of India).

SCOTT–KEMMIS, D. and M. BELL, 1985: 'Technological dynamism and technological content of collaboration: Are Indian firms missing opportunities?', *Economic and Political Weekly* 20 (Nov.): 45–47, pp. 1991–2004.

SENGUPTA, R.P., 1984: 'Technical Change in the Public Sector Steel Industry', *Economic and Political Weekly* 19 (4 Feb.): 5, pp. 206–15.

SIDDHARTHAN, N. S., 1984: 'Industrial Structure, Non-Price Competition and Industrial Development', *Economic and Political Weekly* 19 (Aug.): 31–3, pp. 1307–10.

SRIDHARAN E., 1989: The Political Economy of Industrial Policy for Competitiveness in the Third World: The Case of the Electronics Industry in South Korea, Brazil and India. Unpublished PhD dissertation, University of Pennsylvania.

————, 1996: *The Political Economy of Industrial Promotion: Indian, Brazilian and Korean Electronics in Comparative Perspective, 1969–1984* (Westport, CT: Praeger).

THOMAS, R.G.C., 1990: 'U.S.: Transfers of Dual Use Technologies to India', *Asian Survey* 30(Sept.): 9, pp. 825–45.

VARSHNEY, A., 1984: 'Political Economy of Slow Industrial Growth in India', *Economic and Political Weekly* 19 (1 Sept.): 35, pp. 1511–17.

Regional Differentiation in India[1] • 6

KRISHNA BHARADWAJ

An attempt is made here to identify certain general questions and key relationships to provide a setting for a discussion on the determinants of the aggregate level and composition of economic activity, enabling us to understand the factors influencing the process of accumulation. The following questions I believe are relevant:

1. What is the size of the re-investible surplus, and how and in what activities is it generated?

2. How is the surplus distributed among the productive classes and the revenue sharers?

[1] *Editor's Note*:

During the early stages of the project, Krishna Bharadwaj was invited to write a paper on regional differentiation in the development of the Indian economy. However, due to illness, she could not produce a new paper. In the event, a decision was taken to reproduce her paper entitled 'Regional Differentiation in India: A Note' (*Economic and Political Weekly* 17 (Apr. 1982): 14–16, pp. 605–14) in this volume with the permission of the publisher and the author. The last two sections of this chapter (and Tables 6.13 and 6.14) were added to the original text by the author in January 1991.

Sadly, Krishna Bharadwaj died in March 1992 (just as this volume was being readied for press).

3. How is this surplus utilized; in particular, what channels is it invested in?

4. Are the conditions for productive accumulation (in terms of adequate effective demand, adequate return in productive activities, etc.) maintained?

Further, a sustainable process of accumulation leading to industrialization in a primarily agricultural country such as India would, broadly speaking, require the following conditions to be fulfilled:

(i) Adequate agricultural surpluses would need to be generated and harnessed for sustaining productive accumulation.

(ii) Effective and adequately strong linkages would have to develop between industry and agriculture to enable labour to move into productive industry.

(iii) Industrial expansion would have to be viable with adequate stimulation of effective demand for its product, on the one hand, and on the other, provision of the necessary supplies of wage goods and raw materials for non-inflationary growth.

The key relationships on which I shall focus in the following discussion, therefore, are: the conditions of production, both technological and institutional, in which agricultural activity takes place; the conditions for reinvestment of the surplus into productive channels; the interrelation between agriculture, household industry, and non-household (i.e., manufacturing) industry, and the role of public investment.

Historical background

India's developmental experience cannot be studied without referring to the historical fact of protracted colonial domination and the way the mechanism of imperial exploitation affected the different segments and regions of the economy. The imperialist strategy itself altered from period to period, reflecting partly the varying requirements of the growth of the metropolitan economy, and partly the exigencies of the world political scene. It also differed as among the separate regions of India, depending upon the locally specific economic conditions and upon the position of the region in the scheme of colonial exploitation. Even after Independence, while the strategies of planned development themselves altered, they had diverse incidence on different regions.

For our purpose, I will adopt roughly the following periodization, conforming to qualitative changes in the overall growth situation:

1. The colonial period until World War I;
2. the inter-war period and after, until Independence;
3. the post-Independence period, until the mid-'60s; and
4. the period after the mid-'60s.

The regional dimension of the problem is discussed within each period.

The first colonial phase, until World War I, was marked by India remaining an agrarian–plantation economy with very little of manufacturing industry. During the inter-war period, however, modern industry had its first beginnings in India. The '50s and '60s were marked by active public investment through the Five Year Plans. The '70s and '80s were marked by general industrial stagnation coupled with inflation. Inevitably, my review of the events during these periods will be synoptic.

Colonial period until First World War

Let us first look at the conditions of production in agriculture and the utilization of surplus. As is well known, imperial interest in agriculture was principally as a provider of revenue. The surplus thus exploited was partly transferred to the metropolitan country, invested partly in the military and bureaucratic machinery in order to sustain and extend territorial aggrandizement, and partly in order to strengthen the sources of revenue through public investments in railways, canals, etc. The strategy of imperialist exploitation changed with the gradual establishment of political supremacy and with Britain's own advances in capitalist development. Hence, major differences arose in the way the colonial mechanism operated in the eastern region (Bengal, Bihar, Orissa), the western region of Bombay and Gujarat, the southern region of Karnataka and Madras, and the north-western delta of the Punjab.

The eastern region, which was the first to come under complete colonial domination, constituted the base from which further territorial expansion took place. It thus provided the first source of wherewithal for military conquests. Second, it was the prime experimental base for forced commercialization; the fertile hinterlands were converted into providers of cash and plantation crops in order to feed exports and export-based industries. The latter were predominantly under the control of European capitalists. This is reflected in the fact that Bengal was the principal region where export surpluses originated (occasionally making up for deficiencies in the Madras and Bombay Presidencies). Although treating regional export surpluses as an indication of equivalent drain from the region raises a number of disputable issues, it nevertheless indicates a substantive transfer of resources which is significant––particularly since the benefit of export surpluses did not accrue to the region concerned. In fact, the process of industrialization (based on smaller-scale manufacture, household industries, and crafts) then prevailing in the region was thwarted. And such industries as were newly established or allowed to grow in the region, were mainly intended to serve imperial interests.

Undoubtedly, elsewhere too, the colonial government's prime concern lay in the extraction of surplus from agriculture through land revenue. Apart from the tax burden being heavy, its severity was intensified because it was collected before the harvest and that too with a ruthlessness that was impervious to the plight of producers or to the problems faced by their productive enterprise. However, imperial policy differed considerably from region to region, in terms of the types of land settlement and the extent of commercial penetration into the agrarian economy. This, coupled with the differential policy towards public investment (in irrigation and railways), had a diversified impact on the various regions. These differences were important in shaping the future pattern and rate of industrialization during the inter-war and post-Independence years; they also constituted important factors underlying the rise of the capitalist and financier classes in specific regions. Even after Independence, this legacy has continued to play an important role. Hence the emphasis on the historical background of colonial rule.

With the adoption of the *zamindari* system in some parts, such as Greater Bengal, a number of consequences followed. A class of revenue collectors came up whose primary interest was not in productive cultivation. With growing sub-infeudation, usury, rising ground rents, consequent upon intensified pressure on land of population and pauperized artisans, the surplus in the region was diverted from the direct producers to the traders, moneylenders, and, feudal landlords. This meant that productive investment in agriculture did not grow. On the contrary, it resulted in the parasitical growth of unproductive consumption and the proliferation of unproductive commerce. This is reflected in the relatively poorer growth of agricultural output, and of yield per acre in Greater Bengal as compared to other regions (*see* Tables 6.1 and 6.2). There was very little incentive for the *rentier* lords or for the impoverished, over-exploited tenants, to invest in capital assets and in productive activity.

Table 6.1

Percentage change in acreage and yield per acre in different regions of India

	Acreage			*Yield Per Acre*		
	All Crops	Foodgrains	Non-Food-grains	All Crops	Foodgrains	Non-Food-grains
Total Change 1891–5 to 1941–6						
British India	+16.27	+25.55	+19.84	+ 6.45	− 7.31	+52.96
Greater Bengal	− 3.37	+ 0.79	−29.54	−12.85	−22.17	+54.55
Madras	+16.45	+ 3.89	+82.95	+46.02	+30.35	+79.84
Greater Punjab (including NWFP)	+51.69	+46.42	+94.38	+35.60	+19.24	+74.20
United Provinces	+28.18	+27.93	+29.12	+ 2.62	− 8.63	+29.32
Total Change between 1921–6 to 1941–6						
British India	+ 5.54	+ 7.08	− 0.87	− 1.45	− 9.69	+25.24
Greater Bengal	+ 4.87	+ 6.47	− 9.71	− 7.86	−12.96	+24.45
Madras	+ 3.51	− 1.66	+23.61	+ 0.92	− 4.99	+ 9.45
Greater Punjab (including NWFP)	+ 7.25	+ 7.09	+ 8.48	+13.58	+ 4.91	+39.64
United Provinces	+ 5.00	+ 4.39	+ 7.41	− 2.10	−14.47	+28.88

Source: Compiled from G Blyn, Table 4C of *Agricultural Trends in India* (cited in Bagchi, 1978).

Table 6.2

Yield of crops in Greater Bengal

(Yield in Rupees/acre)

	1891–2 to 1895–6	1921–2 to 1925–6	1941–2 to 1945–6	Change between 1891–6 and 1941–6 (Percentage)		Change between 1921–6 and 1941–6 (Percentage)	
				Average	Yield	Average	Yield
Greater Bengal							
All crops	371.2	351.1	323.5	− 3.37	− 12.85	+ 4.87	− 7.86
Foodgrains	353.7	316.5	275.3	+ 0.79	− 22.17[2]	+ 6.47	−12.96[2]
Non-foodgrains	470.0	583.7	726.4	− 29.54	+ 54.55[1]	− 9.71	24.45
Tea	1571.0	2127.0	3638.0[3]	+106.04	+137.57	11.54	71.04
Sugarcane	9666.0	11592.0	13689.0[3]	− 29.79	+ 41.62	42.60	18.09
British India							
Jute	4885.0	5784.0	5664.0	21.78	15.95	4.91	− 2.08

Note:

1. Acreage shift to non-food crops.
2. Negative yield growth of food crops.
3. Steep rise in yield of plantation crops.

Source: G. Blyn, Tables 3A and 4C of *Agricultural Trends in India* (cited in Bagchi, 1978).

In the *ryotwari* areas, although it is true that the revenue suction mechanism was no less severe than in the *zamindari* areas, some mitigating factors may have been in evidence: First, there was a more direct connection between the producers and the state (as revenue collector) because the practice of revenue farming was not widespread. Second, there was also a better recognition in these areas of legal occupancy rights as well as better records of them. Third, in some regions, legal expropriation and transfer of land rights were restricted, so that the transfer of land from the indebted small operator to moneylender/trader was somewhat more restricted. It would seem then that productive investment was more attractive in these areas (for instance, the Punjab, Madras, Bombay hinterlands), than in the predominantly *zamindari* areas in the eastern region (Bengal, Bihar, Orissa, and Uttar Pradesh). It is worth pointing out, too, that public investment in canal irrigation constituted an additional major discriminating factor.

Public investment during this early colonial period was directed mainly into the railways and irrigation. A high enough rate of return was imperative, particularly in irrigation works. Railways had a priority as much due to military–strategic considerations as due to their being the means of providing a feeder transport network for exports as well as for the European capital-dominated export-oriented enterprises in the port enclaves.

Colonial irrigation policy was also dominated by political and strategic considerations. Its prime motive was to stabilize and increase the revenues from agriculture, with less consideration for some of its consequences on the property structure. The regional distribution of public irrigation was markedly uneven. In the Punjab where, for strategic reasons, canal colonies were set up, there was considerable investment in irrigation. In the *zamindari* areas of the Eastern region, on the other hand, the mechanism of revenue extraction operated through the fixed revenue commitments of the *zamindar*, and less attention was paid to the growth of productivity. In fact, consequent upon increasing sub-infeudation in the region and the growth of an apathetic *rentier* class, even the privately owned minor irrigation works were neglected with untoward consequences.

In the case of UP, Whitcombe (1971) has brought out the ill-effects of such neglect on agricultural productivity, as well as the pernicious consequences of negligent management of public irrigation (e.g. increasing salinity, waterlogging, over-cropping, and denudation of soil nutrients). The most striking consequence of the colonial policy on irrigation was the worsening of the state of the poorer sections of peasantry and tenantry. They were most affected by the deteriorating productivity of the land, on the one hand, and by the increasing hold of usury and debt bondage, on the other. Table 6.3 shows the regional distribution of public investment in irrigation.

Table 6.3

Province-wise distribution of public investment in irrigation during 1860–1947

(Rupees million)

Provinces Period	Punjab	Bihar	Madras	Bombay and Sind	UP	Others	Total
1860–1 to 1897–8	135 (24.2)	34.48 (6.1)	112 (19.7)	50 (9.0)	135 (24.52)	94.52 (16.28)	558 (100)
1898–9 to 1918–9	315 (47.0)	28.09 (4.2)	119 (18.0)	96 (14.4)	58 (8.7)	529 (7.9)	669 (100)
1919–20 to 1946–7	590 (30.0)	46.20 (2.3)	284 (14.8)	514 (26.1)	360 (18.3)	170.8 (8.6)	1968 (100)

SOURCE: M. J. K. Thavraj (1972: 5).
NOTE: Figures in brackets are percentages.

Furthermore, the process of de-industrialization, i.e. the slow decline of household industries and crafts, advanced much faster in the eastern region which, prior to colonial domination, had attained a degree of prosperity in this respect. The imperial connection was the stronger in the eastern region, and European capital and commerce had entrenched itself there, effectively subordinating the indigenous trading classes and financiers. On the one hand, the indigenous industry in the region was faced with a shrinking domestic market as a result of the declining prosperity of the average consumer; on the other, it had no easy access to markets or to finance since finance and commerce were geared to export-oriented activities and markets were flooded with imported goods from Britain. The region became increasingly dominated by the port city, growing as an enclave.

By contrast, the hinterland of Bombay, where *ryotwari* prevailed, presented a different picture. It was characterized by a significant survival of handicrafts and small-scale industries, and local merchants and financiers were in better control of commerce and finance. In addition, the hinterland region was economically relatively better off than the eastern region with the result that the demand for indigenous wares had not yet been stifled. Indeed, it has been noted by economic historians that the class of industrial capitalists/financiers (e.g. Parsis and Marwaris) first arose prominently in this region. In the *ryotwari* region of Madras too, which also had a fair share of canal irrigation and had an ancient tradition of trade and industry, handicrafts and small-scale industries resisted destruction. The Chettiars, for example, were a community of financiers and traders who had ancient connections with other parts of Asia.

The Punjab had a number of factors in favour of its agriculture. First, the average size of the cultivating unit in the canal colonies, where land was *de novo* parcelled out into ownership plots, was much larger, and (in contrast to the eastern region) there was a consequent absence of a mass of impoverished peasantry at the lower levels of holdings. The region, moreover, was favoured with massive public investment in canal irrigation. And, being a military base, a number of small-scale enterprises had sprung up (as in cantonment townships elsewhere such as Meerut). There was, thus, an economic base for generating effective demand for the region. The region had no cities, however, comparable to the port-cities of Bombay, Calcutta, and Madras, or even large traditional metropolitan centres.

The inter-war years

During the first decades of the twentieth century, under the conditions of war prevailing in Europe, and with the development of the phase of high finance capital in Britain, Indian industrialization received some protection from the colonial power. The pace and pattern of development were, however, still under imperial control and subject to the priorities of the metropolitan power. While the establishment of basic intermediate goods industries (e.g. steel) was haltingly approved, hardly any capital goods industries were allowed. Consumer goods industries which had a traditional home market were allowed to grow. Their competitiveness, too, was hampered by trade restrictions. It is not surprising that the regions surrounding Bombay and Madras benefited the most in terms of the rise of manufacturing units. These regions had maintained a better organized indigenous class of capitalists and financiers who had a firm hold over the domestic markets. In contrast to Bombay and Madras, the eastern region suffered more from imperialist penetration and the control exercised by European capital and trade.

While new townships came up (such as Ahmedabad, where the textile industry flourished, and Jamshedpur, where the first steel complex was built), development remained basically of the port-enclave type on the periphery of the Presidencies of Madras, Bengal, and Bombay. The decay of traditional industries, especially in the eastern belt, led to the slow decline of urban settlements such as Dacca, Murshidabad, and Surat. Industry did not spread out from the port-enclaves, or even from the nodal points in the sparse interior centres, and failed to attract labour away from agriculture to any substantial extent.

Thus, the process of industrialization during this period was not only weak but also distorted. There were no capital goods of basic industries worth the name, and industry was severed from its linkages with agriculture. Neither 'backward linkages' nor 'forward linkages' were powerful in terms of diversification of consumer goods industries. India was dependent on imports, particularly imports from Britain of both basic capital goods and a number of consumer goods. A 'dualism', introduced between the traditional agriculture and the islands of manufacturing centres, has continued ever since.

In regional terms, a pattern of 'agglomerated' growth emerged, with islands of concentrated growth but having very weak dispersal effects. This resulted in a marked dichotomy reflecting a disjunction between agriculture and industry. The port-enclave manufacturing centres (e.g. Calcutta) were growing fast even as the hinterland agrarian and traditional industry were deteriorating. On the other hand, regions with relatively prosperous agricultural growth such as the Punjab had no major industrial centres. As late as 1948, the Presidency States (Bombay, Madras, and Bengal) accounted for 76.7 per cent of the total industrial workers, 77 per cent of the total value of industrial production, 82.2 per cent of the value of engineering and electrical goods, and 87.2 per cent of the value of chemical goods in the country. The corresponding figures, for the mineral rich Provinces/States of Bihar, Orissa, the Central Provinces and Berar were 9.61, 9.54, and 5.1, respectively, showing how little the 'natural endowments' of the region mattered in this respect (Raza and Kundu, 1980). The southern and western regions around Madras and Bombay (and especially those parts of the latter which were to constitute the State of Gujarat) were better placed and had a more advantageous start in terms of both agriculture and industry.

Post-Independence period

The post-Independence period divides into two sub-periods: (i) 1947 to the mid-'60s, during which development was guided by a more or less vigorous planning process and public investment played an active role; and (ii) the period since the mid-'60s, when public investment lost its thrust and planning proceeded haltingly. Table 6.4 provides an overall view of the economy's rather uneven performance over the various plan periods (and during the years during which the plan was suspended).

Over the plan periods, the strategy of industrialization varied. A most important change, in terms of industrial breakthrough, took place during the second plan period, when a concerted move was made to build up basic industries and to locate them in the interior. It was hoped that these major industrial complexes would act as nodal points from where growth would spread. However, industrial development remained concentrated in the traditional port-cities and their periphery. It is interesting that industrial complexes (such as Bhilai and Durgapur) appear to have generated little development of subsidiaries in the vicinity and that engineering industries still tend to concentrate in the areas near Calcutta. Of course, some new interior centres have emerged, but the dominance of Madras, Bombay, and Calcutta, and of their hinterlands has not been effectively reduced.

Table 6.4

Annual growth-rates

(percentages)

	National Income	Per Capita Income	Agricultural Production	Industrial Production	Foodgrain Productions
First plan (1951–6)	3.6	1.7	4.3	7.4	5.2
Second plan (1956–61)	4.1	2.0	4.1	6.6	3.9
Third plan (1961–6)	2.4	0.1	-1.0	9.0	-2.0
3 Annual plans (1966–9)	4.1	1.8	7.0	2.6	9.4
Fourth plan (1969–74)	3.5	1.2	3.1	4.0	2.8
1974–5	1.0	-1.0	-3.2	2.6	-5.4
1975–6	9.2	6.8	15.2	6.0	22.0
1976–7	1.4	-0.6	-7.0	9.5	-9.0
1977–8	7.4	5.2	13.9	3.9	14.8
1978–9	(4.3)	(2.3)	(4.0)	(7.6)	(4.0)
1979–80	(-1.5)	(-3.5)	(-7.5)	(2.0)	(-7.1)

NOTES: Bracketed figures are provisional; at 1970–1 prices; based on Index Numbers; Calendar years up to the Third Five Year plan and financial years thereafter. First plan figures are for 1952–5.

Looking aggregatively at State-level statistics during the periods 1950–1, 1964–5, 1969–70 (*see* Tables 6.5 and 6.6), the first four States, in terms of the highest per capita net domestic product, continue to be West Bengal, Punjab, Gujarat, and Maharashtra. Assam, which was in the first five states in 1950–1 and 1964–5, lost its rank to Kerala in 1969–70. Table 6.8 shows that, since the '70s, Punjab, Haryana, Gujarat, and Maharashtra have come to occupy the first four places while West Bengal has been losing in rank continuously, Bihar continues to remain at the bottom of the scale, and Rajasthan has been pushed to lower and lower ranks over time. Tamil Nadu appears to have been gaining ground since the '60s.

Comparing the figures for per capita consumption levels for 1957–8 with those for 1960–1 and 1963–4 (*see* Table 6.7), we note that Punjab occupies the first place; West Bengal, Assam and Rajasthan remain in the top five brackets, while Orissa lurks at the bottom of the rank-order, along with Bihar, Andhra Pradesh, Kerala, and Tamil Nadu. There is, thus, no necessary correspondence between the ranking according to per capita consumption and the ranking according to per capita net domestic product. Even more striking is the lack of correspondence between the rank-ordering of States according to per capita rural and urban levels of consumption—which possibly reflects the disjunction between agrarian and industrial growth. For example, whilst the States of West Bengal and Bombay (later divided into Gujarat and Maharashtra) appear in the top five brackets of urban consumption, they were much further down in rural consumption during the earlier years. The reverse is the case with Jammu and Kashmir which has been slow to urbanize.

Table 6.5

Ranking of States (1950–70)

	Per Capita NDP				
	1950–1	1955–6	1960–1	1964–5	1969–70
Andhra Pradesh (AP)	9	10	7	6	10
Assam	5	5	5	5	7
Bihar	14	14	14	14	14
Gujarat	3	4	3	3	2
Karnataka	7	7	8	8	8
Kerala	6	6	9	9	5
Madhya Pradesh (MP)	13	9	11	11	12
Maharashtra	4	2	1	2	3
Orissa	11	13	13	13	9
Punjab	2	3	4	1	1
Rajasthan	10	11	10	12	13
Tamil Nadu (TN)	12	8	6	7	6
Uttar Pradesh (UP)	8	12	12	10	11
West Bengal (WB)	1	1	2	4	4

Table 6.6

Group ranking of States (1960–70)

High Income States (first five) (descending order)			Lowest Income (five states) (ascending order)		
1960–1	1964–5	1969–70	1960–1	1964–5	1969–70
WB	Punjab	Punjab	Bihar	Bihar	Bihar
Punjab	Maharashtra	Gujarat	MP	Orissa	Rajasthan
Gujarat	Gujarat	Maharashtra	TN	Rajasthan	MP
Maharashtra	WB	WB	Orissa	MP	UP
Assam	Assam	Kerala	Rajasthan	UP	AP

Table 6.7

Per capita consumption (rural and urban)

Highest five (in descending order)

1957–8			1960–1			1963–4		
A	R	U	A	R	U	A	R	U
Punjab	Punjab	Assam	Punjab	Punjab	Assam	Punjab	Punjab	Assam
Assam	Assam	WB	WB	J&K	WB	WB	J&K	WB
Rajasthan	Rajasthan	Bombay	Assam	Bihar	Punjab	J&K	Assam	Bombay
Jammu & Kashmir (J&K)	J&K	MP	Bihar	Assam .	Bombay	Assam	Rajasthan	Punjab
WB	Karnataka	Punjab	J&K	Rajasthan	Karnataka	Rajasthan	WB	Rajasthan
Lowest five (in ascending order)								
Orissa	Orissa	Karnataka	Orissa	Orissa	Orissa	Orissa	Orissa	Karantaka
TN	TN	UP	Kerala	Kerala	Kerala	Kerala	Kerala	Kerala
Kerala	Kerala	J&K	MP	MP	J&K	Bihar	AP	AP
AP	AP	AP	TN	TN	Rajasthan	AP	Karnataka	UP
Bihar	Bombay	Kerala	AP	AP	AP	Karnataka	Bihar	Bihar

A: Aggregative R: Rural U: Urban
SOURCE: For Tables 6.5, 6.6, and 6.7: K. R. G. Nair (1986).

Table 6.8

Per capita income by States

(Rupees at current prices)

	1970–1		1975–6	
Rs 1000 and above	Punjab	(1067)	Punjab	(1688)
			Haryana	(1514)
			Maharashtra	(1455)
			Gujarat	(1236)
			Himachal Pradesh (1+P)	(1165)
			WB	(1100)
			Karnataka	(1038)
			Kerala	(1000)
Rs 750–1000	Haryana	(932)	Tamil Nadu	(997)
	Gujarat	(845)	Nagaland	(949)
	Maharashtra	(811)	Manipur	(904)
			Meghalaya	(899)
			AP	(897)
			Rajasthan	(873)
			Tripura	(872)
			Assam	(848)
			Orissa	(834)
			J&K	(825)
			MP	(790)
Rs 650–750	WB	(729)	UP	(727)
	HP	(676)	Bihar	(669)
	Karnataka	(675)		
Below Rs 650	Meghalaya	(644)		
	Kerala	(636)		
	Rajasthan	(629)		
	TN	(616)		
	AP	(586)		
	Assam	(570)		
	Tripura	(563)		
	J&K	(557)		
	Orissa	(541)		
	Nagaland	(508)		
	UP	(493)		
	MP	(489)		
	Bihar	(418)		
	Manipur	(408)		

SOURCE: Compiled from Central Statistical Organization data.

These aggregative State level data are only indicative, since the pattern of growth has been uneven among different districts of the same State—a classic case being that of western and eastern UP. Differentiation has occurred in terms of both agricultural and industrial growth. A study of regional disparities (Nair, 1986), maintains that, while there were some tendencies towards narrowing disparities between 1950–1 and 1960–1, the same cannot be said for the decade between 1960–1 and 1970–1.

This is not surprising if we notice that public investment had slackened after the mid-'60s and that private investment has also been tardy. The tardiness of private investment has been ascribed to various causes: e.g. infrastructural bottlenecks, deficiency of effective demand, deterioration in world markets, and a complementary slackening of public investment. The strategy of export-led growth which has been advocated in the face of tardy growth of home markets is frustrated by such factors as shrinking world trade, intensifying trade barriers, increasing cost of crucial imports, etc.

Differentiation among regions has also occurred in agriculture with the uneven impact of the 'Green Revolution' after the mid-'60s. First, the 'Green Revolution', which made an initial impact by generating a high level of growth, would appear to have tapered off and its tempo has not been sustained. Second, its impact was felt mainly on wheat and less so on rice. Bhalla and Alagh (1979) have shown that Punjab and Haryana recorded a remarkable growth of output and yield per hectare while Orissa, Maharashtra, and Andhra Pradesh registered negative growth-rates (Table 6.9). Its benefits appear to have accrued to the districts within States where there was either good rainfall or already existent irrigation or that created by private investment. The new technology thus appears to have been adopted and more widely used in regions where the general level of well-being of producers was already relatively high.

Table 6.9

Annual compound growth-rate: 1970–3 over 1962–5

	Area	Output	Rank	Yield	Rank
AP	–1.11	–0.98	16	0.13	15
Assam	1.90	2.23	11	0.31	14
Bihar	0.52	1.52	14	1.47	11
Gujarat	0.22	4.50	5	4.25	5
Haryana	1.22	7.88	2	6.58	2
HP	0.46	3.35	7	2.87	8.5
J&K	0.32	5.80	3	5.45	3

	Area	Output	Rank	Yield	Rank
Karnataka	–1.88	2.45	10	4.41	4
Kerala	0.90	2.15	12	1.24	12
MP	0.68	1.63	13	0.94	13
Maharashtra	–0.97	–4.89	17	–3.96	17
Orissa	0.63	–0.61	15	–1.23	16
Punjab	2.67	10.42	1	7.54	1
Rajasthan	1.33	4.72	4	3.37	6
Tamil Nadu	–0.01	2.87	8	2.87	8.5
UP	1.10	4.30	6	3.15	7
W. Bengal	0.42	2.74	9	2.33	10

SOURCE: Bhalla and Alagh (1979).

Regional differentiation in agriculture

The two-way Table 6.10 shows the distribution of districts (in numbers), according to the initial level of productivity per hectare, as in 1962–5, and the rate of growth annually compounded between 1962–5 and 1970–3. It shows the unevenness of growth and of initial levels of productivity among districts within the same State. Looking at levels of productivity, the same study shows that the highest levels were to be found in the coastal plains, including the interior of Tamil Nadu and some small areas in West Bengal, Assam, and in the north-western parts of Punjab and Haryana. The lowest productivity areas belong to the western, central, and southern arid zones of India, namely western Rajasthan, some parts of Gujarat, Madhya Pradesh, and Maharashtra, some parts of Karnataka and Andhra Pradesh. These are areas of uncertain rainfall and no assured irrigation. However, the highest rates of growth have occurred in dry areas (e.g. Rajasthan, Haryana, and Gujarat) where irrigation has been introduced on a major scale. The other fast-growing regions are the irrigated districts of Punjab and western UP. Leaving these regions aside, a large part of India is characterized by medium, low, or even negative rates of growth. A large number of districts located in the Gangetic plains of eastern UP, Bihar, and West Bengal fall into this category. The high productivity regions in the south have also tapered off in growth-rate, whilst some parts of Bihar and Orissa show actual deceleration.

Table 6.10

Distribution of districts in various States according to yield level in 1962–5 and growth rate between 1962–5 and 1970–3.

(Number of districts shown in brackets)

Yield Level/ Growth Rate	4.5 per cent		1.5–4.5 per cent		0.0–1.5 per cent		Less than 0.0 per cent	
Above Rs 1300/hectare	Karnataka	(8)	AP	(2)	AP	(1)	AP	(3)
	Punjab	(2)	Assam	(1)	Assam	(1)	Karnataka	(3)
	Tamil Nadu	(2)	Gujarat	(1)	Kerala	(3)	Maharashtra	(2)
	UP	(3)	Karnataka	(8)	Maharashtra	(2)	West Bengal	(1)
			Kerala	(4)	Tamil Nadu	(3)		
			Tamil Nadu	(4)	West Bengal	(1)		
			UP	(1)				
			West Bengal	(6)				
Rs 700-1300/ hectare	Assam	(1)	Assam	(3)	AP	(2)	AP	(5)
	Haryana	(3)	Bihar	(5)	Assam	(2)	Bihar	(6)
	J&K	(1)	Gujarat	(2)	Bihar	(3)	Gujarat	(3)
	Punjab	(9)	Haryana	(2)	Gujarat	(4)	MP	(2)
	UP	(6)	HP	(1)	Karnataka	(4)	Maharashtra	(4)
			Karnataka	(3)	MP	(3)	Orissa	(8)
			MP	(3)	Maharashtra	(1)	UP	(1)
			Rajasthan	(2)	Orissa	(3)		
			Tamil Nadu	(1)	Rajasthan	(1)		
			UP	(22)	Tamil Nadu	(2)		
			West Bengal	(7)	UP	(10)		
Below Rs 700/hectare	Gujarat	(4)	Gujarat	(2)	AP	(1)	AP	(3)
	Haryana	(2)	MP	(22)	Gujarat	(1)	Bihar	(1)
	J&K	(1)	Rajasthan	(6)	Karnataka	(1)	Gujarat	(1)
	Karnataka	(3)	UP	(5)	Rajasthan	(3)	Karnataka	(3)
	Rajasthan	(12)			UP	(1)	J&K	(1)
							MP	(5)
							Maharashtra	(17)
							Rajasthan	(2)

SOURCE: Bhalla and Alagh (1979).

The agricultural performance and the general state of welfare in the regions is reflected in the per capita incomes and consumption levels. Thus, Punjab and Haryana—regions of high productivity and high growth, also show high per capita net product and consumption, while West Bengal, Bihar, Orissa, with their deteriorating conditions in agriculture, show poor levels of income consumption. The differences in the spectrum of consumption levels within the region are also of considerable interest. The regions where the overall rate of growth of agriculture is low are precisely those with greater disparities in consumption levels and with a greater mass of people in the lowest bracket of consumption expenditures. Gujarat, Haryana, and Punjab have relatively smaller proportions of their population in the lowest consumption expenditure brackets (*see* Table 6.11).

Table 6.11

Percentage distribution of households by monthly per capita expenditure bracket—October 1973–June 1974 (rural).

(in rupees)

	0–21	21–43	43 and above
Gujarat	1.3	31.7	67.0
Haryana	0.7	20.7	78.6
Punjab	0.5	12.5	87.0
All-India	3.7	37.9	58.4
West Bengal	8.9	45.9	45.1
Bihar	4.2	32.2	63.3
Orissa	7.6	53.4	31.0

SOURCE: *National Sample Survey 1974.*

The uneven nature of agricultural development is also evident from other indicators, such as fertilizer consumption, percentage area irrigated (*see* Table 6.12). It is evident that Punjab and Haryana have shown a remarkable up-thrust with all-round development in the rural sector. Not only were larger agricultural surpluses generated, but they were directed into productive investment. It would seem that the facts of the relatively higher average level of living and a higher average size of the cultivating unit with smaller proportion of extremely poor cultivators have helped to channel the agricultural surplus into more productive avenues. As against this, it should be noted that some parts of this prosperous zone have been characterized by a persistence of usury and reappearance of bondages in different forms. Statements concerning the 'bright' record of prosperous regions should be understood entirely in relativist terms.

Relatively well off regions shine only by comparison with the utterly dismal record of regions such as eastern UP, Bihar, and parts of Orissa and West Bengal.

Table 6.12

Industrial indices of States

State	Daily Factory Employment per 1000 Population (1977)	Gross Factory Output per Capita (1976–7)	Irrigated Area as % of Cropped Area (1975–6)	Fertilizer Consumption Per Hectare 1978–9 (Kilograms)	Electricity Consumption Per Capita (1978–9) (Kilowatts)
Adhra Pradesh	8.2	397	34.9	46.3	112
Assam	4.0	310	18.0	2.4	42
Bihar	5.4	323	29.8	17.2	69
Gujarat	18.2	1187	15.0	31.4	264
Haryana	13.2	778	50.3	37.4	265
Himachal Pradesh	4.0	166	16.6	13.0	81
Jammu & Kashmir	3.3	90	40.2	20.7	125
Karnataka	14.6	423	15.3	32.3	189
Kerala	11.7	425	9.2	33.4	122
Madhya Pradesh	7.6	317	8.8	9.0	106
Maharashtra	19.3	1486	11.0	19.4	268
Orissa	3.3	241	19.2	9.3	117
Punjab	9.8	800	73.8	94.8	373
Rajasthan	3.6	258	17.1	7.7	119
Tamil Nadu	10.8	772	46.7	68.1	219
Uttar Pradesh	5.1	247	40.2	45.5	108
West Bengal	16.2	741	19.4	30.6	147

Regional differentiation in industry

How does the process of industrialization appear from a regional perspective? Reference has already been made to the unsuccessful attempt at generating growth in the backward regions through the establishment of industrial complexes. Raza and Kundu (1980) have made some revealing observations on the pattern of growth that emerged during 1961–71. Using the National Sample Survey's regional demarcations, they have found that, of the 58 regions, only two, namely coastal Maharashtra and the central plains of West Bengal, accounted for about 32 per cent of the total non-household manufacturing employment. These were followed by the following five regions in descending order: Gujarat plains, southern Kerala, coastal and inland Tamil Nadu, and western UP, each accounting for more than 4 per cent of the total manufacturing employment. Even more striking is the fact that 89 and 75 per cent of the non-household industrial workers respectively in the Maharashtra and West Bengal regions lived in class I towns (i.e. exceeding 1,00,000 in population). More than 17 per cent of the total non-household manufacturing employment was thus concentrated on the peripheries of Calcutta and Bombay. A disconcerting fact is that the share of non-household employment in the towns went down (during the interval between 1961 and 1971) in relation to that of cities. Also, while the cities of Saurashtra, inland Karnataka, and the central plains of West Bengal show a rise in the share of non-household employment, those in the western plains of West Bengal, southern UP, and northern Bihar have experienced a decline.

The percentage of workers employed in household industries declined from 6.4 to 3.5 between 1961 and 1971. It is noteworthy that regions with a high percentage of workers in non-household industries (such as in the Gujarat plains, coastal Maharashtra, northern Orissa, and inland eastern Karnataka) constituted only a small fraction of the workforce in household employment. The reverse was the case in regions such as eastern UP, some parts of Tamil Nadu, and Madhya Pradesh, where a relatively active household sector exists but where there is hardly any significant non-household manufacturing activity. It is, however, alarming that an absolute decline in household employment occurred between 1961 and 1971 in regions covering large parts of Bihar, Orissa, and the eastern region of West Bengal where the non-household manufacturing sector is weak. It is in the villages and towns of these regions that the household sector has been losing out.

Raza and Kundu (1980) have also thrown light on two broad patterns that appear to be emerging:

1. the pattern of agglomerated growth, based on generally capital-intensive, large-scale manufacturing units located in a few cities; and
2. the 'dispersed' pattern, where growth is relatively dispersed in small towns and rural settlements.

Suitable indices are compositely constructed to measure these. According to this study, city-based industrial development was found to be localized in the following six regions: (1) the Calcutta conurbation; (2) the Madras conurbation; (3) the Bombay–Gujarat conurbation; (4) the Delhi metropolitan region; (5) the Jamshedpur–Dhanbad–Bokaro complex; and (6) the Ludhiana–Jalandhar complex. The first three reflect a continuation of the historical legacy of port–enclave development. An additional feature of this identification consists of the absorption of the Gujarat region within the shadow of Bombay. The Jamshedpur–Dhanbad–Bokaro complex, too, has historical roots in the colonial period that are traceable to the establishment of the steel mills of Jamshedpur. But this complex has been widened and strengthened with the massive public investments in Bokaro and Dhanbad. Delhi owes its growth mostly to public investment and is basically a tertiary-activity complex, an outgrowth of the services sector feeding the capital. The Ludhiana–Jalandhar region is a relatively new development, although the region was known even earlier for its army-based small sector. It is now situated in the heartland of the Green Revolution and draws its sustenance largely from the surrounding agricultural prosperity.

As distinguished from these agglomerated growth regions, certain parts of AP, Kerala, and Karnataka in the south, Haryana in the north, and Gujarat and parts of Maharashtra, have shown prominently dispersed growth. Kerala, with its continuous settlements, has come up particularly with dispersed industries and its advance in this matter is reflected in its improved position in recent years in terms of per capita output and consumption levels. Gujarat provides a remarkable instance of growth patterns of both kinds: the city-based agglomerated growth as well as dispersed growth in smaller towns and villages.

Bringing together our observations[2] on agricultural and industrial growth in its regional perspective, what broad generalizations can be offered? First, we note the importance of agricultural surpluses in adequate quanta to sustain industrial expansion. The regions sharing some industrial vitality appear to be the ones where agricultural growth has also been promising. We further note that an important precondition to growth is that surpluses are productively reinvested. Public investment appears to have played an important role historically in supporting agricultural growth and in generally supporting an environment for productive channelling of surplus.

An even more important precondition for productive utilization of surplus and forging a link between agriculture and industry would appear to be a general level of well-being, which is roughly indicated in per capita consumption levels and by a smaller proportion of households in the lowest brackets of household consumption. In those areas where both the agricultural growth-rate has been favourable and the 'general level of well-being' has been relatively higher (e.g. Gujarat and Haryana), higher industrial activity is observable.

This association between better agricultural performance and

[2] In this sketchy treatment, no attempt has been made to scrutinize closely the comparability of, or probe deeper into, the statistical evidence put together from diverse sources.

higher level of well-being, on the one hand, and higher industrial performance, on the other, may be interpreted in the following way. If the region is afflicted by the deadweight of utter poverty, the situation offers a breeding ground for exploitation through usury, speculative trading, commercial exploitation, etc. All these divert surpluses into unproductive channels and create relative advantage favouring such operations. The preponderance of such 'unproductive' uses of surplus implies a perverse system of 'social subsidization'[3] under which the process of pauperization continues even as the surplus is diverted into unproductive channels. Second, if the general level of living is reasonably high, an effective demand for the products of the industry is created. The emergence of Haryana–Gujarat on the industrial map, particularly in their 'dispersed' pattern of industrialization, would appear to reflect this situation, at least in part. By contrast, Bihar, Orissa, eastern UP, parts of Rajasthan, and MP, would appear to illustrate the contrary case of retardation.

Dichotomous development in the '80s:[4] some significant macro-economic trends

The '80s have been a dramatic decade for India, arousing major contradictions regarding the performance of the economy. India has entered into the '90s with an engulfing crisis in its international finance along with instability and crisis in the governance of the country. I deal here with the critical features of the macroeconomic trends in the '80s in comparison with past trends in order to indicate the dynamic movements of the economy that have given rise to major controversies in respect of development strategies. Let us identify some critical features of the macroeconomic trends that have evoked deep concern and sharp disagreements.

In the realm of output, i.e. levels and composition of national product, India, during the '80s performed creditably in terms of widening its range of products and services and extending their

[3] The term 'social subsidization' is used here in a specific sense to signify the form of consumption loans. From the viewpoint of macro-level real flows, a part of the surplus is diverted to barely sustain a section of the population. However, the system through which these flows operate is doubly pernicious: first, it does not usually work towards enhancing the productive potential of the population so sustained and, probably, even works in the reverse; second, it diverts re-investible surpluses away from directly productive uses.

[4] This section, covering the '80s, was completed in January 1991 before the events leading to considerable changes in government as well as in political and economic conditions in India and in the outside world took place.

volume in industry, agriculture, and services. There has been a substantial growth in product terms: the record of economic performance shows a GDP, at constant 1980–1 prices, of around 5 per cent per annum up to 1987–8, with the value added in agriculture growing at around 2 per cent per annum, in manufacturing at around 8 per cent per annum, and in services around 6 per cent per annum in the same period. Remarkable also was the resilience shown by the economy in recovering sharply from the severe and widespread drought of 1987–8 following in succession the two previous lean years. In 1988–9 the GDP increased by 9 per cent and, more remarkably, agricultural production showed a rise of around 20 per cent. Industrial production which, it was generally expected, would be severely affected by the depression brought about by the drought, recorded a growth of 8.8 per cent. In all these, the performance in 1989–90 was poorer, the GDP growth being no more than 4 per cent and, even with a normal monsoon, agricultural production rose by only about one per cent over the high level of 1988–9.

Even more remarkable was the fact that, contrary to the fears generated by the severity of the drought, the pace of reported inflation in 1989–90 was also restrained. Indeed, a point to note is that the critical drought situation was met not with the traditional textbook wisdom of curtailing public expenditure, but rather was countered by expanding public investment, supported partly by the drawals upon accumulated food stocks and on foreign exchange reserves to finance heavy imports. It was, however, the manner of financing this high-speed growth (further reinforcing the past fiscal trends) through deficit financing and massive borrowing from internal and external sources, as well as the pattern of growth adopted, that has led to a continuing grave fiscal crisis. The hectic access to borrowing and spending of foreign exchange under the new policies drove the economy into a foreign exchange crisis situation which was to engulf the economy and the planning process entirely by the beginning of the '90s.

A disturbing feature of the industrial growth of the '80s was the wide year-to-year fluctuations in growth-rates as well as wide divergences in the growth performance of different sectors. Whilst growth-rates have been considerably lower for the intermediate goods and non-durables among consumer goods, the consumer goods sectors (particularly electric and electronic appliances) have forged ahead, reflecting the new high-style consumption patterns becoming entrenched amongst the richer sections of society. The latter have also proved to be import-intensive (both, in terms of import of technology and of inputs) and have evoked powerful and pervasive demonstration effects on life-styles, particularly of the middle classes.

The technological advances in the capital and intermediate goods sectors have been slow with a very tardy process of indigenization of imported know-how. There has been growing sickness in older establishments and low profitability in traditional industries. Thus the industrial growth scenario is heterogeneously structured, although showing some advances.

In agriculture, too, despite the significant achievement of self-sufficiency, in the sense of eliminating the regular and critical dependence on imports of the earlier years, the per capita net availability has not shown much sustained improvement. The volatility of production prevails particularly in foodgrains. The production of pulses, coarse grains, and oilseeds show very low rates of growth despite rising prices. The oscillations in the rate of growth of foodgrains production and escalating per unit costs (which have resulted in continuous pressures on Agricultural Costs and Prices Commission (ACPC) to raise support prices) indeed constitute a cause for concern. This is so, notwithstanding the mounting fertilizer and irrigation subsidies. It is also worth noting that marginal returns on the modern inputs are showing a declining trend, particularly in the high growth regions. On the other hand, the elasticity of labour demand which measures labour absorption, is showing a decline, with negative signs in some cases. This picture contrasts with that in the '70s, when it was hoped that labour absorption in agriculture would be maximized through increased productivity of land.

A striking feature of the changed structure of growth is also the enormous expansion of services, part of which reflects infrastructural growth, but a percentage of which is also connected with the burgeoning public administration, the banking and insurance sectors, and commercial services such as hotels, restaurants, trade, and transport. The expansion of the service sector has two features: 1. the positive, dynamic feature refers to the emergence of the service sector fulfilling the demands of the consumers and supporting and extending the lines of activities of the primary and secondary sectors; and 2. it is a form of 'disguised employment' in which the unemployed enter into small trading activities, spreading the distributive networks, and sharing the actual margins earned in trade.

A general feature of such growth is the increasing fluctuations in the aggregate product, uneven growth of sectors and regions, the expanding sectors being increasingly oriented towards the modern or 'leading' sectors and its beneficiaries—the rich and middle classes. With the wave of modernization and easing of import controls, new commodities, wants, 'needs', and consumption styles have been introduced. Many of the enterprises producing the consumption goods and services are operating either as small-scale ancillary units of the large-scale manufacturing sector or are displacing traditional small-scale production activities in competing areas, particularly in food-processing, textiles, clothing, etc.

While there have been considerable advances in output, there have been no corresponding or comparable gains in employment. The manufacturing sector absorbed barely 10 per cent of the labour force even in 1981 and the dependence on agriculture as a precarious means of livelihood has continued to persist, over several decades, at around 70 per cent of the workforce. The 1981 census reported some increase in employment in the manufacturing household industry which was recorded as higher than in services. However, it is unevenly spread over regions and shows very low productivity. Even in the prosperous High Yield Variety (HYV) seeds belt of Punjab and Haryana, employment in the manufacturing household industry is principally concentrated in towns.

At the same time, as rural industrialization has failed to advance in new activities, the traditional artisan and household industry has also been losing ground. The pace of urbanization and temporary movements of labour have been greatly accentuated. The urban manufacturing households that have mushroomed are known for their demand for casual and sweated labour. In a few relatively more productive rural areas, they operate predominatly as seasonal sideline activities, mainly providing a supplementary source of income. No sustained non-agricultural activity yielding secure livelihood appears to have emerged in the rural areas. There is no evidence to suggest that substantial occupational diversification has taken place during the last decade.

These uneven patterns of output and consumption growth over regions and sectors of the economy have been accompanied not only by a growing differentiation of technological and organizational processes in production but also by differential patterns of employment growth and conditions of labour. Tenurial conditions as well as forms of credit and of labour, and systems of rent and wage payments are changing radically with the uneven spread of commercialization in agriculture. However, no strikingly significant impact on occupational diversification or on livelihood opportunities or on stable living standards is evident (even though there may be a shift in the products that enter into this standard). While overall employment (in terms of person-hours) does not show striking growth, remarkable wage and income differentials are emerging with the opening of additional employment opportunities mainly in the tertiary sector and among the white-and-blue collar categories. Their salaries are also increasing in relation to those of unorganized labour due to wage indexation and negotiated wage settlements. At the same time, there is also growing differentiation in the conditions of labour in the organized and unorganized sectors.

These structural imbalances in the processes of income generation and income formation are also reflected in consumption styles and demand patterns. The concentration of incomes in the rich and middle classes has provided a booming market for consumer durables which, under the liberalized regime has shown considerable buoyancy in production. A dichotomy appears to be emerging between the sections of the population which are beneficiaries and receive their incomes from the 'modern' sector, on the one hand, and those which are peripherally drawn into the commercialization process, on the other. The latter have to function in adverse conditions of both inadequacy of means of livelihood and rising costs of living with changing needs of survival. With the shrinking of inadequately expanding work opportunities and hence low and uncertain income, they are disenfranchised of their economic votes on the market, on the one hand, while on the other hand, they have to pay the full consequences of a pattern of growth in which production is geared to the new consumption styles under escalating cost conditions.

Increasingly, the traditional artisan and household industries in rural areas and small towns (which are producing either export or luxury-oriented skill-intensive products) survive, whilst their own subsistence goods (wage goods) are being provided through the large-scale sector, often via its small-scale ancillary units. The pricing of these wage goods follows the oligopolist market norms, with hefty trade and distribution margins, including sales and advertising costs. Although wage incomes increase, they are associated with the emergence of casual labour or with contractual labour. We find various changes in the forms of employment, such as a shift to money wages or as fixed period payments, as well as changes associated with changing patterns of leasing.

Non-agricultural activities, particularly in the tertiary sector, are also increasing. With the growth of transport and communication facilities and with the rise of the activities of the state, the process of commercialization is displacing the earlier traditional communitarian system of security and of family relations. The poorer sections of the society are in a state of flux. This may appear to be a sign of dynamism or a shift in the direction of greater 'modernization', but it is also an indicator of the dangerous instability, taking the form of more and more overt political conflicts. If more regular patterns of occupational differentiation and stability of livelihood do not emerge, social instability is likely to become endemic.

A remarkable feature of the development during the '80s was the emergence of the balance of payments crisis which accentuated in its severity to the extent of dominating the attention of the policy-makers. While there has been a recurrence of the balance of payments problem since 1956 when it first surfaced, there is a qualitative difference between the crisis of the '80s and that of earlier decades. Until 1975–6, with tight controls on imports through quantitative restrictions and strict foreign exchange regulations, the current account deficits were maintained at a low level and were financed by inflows of external assistance on highly concessional terms. After 1976–7, however, the balance of payments situation did improve. Apart from some advances in exports, the sudden rise in net invisibles, mainly on account of remittances, brought about a striking improvement in the balance of payments. This period witnessed a process of industrial and import liberalization and, particularly in 1978–9, many items were put under an Open General Licence, to be imported mainly for domestic production. In 1979–80, the payments situation was adversely affected as a consequence of the second international oil crisis and led to a deterioration of the terms of trade.

Since 1980, the crisis has assumed an altogether different dimension. In 1981, India entered into an IMF loan of SDR US dollars 5 billion; in the event, only SDR US dollars 3.9 billion was drawn before the arrangement was terminated in 1984. A few favourable features that appeared on the economic horizon included an increase in oil production. The gradual decline in net receipts from invisibles and a reduction in flows of concessional assistance to India from the World Bank and the International Development Association (IDA) led to a hardening of credit terms. India had to finance the current account deficit by means of substantial inflows of capital through commercial borrowing and deposits of non-resident Indians (NRIs). The debt–service ratio as a percentage of export of goods and services increased from 9.1 per cent in 1980 to 18.1 per cent in 1984, and further to 21.3 per cent in 1989.

Thus, it becomes clear that both the course of financing and the mode of its utilization since the '80s have been qualitatively different from the past. The balance of payments crisis in the '80s was accompanied by a process of industrial growth within the economy. The pattern of industrial growth in the '80s was such as to derive savings mainly from the domestic households sector (*see* Table 6.13) and borrowings from abroad. Since the '80s, the economy has been characterized by a rapidly intensifying foreign exchange crisis.

The dichotomous dynamics of growth

Certain asymmetries and dichotomies have emerged in the economy in the growth process. It would seem that a part of the economy— which, for the lack of a better term, may be called the 'capitalist', 'modern' or 'leading' sector— shows greater rates of capital accumulation, sectoral growth-rates, and greater technological dynamism, and also generates a relatively rapid growth of incomes. This growth of incomes through product expansion is also aided by the wage policies pursued by certain small segments of employment such as the services and the corporate sectors with higher structures of wage emoluments and attached perquisites and incentives. Not only do these incomes form the main and growing source of savings (*see* Table 6.14), but they also benefit from the interest rate and incentive-based fiscal policies of the government. At the same time, they provide the effective demand for consumer goods and services which have given rise to new consumption styles and also sustain the pattern of growth, albeit uneven and unstable.

Linked crucially to this sector by the advancing forces of commercialization and through the fiscal policies and development strategies of the state, is the vast majority of those who do not have a regular and assured means of livelihood. Some of them cannot raise, through their labour, even enough for subsistence and are 'unproductively' employed in the social sense, i.e. they cannot produce even for their own maintenance and have to find subsistence through networks of usury or from the relief programmes of the state, The process of growth has generated neither sufficient employment nor income in this 'dependent' sector. To satisfy its consumption needs (particularly in the case of non-agricultural goods, i.e. wage goods other than food), this sector is increasingly dependent on the produce of the 'leading sector'. More and more rural industries have been eliminated or competed out of the market. Even the rural industries which produce specialized luxury items for export or the fanciful consumption of the rich and middle classes have tended to show only limited buoyancy. The stories of success of small-scale units are mostly connected with the ancillarization of small units as an extended activity of larger units. Small-scale units, independently run by small producers, have to face a market which is highly monopolized through product differentiation and sales promotion.

The dichotomous pattern of development, which has been intensified since the '80s, raises many questions on the effects that an advancing process of commercialization generates on the production and distribution of products. However, the question of providing productive employment to vast sections of the population continues to persist.

Table 6.13

Gross Domestic Saving and Gross Domestic Capital Formation

(as percentage of GDP at current prices)

	Gross Domestic Savings				*Gross Domestic Capital Formation*		
	Household Sector	Private Corporate Sector	Public Sector	Total	Public Sector	Private Sector	Adjusted Total
1980–1	16.1	1.7	3.4	21.2	10.3	12.5	22.1
1981–2	14.1	1.6	4.6	21.0	11.1	14.5	22.6
1982–3	13.4	1.6	4.4	19.4	11.5	12.0	20.8
1983–4	14.7	1.5	3.3	19.5	10.6	11.6	20.8
1984–5	14.1	1.7	2.8	18.7	11.5	10.9	20.1
1985–6	15.2	2.1	3.2	20.4	11.8	13.7	22.8
1986–7	15.1	1.7	2.7	19.5	12.1	12.2	21.7
1987–8	15.9	1.7	2.1	19.6	10.0	12.2	21.7
1988–9	17.5	1.8	1.8	21.0	10.5	13.9	23.9

SOURCE: Government of India (1990), S–6; S–7; S–8; S–9

Table 6.14

Domestic Savings by Type of Institution

(as percentage share in total)

	1980–1	*1987–8*
I *Gross Domestic Savings*	100.00	100.00
1. *Household Sector*	*75.9*	*82.2*
Financial Savings	29.9	36.6
Savings in Physical Assets	46.0	45.6
2. *Private Corporate Sector*	*7.9*	*8.5*
Joint Stock Companies	7.3	8.1
Corporative Banks & Societies	0.6	0.4
3. *Public Sector*	*16.2*	*9.3*
Administrative Departments	9.7	–6.3
Non-departmental Enterprises	6.4	15.6
II *Consumption of Fixed Capital*	*42.0*	*50.5*
4. Household Sector	19.1	20.3
5. Private Corporate Sector	5.9	7.4
6. Public Sector	17.0	22.4
III *Net Domestic Saving*	*50.5*	*49.5*
7. Household Sector	56.6	61.4
8. Private Corporate Sector	2.0	1.2
9. Public Sector	–0.6	–13.1

REFERENCES

BAGCHI, A.K., 1978: 'Reflections on the Pattern of Regional Growth in India during the Period of the British Rule' (Centre for Studies in Social Sciences, Calcutta) (mimeo.).

BHALLA, G.S. and Y.K. ALAGH, 1979: *Performance of Indian Agriculture—A District Study* (Delhi: Sterling).

Government of India, 1990: *Economic Survey 1989–90* (New Delhi: Ministry of Finance (Economic Division)).

NAIR, K.R.G., 1986: *Inter-state Income Disparities in India: An an All India Conference on Regional Disparities in India held in India, April 1979*, 2 edn. (New Delhi: Agricole Publishing Academy).

NATIONAL SAMPLE SURVEY, 1974: *28th Round October 1973–74* (New Delhi: Central Statistical Organization).

RAZA, M. and A. KUNDU, 1980: 'Regional Disparities in India' (Project Report) Indian Council of Social Science Research, Delhi (mimeo.).

THAVARAJ, M.J.K., 1972: 'Regional Imbalance, and Public Investment in India (1860–1947), *Social Scientist* 1 (Nov.): 4, No. 40, pp. 3–24.

WHITCOMBE, E., 1971: *Agrarian Conditions in Northern India: United Provinces under British Rule 1860–1900* (New Delhi: Thomson).

India's Uneven Development and Its Implications for Political Processes: An Analysis of Some Recent Trends • 7

RAVI SRIVASTAVA

In recent years, the Indian political process has been characterized by two significant trends: a growing regionalization of the Indian polity; and the emergence of new social groups as significant actors on the political scene. In this chapter certain features pertaining to the pattern of economic growth, which may throw light on some of the observed trends, are explored.

Some general features

Despite an avowed strategy aimed at reducing disparities between regions, the principal indicators show that the economic distance between the regions has grown in many ways since the inception of planned development. The inequalities in per capita State Domestic

Product show a steady increase. Between 1960–1/1962–3 and 1984-5/1986–7, the coefficient of variation in per capita S.D.P. has grown from 17.95 per cent to 41.02 per cent. Between 1970–1 and 1986–7, the coefficient of variation of sectoral net products in each of the primary, secondary, and tertiary sectors all show an increase. Between 1950–1 and 1984–5, the coefficient of variation in foodgrain production per capita also shows an increase from 38 per cent to 53 per cent.

More significantly, there is a greater regionalization of some of the significant economic problems such as poverty, unemployment, and low wages. The bulk of the poor are now concentrated in the Central and Eastern regions. Between 1977–8 and 1983–4, the percentage of persons below the official poverty line in these regions rose from 57.32 per cent (of the national total) to 62.42 per cent (Rao, 1986; Parthasarathy, 1987; and World Bank, 1989). Moreover, certain other significant aspects of poverty and deprivation, not covered by the data on which the distribution of poverty is based, further contribute to its regionalization. These arise out of the reduced access of the poor to common property and other social resources such as forests; large-scale destruction of environmental and natural resources mainly affecting the poor; and displacement and pauperization of labouring households from the catchment areas of large development projects.

Thus, Indian development during the last few decades, has been characterized by extreme unevenness, not only between regions, but also, as we shall see later, within regions and sectors. This chapter focuses on some aspects of the changing interplay between (emerging) social groups, which are both the cause and effect of this unevenness.

Regional dimensions of change in the agrarian sector

The colonial inheritance

The disparities inherited from the past have been closely associated with the disarticulated and uneven growth imposed by colonial policy. While some economic historians have debated whether the economic distance between regions increased or decreased during the colonial period, the qualitative impact of colonial policy on regions and sectors is amply clear (e.g. Kumar and Krishnamurthy, 1980).

The impact of British policy on agrarian differentiation can be seen as a consequence of several mutually reinforcing policies, which were themselves subject to change in the different phases of British rule.

First, since land revenue was the principal form of surplus, the form of agrarian organization best suited to the maximization of surplus, over a certain period, was crucial to British policy. This underwent changes as the British consolidated their rule, first in the Eastern, then in the Southern, Western, and North-western parts of the country. The traditional proprietary rights were ignored, most of all in the permanently settled areas in the East, where initially high revenue demand and its subsequent limitation led to a complex web of non-cultivating intermediaries, all of whom had a claim upon the meagre surplus of the direct producers. In the *ryotwari* and *mahalwari* regions, the leading proprietors were recognized under the revenue arrangements, though even here the revenue demands were not only heavy, but were also subject to periodic upward revision.

The differences in revenue arrangements implied less sub-infeudation in the *ryotwari* areas. Further, demographic and economic factors, together with the differential impact of rent/revenue on different cultivators, created the ground for regional differences in the pattern of surplus appropriation and the extent of economic differentiation among the peasantry. Tenancy and debt relief legislation provided a further basis for differentiation and creation of a narrow stratum of rich peasantry, to a lesser or greater degree, in different regions (Misra, 1982).

The impact of public investment further heightened the processes of regional (as well as within region) differentiation. Public investment in British India (chiefly in the building of communication networks and irrigation) was guided mainly by military and commercial considerations. The latter, in particular, implied that investment in irrigation was concentrated chiefly in areas where gains in productivity could be skimmed off as additional revenue, viz. the North-West, Bombay, and Madras provinces.

The narrow base of a surplus-owning peasantry in the East, coupled with negligible productivity-raising investment by the state, contrasted with the relatively prosperous base of the middle/rich peasantry in the North-West, reinforced by public investment. The mutually reinforcing elements resulted in widely varying growth dynamics in the different regions, as shown by Blyn (1966) and confirmed in a number of more detailed regional studies.

Dimensions of agricultural change since Independence

The agrarian sector in India has undergone rapid and uneven change since Independence, reflecting the mutually determined interaction between, (1) policy-induced changes in the agrarian sector, investments, and technological change, (2) extant production relations, and (3) region-specific agro-ecological and demographic characteristics. These changes have brought to the fore new *dramatis personae* whose characteristics and mutual strength continues to vary between regions.

i. The initial years

In the early years of planning, agrarian policy was built on two main planks.

First, a series of policy measures to reorganize the agrarian structure and to make it more conducive to growth and equity were undertaken. The outcome of the reforms varied with the ability of the dominant landholders affected by them to dilute their scope and implementation. This varied from State to State, and region to region. For instance, the reform came into effect in several Eastern Indian States (Uttar Pradesh, Bihar, West Bengal) with a delay ranging from 4 to 10 years. The principal outcome of these reforms was an *uneven* shift from tenant-based cultivation to owner-based cultivation.[1]

The impact of the second prong of the planning strategy, namely, increases in yields through increased irrigation, fertilizer, seeds, etc. was also regionally disparate. The use of modern inputs remained low in the '50s. Cumulative expenditure on capital irrigation works was concentrated in Bombay, Uttar Pradesh, Madras, and Punjab. The other States which benefited were Orissa and Rajasthan with 13.2 and 9.6 per cent of the cumulative expenditure during the period between 1946 and 1960–1 (Banerjee and Ghosh, 1988: 143).

The agrarian scene was radically altered by the new agricultural strategy which took shape during the '60s. The wheat-growing areas of the North–West (Punjab, Haryana, western Uttar Pradesh) were the best placed to reap the benefits of the new strategy. Not only did they enjoy a high proportion of area under assured irrigation, but the regions were less burdened by the deadweight of pre-capitalist exploitative relationships, the land base of the well-to-do peasantry was relatively more adequate, and government effort, as already stated, was also focused on the relatively developed regions.

Consequently, during 1962–5 and 1970–3, districts experiencing high growth rates for all crops were concentrated in Punjab, Haryana, Uttar Pradesh, and Rajasthan, with these States accounting for 36 of the 50 high growth districts. Gujarat, Karnataka, and Tamil Nadu accounted for another 10 of the high growth districts. Only one out of the 60 high growth districts was located in the central or eastern

[1] In 1954–5, an estimated fifth of the cultivated area was under tenancy, and over one-fourth of the rural households leased in land, but the proportions varied between States. The early record of implementation was noticeably good in Jammu and Kashmir, Karnataka, and Kerala (Walinsky, 1979).

regions. At the other end, 133 out of the 289 districts studied experienced growth rates below 1.5 per cent per annum. Most of these were located in Andhra Pradesh, Bihar, Karnataka, Gujarat, Madhya Pradesh, Orissa, and Uttar Pradesh. During the same period, 25 per cent of the growth of incremental output (of 38 major crops) came from as few as 13 districts, which also accounted for 21.8 per cent of incremental irrigated area, 15.7 per cent of fertilizer consumption, 20.8 per cent tractors, and 8.9 per cent tubewells (Bhalla and Alagh, 1979).

Thus, in these and other areas, where appropriate economic, institutional and irrigational infrastructure was already available or was rapidly becoming available, the new strategy accelerated processes of qualitative change, whereby the former landed proprietors or the well-to-do peasantry initiated a process of limited capitalist development (Byres, 1972). These processes implied, on the one hand, changes in the agrarian organization leading to resumption of land, changes in lease relationships and rapid growth in proletarianization and part-proletarianization and, on the other, a further concentration of investible surpluses, assets, and incomes in the hands of these classes. Thus, inter- and intra-regional inequalities appear to have worsened over this period, engendering fresh and dramatic contradictions on the issue of land and wages.[2]

ii. Aspects of agrarian change in the recent period

Agrarian change in India since the '70s marks a strengthening of the trends (towards capitalist penetration) witnessed earlier. A central aspect of this change has been the drawing of new groups in hitherto backward areas into the process of capitalist production and accumulation, a further increase in proletarianization and part-proletarianization (both as a result of demographic as well as economic processes), and greater dependence on the market and market-related processes, implying considerable vulnerability in the face of market fluctuations. However, the nature and the pace of these changes continue to be highly uneven, so that economic parameters suggest a contradictory picture, implying both a widening of the growth process and an increase in the economic distance beween regions.

Below are presented some stylized facts regarding agrarian change in the recent period. These are then related to factors influencing the emergence of capitalism in agriculture in the subsequent section.

[2] The evidence is reviewed in detail in Byres (1981).

(a) Agricultural growth

Between 1967–72 and 1981–4, State level results show sustained high growth-rates (over 3 per cent) in Punjab and Haryana, and an acceleration to high growth-rates in a few States in the North, West, and South (Maharashtra, Gujarat, Uttar Pradesh, and Andhra Pradesh). Growth-rates in Karnataka and Rajasthan show some deceleration compared to the '60s, but remained above average. On the other hand, agricultural growth-rates remained low or decelerated even further in Tamil Nadu, Kerala, Madhya Pradesh, Orissa, West Bengal, and Bihar.

District level analysis (Bhalla and Tyagi, 1989) for a similar period (1970–3 to 1980–3) indicates that in the decade ending in the triennium centred on 1971–2, only 71 districts (out of a total of 284) in India, accounting for roughly 25 per cent of gross cropped area, experienced agricultural output growth-rates above 3.5 per cent. By the end of the subsequent decade, the number had shifted up to 95 districts, covering nearly 40 per cent of gross cropped area. At the same time, a large set of districts remained in the low growth ranges. Districts with output growth rates at or below the population growth rate accounted for 61.5 per cent of the gross cropped area in 1970–3. By 1980–3, the area under these districts had shrunk, but only by about 10 points to 50.6 per cent. In addition to many of the sparsely populated districts of Madhya Pradesh and Rajasthan, these low (or negative) growth regions included most of the most densely populated areas of the country in Bihar, Tamil Nadu, and West Bengal.

The noteworthy feature of the growth picture during this period was the continuation of very high growth-rates in most of the original Green Revolution areas, the appearance of a large new set of the high growth districts in Uttar Pradesh, Andhra Pradesh, Gujarat, and Maharashtra, together with the unambiguous worsening of conditions in Bihar, Tamil Nadu, and West Bengal, and the downward shift of some districts in Rajasthan and Madhya Pradesh (Bhalla, 1988; Bhalla and Tyagi 1989).

While a larger number of districts showed improved performance during the above period (1970–3 to 1980–3), the concentration of growth appears to be only very marginally mitigated. Thus, 14 districts still contributed 25 per cent of the incremental output and accounted for 17.9 per cent, 22.3 per cent, 28.5 per cent, and 11.2 per cent of the increment in irrigated area, fertilizer consumption, tractors, and tubewells respectively. But the number of districts contributing 50 per cent of the incremental output went up from 37 to 44 (Bhalla and Tyagi, 1989).

The analysis of State level growth of foodgrains during the '80s shows that, in recent years (1977–8/1979–80 to 1986–7/1989–90) growth-rates have been high in some of the 'lagging' states (West Bengal, Orissa, and Madhya Pradesh) and have decelerated in some Southern and Western states (particularly Karnataka and Maharashtra, whereas Kerala and Tamil Nadu continue to register negative growth-rates). As a result, the declining share of the Eastern region in total foodgrain production has been reversed, whereas the North-Western region continues to account for a growing share of foodgrain production (Srivastava, 1994).

(b) Trends in labour productivity and labour use

Changes in the production–organization of agriculture have, in conjunction with demographic forces, led not only to increased proletarianization of rural labour but also to changes in patterns of labour use and in labour absorption. Changes in area sown per worker have overshadowed yield improvements in slow growth areas, leading to growing disparities in labour productivity. This is of considerable concern since labour productivity appears to be a proximate determinant of agricultural wages (Jose, 1988; World Bank, 1989).[3] Between 1962–5 and 1975–8 a number of regions with low initial levels of labour productivity experienced low or medium growth-rates in this variable, or otherwise experienced declining labour productivities (Mahendra Dev, 1986). Thus, several regions, chiefly in the Eastern, Central, and Western parts showed low productivity–low growth.

Between 1970–3 to 1980–3, productivity per agricultural (male) worker rose rapidly. This was also due to slower growth in male agricultural workforce. Agricultural labour productivity rose sharply in Punjab and Haryana but remained stagnant in most Southern (except A.P.) and the Eastern States. Among the Central States, labour productivity increased in Maharashtra and Gujarat, but registered declines in Rajasthan and Madhya Pradesh. The coefficient of variation increased from 34.1 in 1960–3, to 44.1 in 1970–3, and to 55.8 in 1980–3.

The growing disparity in labour productivity implies that regions which have experienced faster growth have experienced increases in yield with smaller declines in the net area sown per worker, whereas in low growth regions, yield increases and changes in the net area sown per worker have largely offset each other (Mahendra Dev, 1986; Kundu and Raza, 1982). Thus the underlying employment elasticities

[3] However, Jose does find a narrowing of wage disparities since the mid '70s despite the contrary evidence on agricultural labour productivity.

and the patterns of labour use generated by the disparate growth processes are also different across regions. At the aggregate level, the elasticity of labour absorption declined from 0.71 in 1960–3 to 0.59 in 1970–3 and further to 0.51 in 1980–3.

(c) Patterns of commercialization, marketed surplus, and terms of trade

In the early phases of colonial rule, the commercialization of agriculture was bolstered by the colonialist's demand for agricultural raw materials and foodgrains. A rigid revenue demand payable in cash made it imperative for the cultivator to enter the market nexus which was further facilitated by the mediation of the merchant–moneylender who was also often the landlord (Chowdhury, 1964; Amin, 1982).

Debt-relations with the merchant–moneylender or a rigid demand for cash still induces the smallholder to grow crops exclusively for the market or to sell crops at terms which may otherwise be disadvantageous. For such cultivators, the strategy of reproduction *via* the market is more viable in regions which are commercialized (so that the market for crops is less imperfect) or where the demand for commercial crops is high. In general, however, the degree of commercialization depends on the landholding structure; on the cropping pattern which has partly evolved from the colonial circumstances; and on the degree of penetration of capitalism in agriculture which determines the market orientation and the marketable surplus of the dominant producers and the region as a whole.[4] Detailed data regarding regionwise trends in commercialization are scanty. The North-west regions in the country are clearly highly commercialized and produce the major part of the marketed surplus of wheat and paddy (Patnaik, 1975). The Western and Southern regions follow, while the Eastern and Central parts of the country are perhaps the least commercialized.

The market susceptibility of agriculture has clearly grown (Chakravarty, 1987). With this, the relative price of input as well as output, and the fluctuations in output price have become very relevant. Further, the welfare of rural households also depends on the price of its consumption basket in relation to income. Not surprisingly, the 'terms of trade' debate figures centrally in both academic as well as farming lexicon. The price of industrial inputs into agriculture increased sharply following the first oil crisis, and was reflected in the

[4] Cf. Patnaik, 1975. For subsistence crops, Chattopadhyay and Sen (1988) reach the conclusion that while the relationship of marketable surplus to landholding size may vary across regions, the relationship between the former and per capita cropped land under subsistence crops is positive.

adverse movement in the barter terms of trade between agriculture and industry (Nadkarni, 1987; Ghosh, 1988). This movement may have implied lower net realization per unit output of specific crops.

A second adverse shift took place following the second oil crisis from which the economy appears to have recovered by the early '80s (Virmani, 1988). The movement in the terms of trade has raised questions regarding, (1) its impact on profitability and agricultural growth, (2) its impact on rural incomes and inter-sectoral flow of resources and, (3) its inter-regional and intra-regional differential impact on rural households. There are strong adherents to the view that movement in the terms of trade has adversely affected profitability and rural incomes, and has led to a flow of resources out of the rural sector.[5] But it has been convincingly argued that the above view is partial as it does not take into account either the government or the external sector, and that both industry and agriculture have been subject to a squeeze as a result of the price rise. It has also been argued that the above view does not distinguish between surplus and deficit rural households and regions in analysing the impact of movements in terms of trade (Mitra, 1977; Janvry and Subba Rao, 1987; Nadkarni, 1987; Ghosh, 1988; Virmani, 1988). Nevertheless, it will be shown below that relative price movements have become part of the political lexicon of farmers' unions all over the country.

(d) *Disparities in investment performance*

Both macro- and micro-evidence reveal significant differences in the assets held and the investment performance of different groups of rural households across regions and across time. The share of productive farm and non-farm equipment has increased and the change has been more significant since 1972 (Krishna and Raychaudhury, 1982; Srivastava (forthcoming)). States with low average value of all assets and farm equipment per rural household show rapid rates of growth in the average value of all farm equipment[6] as well as for individual items of machinery such as tractors and pump-sets.

Interestingly, while, during the '60s, a number of States showed an unambiguous worsening of asset distribution in the rural areas, most States did show some improvement in asset distribution during the '70s (Srivastava (forthcoming)). Thus, intra-State asset inequality has

[5] Different versions of this hypothesis may be found in Shetty, 1978; Patnaik, 1986; Patnaik, 1984, 1985; Bardhan, 1984; and Chakravarty, 1987.

[6] The regressiveness in the States' own resources (per capita State Plan Outlays: per capita Central Plan Assistance) is shown by its positive correlation with per capita income (George, 1985: 115–17).

moved along a different direction during the '70s as a result of the pattern of accumulation or other factors.

An analysis of changes in the asset structure of different groups of households across States, based on Reserve Bank of India (RBI) Debt and Investment Surveys, provides interesting results. First, households below the first decile in terms of asset ownership show relatively greater investment dynamism in most States. Second, the top decile of households shows a slow-down in investment performance, suggestive of technical or market-related constraints. Third, the top groups of households in developed (e.g. Punjab, Haryana) as well as relatively backward States (e.g. Orissa and M.P.) show a relatively greater preference for non-farm assets.

A large number of micro-studies carried out in different regions confirm the disparate inter-regional and inter-group investment performance. They also corroborate some of the conclusions of the macro data pertaining to the relative investment dynamism of the upper peasant groups despite a slow-down in agricultural investment in certain regions, and diversion of surpluses to other avenues.[7]

Regional disparity in agriculture and emergent rural classes

What accounts for the pattern of agrarian disparity discussed above? At a proximate level of analysis, the pattern of growth appears to be related principally to certain factors such as the use of modern inputs or irrigation and government policy. Such an analysis precludes the role of endogenous factors in shaping the pattern of growth.

By contrast, Bhaduri (1981) has provided an alternative explanation of regional differences in accumulation in terms of the heterogeneous nature of the dominant rural classes. A more comprehensive explanation of the pattern of regional differentiation is provided by Patnaik (1976, 1986), who interweaves the colonial pattern of agrarian relations with post-Independence public policy, the specifics of the Green Revolution strategy, and structurally related rural unemployment. Other authors have also suggested that differences in the modes of surplus appropriation between regions determine, in some sense, the attitude of the appropriating classes to technical improvements and investment. Attention has been drawn particularly to interlinked land–labour–credit markets which reduce the feasible

[7] On investment dynamism of large–medium farmers, *see*, for instance, Rutten (1986). On the slowing down of investment, *see* Mody (1983); Bhalla (1976); Patnaik (1976); and Parthasarthy and Pothana (1983).

options for the poor, as well as the strategies open to the rural rich (Bharadwaj, 1979, 1985; Bhaduri, 1984).

I have argued elsewhere that the pattern of uneven development in agriculture reflects its gradual transition from pre-capitalist agriculture (Srivastava, 1994). The principal issue is the determination of relevant conditions under transition which cause different rural classes to shift/reorient their strategies of production, accumulation, and reproduction. Further, under such a reorientation, how do surplus appropriation strategies change, and what factors account both for the resilience and for the change of pre-existing relations? Through such an enquiry, it is possible to focus on the mutual relationship between factors such as public investment, demography, and class relations (Srivastava, forthcoming). Our own results show that, with the penetration of capitalism, the forms of appropriation have undergone subtle but related changes, and bring out the futility of focusing on the relationship between the 'forms' of appropriation and accumulation outside the context of agrarian transition.

I have shown earlier that major differences in agrarian structure continued to persist in the wake of post-Independence land reform. These related mainly to the nature and strength of the major rural classes and the differences between them. In many parts of the country, for instance, in the 'permanently settled' regions, the dominant rural classes had little operational accountability for production. Furthermore, the intermediate rural classes had a weak base. The converse was the case in the north-western region. Since Independence, a number of factors such as land reform, demographic changes (including migration), political processes, public investment, and other strategies, and technological innovations have impinged upon and interacted with class relations in a regionally differentiated fashion. The outcome has been a change in the characteristics and strength, and the emergence of qualitatively different classes, bringing about uneven shifts in the production/appropriation and accumulation strategies of the surplus-appropriating groups, with far-reaching impact on the whole gamut of agrarian organization (Byres, 1981; Srivastava, forthcoming).

These shifts have been most dramatic in cases where pre-existing options are no longer available to the major appropriating classes, and this in turn is accelerated by the existence of alternative subsistence strategies for the rural poor as well as the emergence of the rural proletariat as a 'class for itself'. Thus, alternative employment opportunities and organization among the rural poor accelerate the demise of pre-existing relations, whereas the converse may be the case if no such opportunities are available (Bagchi, 1973; Bharadwaj, 1985; Patnaik, 1986; Chandra, 1974).

There is sufficient evidence from the backward regions to show that shifts symptomatic of uneven transition—though still far from universal—are occurring in some areas under the impact of the kind of factors which are discussed above. This has brought about certain changes in production and investment strategies as well as changes in land–lease–labour relations.

As far as the erstwhile proprietors and other dominant landholders in several regions are concerned, such changes are occurring very slowly. At the same time, however, in regions where feasible opportunities exist (and this is true on a larger scale now, despite the persistence of relative differentials), the dominant landholders consisting of old and new proprietory groups are themselves differentiated. Also, in such cases, the intermediate groups of cultivating peasantry appear to show greater dynamism. This may account both for the apparent resilience to change (because the dominant landholders have a greater weight in the economic process), as well as the somewhat greater dynamism of several backward regions in the recent period. In other words, differences between regions appear to reflect within-region and between-region inter-group differences in production/accumulation.

The above hypothesis is strongly supported by my analysis based on fieldwork in Uttar Pradesh (Srivastava (forthcoming)). This study shows greater dynamism and better agricultural performance, (1) among all classes in the more developed villages relative to the less developed ones; (2) among the rich peasant and upper middle peasants, relative to rich farmers and landlords within the study villages.[8] I have already noted the results of the detailed macro evidence in a preceding section which underscores the relatively dynamic behaviour of the upper-intermediate group of households (defined as those in the first to third decile in terms of assets owned) in several lagging States.

To summarize: recent changes in the agrarian sector are symptomatic of uneven capitalist development and show considerable differences between regions and rural classes in terms of agricultural and investment performance, and the nature and extent of involvement in the market. While hitherto agriculturally stagnant regions show considerable resilience to change (which can be traced to the persistence of pre-capitalist exploitative relationships), in recent years there has been some change. In the transitional situation, it is possible to discern some differences both within the dominant proprietory classes, and between these classes and the intermediate groups (the new rich and middle peasantry) in matters relating to agricultural

[8] The empirical content of the class categories has been defined in Srivastava, 1989.

performance and investment behaviour. Finally, it is apparent that in several relatively backward regions, the upper strata of the peasantry are playing a key role in bolstering investment and output.

The emerging rural scene

As shown above, the land reforms initiated at Independence brought about an uneven shift from rent-based to owner-based cultivation. Dominant landholders exercised considerable influence on policy-making, initially in the States and subsequently at the Centre as well. In certain north-western and southern States, the dominant land-holders were principally self-cultivating and incipiently capitalist. They were also primarily drawn from cultivating castes. Policies during the '60s accelerated the development of capitalism in favourably endowed regions. With the increasing penetration of capitalism, both from below and above, and an increase in the numbers of the surplus-owning peasantry, the social base of the dominant land-holders also increased, as did their influence over regional politics (Nadkarni, 1987).

During the period between the '60s and the early '70s, the base of incipient rural capitalists extended both vertically and horizontally in the developed areas and, to a smaller extent, in the less-developed areas (Walinsky, 1979). This was achieved mainly through policies initiated from above, in which the emerging rural classes themselves played little direct role. These policies were adopted ostensibly in the interest of capitalist development as a whole (cf. Varshney, 1989). However, this period witnessed the emergence of important political formations, notably the Bharatiya Kranti Dal (BKD), which espoused the cause of the peasantry (Brass, 1985; Hasan, 1989; and Frankel, 1989). The BKD drew its base mainly from among the peasants in the northern regions—principally western Uttar Pradesh and Haryana, but also, particularly in the later stages, from Bihar and eastern Uttar Pradesh. The later developments were partly a consequence of the uneven penetration of capitalism in this region, the implications of which are again discussed below.

Simultaneously with the strengthening of the rising rural capitalist class, changes occurred also in the strength and characteristics of the agricultural proletariat. These were impelled by both economic and demographic forces.

The change from rent-based to owner-based cultivation brought in its wake a spate of evictions from land, thus increasing the pace of proletarianization of the peasantry. The New Agricultural Strategy (NAS) further stepped up proletarianization, polarizing the countryside in many regions. During the late '60s and early '70s, a spate of land and wage struggles, led by political formations on the left, erupted in several parts of the country, leading many observers to comment on the possibility of the 'green revolution turning red' (Frankel, 1971; Sharma, 1973). While these struggles, specially those led by the radical left, met with widespread repression, the government also responded by initiating a second series of land reforms, and by launching a number of programmes, among others of asset distribution and transfer to the rural poor aimed at alleviating growing joblessness and impoverishment in the agrarian sector.

It may be said that until the '70s, political action taken by the rural proletariat was far more intense and widespread than that initiated by the nascent rural capitalists. However, the tide had begun to turn by the mid-'70s. The previous sections have traced the process of strengthening capitalism in the last two decades, during which the dependence of agriculture on wage labour has also grown. Further, casualization of agricultural labour has also increased (Jose, 1988; Vaidyanathan, 1986; World Bank, 1989; Bardhan, 1989). Even so, open, organized, and sustained action by the rural poor has been confined to a few regions in recent years even though available evidence, including my own, does suggest that localized resistance by the labour force to specific forms of control in the labour process is far more widespread. These have accelerated the search for new forms of labour control by employers which make labour easier to control and appropriate (Srivastava, forthcoming). The significant theme of why the rural proletariat has been unable to keep its political ascendance since the mid '70s is not pursued here. It has been argued (Bardhan, 1989) that the emerging forms of labour control succeeded in effectively segmenting and making clients of the rural proletariat. This theme needs to be explored in greater detail.

Here I pay greater attention to the developing politics of the cultivating classes. Since the mid '70s, three interrelated developments have accelerated the transformation of the rural capitalist 'class in itself' into a 'class for itself'. First, as discussed earlier, beginning with the first oil crisis, prices of the principal material agricultural inputs rose sharply. Second, agriculture became somewhat demand-constrained and agricultural prices rose relatively slowly. Third, in more capitalistically developed areas, the production of principal crops became subject to declining returns and it became more and more difficult to achieve sustained increases in productivity.

Beginning with Tamil Nadu, farmers in Karnataka, Maharashtra, Punjab, and Uttar Pradesh organized themselves into farmers' organizations. Their principal demands centred on subsidization of inputs (particularly credit and electricity) and raising the procurement prices of agricultural commodities.

These movements have been analysed in some detail in several studies.[9] Nadkarni has convincingly shown that the main bases for these movements are situated in regions which have a high degree of commercialization and which also have a higher than average number of surplus landowners. Conversely, it is difficult to sustain these movements where a substantial rich peasant–rich farmer base does not exist. These features are also corroborated in the other studies.

The analysis of the demands and character of the movements appear to show that these are rich farmer–rich peasant led movements based in areas of well-entrenched agrarian capitalism.[10] Indeed, occasional forays by the leaders of these movements into less developed areas—for instance, Joshi in Nipani and Tikait into eastern Uttar Pradesh—have met with little success. It has also been noted that even though the demands of the farmers' organizations are centred on the sections of farmers producing a net marketed surplus, they are also relevant for other classes of peasants which are deeply involved in the market, particularly in commercialized areas (Nadkarni, 1987; Gill and Singhal, 1984a; Omvedt, 1981; Nathan, 1981). These movements have been careful in avoiding issues which might divide the peasantry in these areas.[11]

As with the political formations representing the peasantry, and given the 'dominant' status of certain castes in the areas of strength, farmers' organizations have also used caste and kinship networks for mobilizational purposes, but their appeal is not restricted to members of particular castes. Another aspect of the farmers' movements which has attracted attention pertains to their professedly 'non-party political' character and to the zeal with which these organizations strive to protect their autonomy *vis-à-vis* existing political formations.

Parallel to the development of independent farmers' movements, there has been a steady increase in the influence of rich farmers–rich peasants over the institutional political process which owes as much to changes occurring in the hitherto agriculturally 'backward' Hindi-

[9] E.g. Alexander (1981), Kerala and Tamil Nadu; Nadkarni (1987), Karnataka and Tamil Nadu; Hasan (1989a) and Gupta (1988), Uttar Pradesh; Omvedt (1981), Maharashtra; Gill and Singhal (1984a), Punjab. The general issues and implications of the farmers' movements are discussed in Nadkarni (1987) and Varshney (1989).

[10] A comparison of studies cited in footnote 9 would bear this out.

[11] According to Nadkarni, the lack of emphasis on foodgrain prices is one such issue. However, Varshney demonstrates that even on the question of higher foodgrain prices, a fairly broad-based unity with small peasants was possible, given their extensive involvement in the market for inputs in the developed areas, and the potential for increased viability on a new production frontier emerging from a greater use of modern inputs.

speaking hinterland as to those taking place in regions where capitalism is relatively well developed. With the entry of Charan Singh as one of the key leaders in the Janata coalition at the Centre in 1977, the politics of rural populism moved centre-stage (Byres, 1988). Charan Singh's main support base was much the same as that of the farmers' movements. He espoused causes which the latter voiced. But, he naturally refrained from taking active part in agitational politics when he was in power.

The late '70s marked another phase of consolidation of power by the new dominant landholders, rich farmers, and rich peasants belonging principally to the intermediate cultivating castes. The growing clout of the newly emerging rural classes has been analysed by a number of authors (e.g. Kohli, 1987; Varshney, 1989; Frankel, 1989; and Hasan, 1989). Some of these studies wrongly characterize the process as one of growing dominance by the 'middle peasantry', a term which apparently covers all self-cultivating peasantry belonging to the intermediate castes (cf. Prasad, 1988; Kohli, 1987). While middle peasants undoubtedly have been beneficiaries in several regions, I have shown above that the recent agrarian changes encompass the entire gamut of rural classes, including middle peasants, rich farmers, and rich peasants belonging to the intermediate social groups. A central aspect of these changes is that categories do not remain static with developing capitalist relations which are characterized by differentiation in both directions (upward and downward). Further, rich farmers and rich peasants have been the principal beneficiaries of recent changes.[12]

As shown earlier, during the '80s the trends in the hitherto backward regions (eastern Uttar Pradesh, Bihar, and to some extent Madhya Pradesh) signify uneven capitalist penetration among different social classes and groups, and this has further strengthened the social and economic base of the rural capitalists in the north-western and other regions of the country. At the same time, these classes and groups are still relatively weak in several regions, including eastern Uttar Pradesh and Bihar, and the fields of option of the dominant landholders amongst them include both capitalist and pre-capitalist modes of reproduction.[13]

These apparently contradictory developments have, however, strengthened the base of rural populism, and signified a further shift in the social bases of the ruling classes towards the rural interests represented by the intermediate groups. The outcome of the 1989 elections in the northern and eastern States, and at the Centre

[12] The categories used here (rich farmers, rich peasants, middle peasants, etc.) are elucidated in Srivastava (1989).

[13] Cf. Chaudhry (1988) for a study of Patna region.

underscored this shift. While the significant 'pro-rural' steps taken by the ruling party at the Centre (in the matter of allocation of resources, pricing, and credit) may not have ensured the interest of the intermediate groups in the less developed regions, the Centre and several States took further steps to ensure the increased participation of these groups in the power structure.[14] However, the latter move provided the principal reason for a split in the Janata Dal, not only due to entrenched upper caste interests, but also opposition from sections of the intermediate groups themselves.

The June 1991 election brought different political parties into power at the Centre and in the States, and implied a considerable setback for Janata Dal, the major agrarian populist party. While the BJP swept into power in Uttar Pradesh on a wave of Hindu resurgence and what may be termed social populism (which also brought into its fold sections of the intermediate cultivating groups and won the BKU's support for it), the Janata Dal was able to retain its hold only in Bihar. Under the impact of the acute economic crisis, the Centre has been compelled to cut back on some elements of the subsidy package evolved for the agrarian classes over the last few years, but progress on this front has been erratic on account of the virtual political consensus favouring the package in the past and the political clout of the farmers themselves.

Thus, the agrarian changes discussed in the preceding section have influenced recent political changes in two principal ways. First, the influence of rich farmers and rich peasants and other rural classes with a relatively greater capitalist orientation has, in general, been strengthened, and the slow and uneven penetration of capitalism in the 'backward' regions has contributed in no small measure to this. Second, the demands of the rural capitalists in developed regions, which arise out of a perception of crisis but which also relate to the consolidation of economic and political gains have been, in the main, incorporated in the national agenda. As a result, despite economic compulsions, the Centre has been able to move haltingly in effecting change in the content of agrarian populism. The implication of these moves is still to be felt, but farmers' organizations and opposition political parties have begun to step up their campaigns. It is evident that these changes will crucially affect the content of economics and politics in the country in the years to come.

[14] While the new policies in themselves should be favourable to emerging capitalism in agriculture, their spatial impact is not clear. But the policies of reservation, first announced on 7 August 1990 and modified by the minority Congress(I) government in 1991, are specifically aimed at providing a greater number of entry points to members of these groups in the organs of state power.

Some changes in the industrial sector

As in agriculture, so too in other spheres of production, changes have been very uneven, and have developed under the mutual interaction of several factors, including the nature and characteristics of emerging social classes. Of several disparate changes, disparities in the industrial sector are analysed below. Once again it is to be noted that this sector too is characterized by a certain widening of the growth process, accompanied by the emergence of a new capitalist class in hitherto backward areas. But, at the same time, the economic distance between regions remains considerable and has been influenced by the weight and character of the capitalist class in already developed regions.

In colonial India, the commercial networks laid by the British served to link the hinterland to the three port-towns of Calcutta, Bombay, and Madras. It was around these centres that the limited enclave-type industrialization of the nineteenth century took place. The indigenous merchant–industrial interests underwent a greater degree of subordination in the Eastern provinces, which also witnessed a more thorough destruction of indigenous handicrafts and industry (Bagchi, 1976; Banerjee and Ghosh, 1988).

In contrast to the Eastern region, the subordination of the merchant–capitalist class was less complete in the Presidencies of Bombay and Madras, as was also the destruction of indigenous crafts and industries. In these regions, the contrast between the port-enclaves and the hinterland was somewhat less marked, and more conducive to growth. Furthermore, when opportunities for limited industrialization did arise, it was the merchant–industrialist–financiers from these regions who played a more prominent role. However, despite some dispersal, industrial development continued to be concentrated in the Bombay–Ahmedabad, Madras, and Calcutta regions. The only two noteworthy exceptions were Jamshedpur and Kanpur in the North. The north-west remained devoid of large industrial centres, though some industrial development did take place around cantonment townships such as Meerut and Jalandhar (Bharadwaj, 1982).

During the '50s and '60s, a number of large public sector plants was established near sources of raw materials in industrially backward States (Bihar, Orissa, Madhya Pradesh, and Andhra Pradesh). A certain measure of success was also achieved in setting up consumer industries (chiefly, sugar and cotton textiles) in backward States. But, in general, the spread effects of industrial growth were confined to regions close to earlier agglomerations, second-level industrial centres (Delhi, Bangalore, Jalandhar–Ludhiana), or areas of heavy public sector investment with some historical growth antecedents (e.g. the

Jamshedpur–Dhanbad–Bokaro region). As the relative importance of secondary centres such as Delhi, Hyderabad, Bangalore, Bhopal, Ludhiana, and Jalandhar increased, a certain measure of dispersal was achieved. Between 1961 and 1971, the coefficient of variation of the share of factory employment to total employment fell from 72.9 per cent to 57.06 per cent at the State level.

The detailed characteristics of the regional pattern of industrial growth between 1961 and 1971 have been studied by Kundu and Raza (1982) at the National Sample Survey region level. The authors note the persistence of a 'disarticulated' pattern of growth, with profitable elements of growth being captured by the advanced regions.

Thus, the main elements of the industrialization pattern during the '50s and '60s suggest a continuation and strengthening of the principal agglomerating factors, continued disjuncture between different forms of industry and industry/agriculture, and limited impact of large public sector projects on the spatial economies.

The promotional policy phase in the industrialization of backward areas in operation since the late '60s again appears to have achieved a dispersal of industrial growth to regions/States outlying the areas of agglomeration, chiefly in the north-western, western and southern parts of the country, but with very limited impact on the eastern States.

The principal impact of government policy on industrial dispersal so far has been to strengthen the 'spread' effects of industrialization to regions, States and districts in proximity to the main industrial centres, or to accelerate the growth of second-level industrial centres. The principal foci of such growth are either in the north-west (Punjab, Haryana, Delhi, western Uttar Pradesh), or in the western and southern States. Industrial growth in Bengal has stagnated, while the eastern States, in general, have lagged behind.

It has been pointed out already that considerable differences emerged in the nature of the regional capitalist class under the impact of colonial policy. The pattern of growth and dispersal since Independence, to which we have referred, has led to the emergence of a nascent capitalist class in backward regions and, in turn, reflects the pre-existing differences in the strength of the regional capitalist classes. Banerjee and Ghosh (1988) have suggested that State and Central governments have intervened more effectively in favour of industrialization and its related requirements, where the regional capitalist class has been strong (as in the north-west, western and southern parts of the country), and where it has traditionally played a more important political role along with the rich peasantry. Obversely, support given to industrialization has been weak in areas (such as the Eastern States) with a weak indigenous capitalist class. It could further be argued that capitalists (both regional and monopoly–national) tend to concentrate in regions in which they can more effectively influence policies and decisions in favour of their development. There is, thus, clearly a reciprocal relationship between the state and the capitalist class, buttressed by historical and economic reasons. This relationship is notably weak in the backward regions.

Centre–state financial transfers: limits to accommodation?

Government policy has attempted to keep in check the growing disparity between regions, as outlined in earlier sections, through policies which bias public and private investment towards less-developed regions; and policies of transfer of resources to less-developed regions and States.

The net outcome of these measures in the past has been a small diminution in disparity between regions in the availability of infrastructure and occasionally in other economic parameters as well. This impact is likely to be further drastically reduced following severe manifestations of economic crisis and policy shifts induced by IMF conditionality. The overall tempo of public investment maintained by the state (both at the Centre and in the States) is severely limited by its capacity to raise resources in a non-inflationary way, by taxing the propertied classes.[15] The criticality of a 'resource contraint' has been underscored in a series of planning and budgetary exercises and has now reached crisis proportions (e.g. Government of India, 1991).

Within the overall constraint, the division of resources between backward and developed States and the pattern of investment in the backward regions assumes a critical role, since a higher rate of public investment in the underdeveloped regions is an important prerequisite for the reduction of disparities (Chakravarty, 1987; Prasad, 1988).

The States' own planned investment is likely to correspond to their resources. These are likely to be very meagre in the backward States.[16] Hence, if the Central component in State plans is not sufficiently progressive, there will be little or no progressiveness in these plans. However, the correlation between per capita Central Plan Assistance

[15] Different versions of this hypothesis may be found in Shetty, 1978; Patnaik, 1984; 1985, 1986; Bardhan, 1984; and Chakravarty, 1987.

[16] The regressiveness in the States' own resources (per capita State Plan Outlays: per capita Central Plan Assistance) is shown by its positive correlation with per capita income (George, 1985: 115–17).

and per capita income of States, has been low and statistically insignificant for all Plan periods, including those covered by the Gadgil formula. It is the middle and not the low income States (excluding the special category ones) which have received higher per capita central Plan assistance in all Plan periods (except the Fourth) (George, 1985: 115–17). The net outcome is that State Plan outlays have maintained their regressive character throughout.

While an explanation for the lack of progressiveness in Central plan assistance can be sought in the underlying criteria, the main explanation appears to rest on the nature of the bargaining processes between the States and the Centre, and the effectiveness with which the States project the interests of the dominant classes. As one analyst has noted,

> the conflict of interests between the rich and the poor states, represented by the National Development Council, could have been resolved only by political compromises, as it happens, rather than on the principle of equity [George, 1985: 130].

The second major component of transfer of resources to States is through direct investment in non-departmental undertakings. Despite large investments in several backward States rich in natural resources, the overall correlation between per capita (gross, up to 1979) investment and per capita income is weakly positive (r = +0.24; *vide* Thimmaiah, 1985: 103). The transfer of resources can also be promoted through the Central Financial Institutions (CFIs), but this is certainly regressive, according to most indicators (Thimmaiah, 1985; Prasad, 1988; Asthana, 1985).

Furthermore, from an analysis of investment premium, credit-deposit, and investment deposit ratios, it has been concluded that

> these ratios indicate that Central Government institutions mobilise a lot of savings from even the less developed states, but divert them to less developed states in response to higher profitability [Thimmaiah, 1985: 99].

It, therefore, appears evident that there are inherent, and sharply escalating, politico–economic constraints on the Centre's ability to impart significant progressiveness to its investment or transfer to the backward States. Corresponding to this is the inescapable conclusion that the Centre has limited autonomy in mitigating uneven growth or its several consequences.

Some aspects of regional movements in the '70s and '80s

The '70s and '80s have witnessed a number of sustained regional and ethnic movements (Punjab, Assam, Jharkhand, etc.) which bear analysis. These movements have diverse social and historical origins, are hegemonized by different social groups and classes, and diverse social ideologies, and have to some extent acquired a self-fulfulling dynamic, all crucial in explaining their origin and the nature of their development. However, there is no intention of providing a full-scale analysis of this phenomenon in this chapter. Rather, I draw attention here briefly to some features of these movements which correspond to the specific and uneven nature of capitalist development of the Indian economy.

The regional and ethnic movements referred to above have arisen in dramatically different regional economic environments, which nevertheless relate to different aspects of a single economic dynamic. There is also a need to relate these movements to the specific nature of emerging classes and social groups in these regions.

The principal paradox concerns the development of a strong regional movement in Punjab, which has been an important beneficiary of direct and indirect resource flows during the colonial period and since (Puri, 1983).

First, the specific nature of development of capitalism in agriculture has led to a situation in which the rural classes in developed regions do not find it easy either to sustain productivity improvements or to sustain rates of return in the face of an imminent realization crisis, and rising input costs. Where agricultural capital confronts merchants and industrial capital (rendering its transition into the latter forms difficult, as in Punjab), it becomes increasingly difficult to augment agricultural productivity (Alam, 1986).

Second, the labour-saving character of agricultural growth in the developed regions, combined with inadequate labour absorption in the secondary and tertiary sectors, and mismatched educational planning, makes for more rapid growth in open or undisguised unemployment (Gill and Singhal, 1984; Puri, 1983).

Third, the weak egalitarianism, implicit in the modified Gadgil formula, conveys the impression that Punjab is subsidizing the development of other States, which it can ill afford to do.

The other regions in which strong movements have arisen in recent years (Assam, Jharkhand, Kashmir) fall at nearly the opposite end of the spectrum of economic development to that where Punjab stands.

Heavy State investments in Assam and Jharkhand have created no significant forward or backward linkages of an intra-regional character.[17] Such investments have also created little local employment. The

[17] The nature of the agrarian problem in Punjab is discussed in a number of studies (e.g. Puri, 1983; Gill and Singhal, 1984 and 1984a; Alam, 1986).

economies of these regions are dominated by strong monopoly—industrial and mercantile interests which have roots outside the region. The fact that Assam and Jharkhand are resource-rich has given rise to the feeling that the exploitation of resources has not benefited the regional or local economies and that they are on the fringes of a 'core–periphery' relationship (cf. Misra, 1980).[18]

The primary and agricultural sectors in all three regions are largely stagnant. In the case of Assam, this aspect of the agrarian economy is further compounded by the fact that no increases in production/incomes can be effected through an expansion of the area under cultivation. Again, the situation provides little scope for expansion in employment in the primary, secondary, or tertiary sectors. Both Kashmir and Assam are 'special category states' whilst Jharkhand is a 'designated backward tribal area'. These regions are therefore entitled to a special flow of resources. But the pattern of utilization of these resources has not given rise to conditions under which these regions could develop economically.

The nature and degree of underdevelopment described above provides the economic underpinnings of discontentment among broad sections of the people. This has been compounded by social, historical, and political factors. It would, however, be a fallacy to regard ethnic and regional movements as resulting exclusively from economic stagnation, since the limited changes in these regions have given rise to a nascent capitalist class (principally consisting of small and medium capitalists, both in agriculture and in industry) which feels thwarted by the pattern of ensuing (under)development. There is, moreover, a new rural élite which has benefited from the siphoning off of the increased flow of resources from the state. The minuscule opportunities of employment in the organized sector also provide little room for the rapidly growing numbers of the educated unemployed. There is, thus, an objective basis for a sharpening of the contradictions between the interests of the regional bourgeois–landlord classes and the aspiring sections of the petty-bourgeoisie in the States, on the one hand, and the interests represented by the Centre, on the other.

It is worth emphasizing that increased investment, or flow of resources from the state to these regions—even if this was within the capacity of the state—would merely replicate the pattern of development described above.

[18] This point is readily conceded in the policy documents issued by the Government of India (e.g. the draft Seventh Five Year Plan).

Conclusion

This chapter presents some facts about recent changes in the Indian economy, focusing on changes in the agrarian and industrial sectors. These have been used to argue that the economy has been subject to the uneven penetration of capitalist relations, and consequently the strength of different classes varies according to the economic characteristics of the diverse regions. I have argued that a central feature of recent changes has been an increase in the influence of the newly emergent classes on the state at the national as well as regional levels. Certain crucial elements in government policy are used to highlight the conclusion that public policy provides limited scope for mitigating existing disparities. Finally, the chapter also presents certain facts and tentative conclusions derived from a brief discussion of ethnic and regional movements, in order to advance the argument that an understanding of their political dynamic can be gained only by situating them in the complex and uneven changes taking place in the regions. These changes have a profound impact on the precise direction assumed by the processes inherent in class formation which are discussed as a separate theme in Volume 4.

REFERENCES

ALAM, J., 1986: 'Political Implications of Economic Contradictions in Punjab', *Social Scientist* 15 (Oct.): 10, No. 161, pp.3–26.

ALEXANDER, K.C,. 1981: *Peasant Organisations in South India* (New Delhi: Indian Social Institute).

AMIN, S., 1982: 'Small Peasant Commodity Production and Rural Indebtedness: the Cultivation of Sugarcane in Eastern U.P., 1880–1920' in GUHA (Ed.), 1982: pp.88–142.

ASTHANA, P., 1985: 'Institutional Finance and Industries: Inter-regional Disparities' in MISHRA (Ed.), 1985: Vol. 1, pp.351–62.

BAGCHI, A.K., 1973: 'Some Implications of Unemployment in Rural Areas', *Economic and Political Weekly* 8 (Aug.): 31–3, pp.1501–10.

———, 1976: 'Reflections on the Pattern of Regional Growth in India during the period of British Rule', *Bengal Past and Present* 95 (Jan.–June): 1, pp.247–89.

———, (Ed.), 1988: *Economy, Society and Polity: Essays in the Political Economy of Planning* (Calcutta: Oxford University Press).

BAIROCH, P., (Ed.), 1980: *Disparities in Economic Development Since the Industrial Revolution* (London: Macmillan).

BANERJEE, D., and A. GHOSH, 1988: 'Indian Planning and Regional Disparity in Growth', in BAGCHI (Ed.), 1988: pp.104–65.

BARDHAN, P.K., 1984: *The Political Economy of Development in India* (Delhi: Oxford University Press).

BARDHAN, K., 1989: 'Poverty, Growth and Rural Labour Markets in India', *Economic and Political Weekly* 24 (25 March): 12, pp.A21–38.

BHADURI, A., 1981: 'Class Relations and the Pattern of Accumulation in an Agrarian Economy', *Cambridge Journal of Economics* 5 (March): 1, pp.33–46.

——————, 1984: *The Economic Structure of Backward Agriculture* (Delhi: Macmillan).

BHALLA, G.S. and Y.K. ALAGH, 1979: *Performance of Indian Agriculture: A District-Wise Study* (New Delhi: Sterling).

BHALLA, G.S. and D.S. TYAGI, 1989: *Patterns in Indian Agricultural Development: A District Level Study* (New Delhi: Institute for Studies in Industrial Development).

BHALLA, S., 1976: 'New Relations of Production in Haryana Agriculture', *Economic and Political Weekly* 11 (27 March): 13, pp. A23–30.

——————, 1988: 'Regional Differentiation and Labour Use in Indian Agriculture and Some Implications for Development Policy' (Patna: A.N. Sinha Institute of Social Studies) (mimeo.).

BHARADWAJ, K., 1979: 'Towards a Macro-economic Framework for a Developing Economy: The Indian Case', *The Manchester School of Economic and Political Studies* 47 (Sept.): 3, pp.270–302.

——————, 1982: 'Regional Differentiation in India: A Note', *Economic and Political Weekly* 17 (April): 14–16, pp. 605–14.

——————, 1985: 'A View on Commercialisation in Indian Agriculture and the Development of Capitalism', *Journal of Peasant Studies* 12 (July): 4, pp. 7–26.

BLYN, G., 1966: *Agricultural Trends in India, 1891–1947: Output, Availability and Productivity* (Philadelphia: University of Pennsylvania Press).

BRAHMANANDA, P.R. and V.R. PANCHAMUKHI, 1987: *The Development Process of the Indian Economy* (Bombay: Himalaya).

BRASS, P.R., 1985: *Caste, Faction and Party in Indian Politics* (Delhi: Chanakya).

BYRES, T.J., 1972: 'The Dialectic of India's Green Revolution', *South Asian Review* 5 (Jan.): 2, pp. 99–111.

——————, 1981: 'The New Technology, Class Formation and Class Action in the Indian Countryside', *Journal of Peasant Studies* 8 (July): 4, pp. 405–54.

——————, 1988: 'Charan Singh 1902–1987: An Assessment', *Journal of Peasant Studies* 15 (Jan.): 2, pp. 139–89.

——————, (Ed.) 1994: *The State and Development Planning in India* (Delhi: Oxford University Press).

CHAKRAVARTY, S., 1987: *Development Planning: The Indian Experience* (Oxford: Clarendon Press).

CHANDRA, N.K., 1974: 'Farm Efficiency under Semi-feudalism: A Critique of

Marginalist Theories and Some Marginalist Formulations', *Economic and Political Weekly* 9 (Aug.): 32–4, pp. 1309–32.

CHATTOPADHYAY, M. and I. SEN, 1988: 'Marketable Surplus and Size-Classes of Holdings', *Economic and Political Weekly* 23 (24–31 Dec.): 52–53 pp. A151–6.

CHAUDHRY, P.K., 1988: 'Agrarian Unrest in Bihar, A Case Study of Patna District, 1960–1984', *Economic and Political Weekly* 23 (2–9 Jan.): 1–2, pp. 51–6.

CHOWDHURY, B., 1964: *Growth of Commercial Agriculture in Bengal (1757–1900)* (Calcutta: Indian Studies Past and Present).

DUNCAN, I., 1988: 'Party politics and the North Indian Peasantry: The Rise of the Bharatiya Kranti Dal in Uttar Pradesh', *Journal of Peasant Studies* 16 (Oct.): 1, pp. 40–76.

FRANKEL, F., 1971: *India's Green Revolution: Economic Gains and Political Costs* (Bombay: Oxford University Press).

————, 1989: 'Caste, Land and Dominance in Bihar: Breakdown of the Brahmanical Social Order', in FRANKEL and RAO (Eds.), Vol. 1, 1989: pp. 46–132.

FRANKEL, F. and M.S.A. RAO, (Eds.), 1989: *Dominance and State Power in Modern India: Decline of a Social Order*, Vol. 1 (Delhi: Oxford University Press).

GEORGE, K.K., 1985: 'Inter-state Disparities in Plan Outlays', in MISHRA (Ed.), pp. 106–31.

GHOSH, J., 1988: 'Inter-sectoral Terms of Trade, Agricultural growth and the Pattern of Demand', *Social Scientist* 16 (April): 4, No. 179, pp. 9–27.

GILL, S.S., 1989: 'Changing Land Relations in Punjab and Implications for Land Reforms', *Economic and Political Weekly* 24 (24 June): 25, pp. A79–85.

GILL, S.S. and K.C. SINGHAL, 1984: 'The Punjab Problem: Its Historical Roots', *Economic and Political Weekly* 19 (7 April): 14, pp. 603–8.

————, 1984a: 'Punjab: Farmers' Agitation: Response to Development Crisis of Agriculture' *Economic and Political Weekly* 19 (6 Oct.): 40, pp. 1728–32.

GOUGH, K. and H.P. SHARMA, (Eds.), 1973: *Imperialism and Revolution in South Asia* (New York: Monthly Review Press).

GOVERNMENT OF INDIA, 1990: *Approach to the Eighth Five Year Plan* (New Delhi: Planning Commission, Government of India).

————, 1991: *Economic Survey 1990–91* (New Delhi: Ministry of Finance, Government of India).

GUHA, A., 1980: 'Little Nationalism Turned Chauvinist: Assam's Anti-foreigner Upsurge, 1979–80', *Economic and Political Weekly* 15 (Oct.): 41–3, pp. 1701–20.

GUHA, R., (Ed.), 1982: *Subaltern Studies I: Writings on South Asian History and Society* (Delhi: Oxford University Press).

GUPTA, D., 1988: 'Country–Town Nexus and Agrarian Mobilisation: Bharatiya Kisan Union', *Economic and Political Weekly* 23 (17 Dec.): 51, pp. 2688–96.

HASAN, Z., 1989: 'Power and Mobilization: Patterns of Resilience and Change in Uttar Pradesh Politics', *in* FRANKEL and RAO (Eds.), 1989, Vol. 1, pp. 133–203.

————, 1989a: 'Self-serving Guardians: Formation and Strategy of the Bharatiya Kisan Union', *Economic and Political Weekly* 24 (2 Dec.): 48, pp. 2663–70.

JANVRY, A. de and K. SUBBA RAO, 1987: *Agricultural Price Policy and Income Distribution in India* (Delhi: Oxford University Press).

JOSE, A.V., 1988: 'Agricultural Wages in India', *Economic and Political Weekly* 23 (25 June): 26, pp. A46–58.

KAHLON, A.S. and D.S. TYAGI, 1980: 'Inter-sectoral Terms of Trade', *Economic and Political Weekly* 15 (27 Dec.): 52, pp. A173–184.

————, 1983: *Agricultural Price Policy in India* (New Delhi: Allied).

KOHLI, A., 1987: *The State and Poverty in India* (Cambridge: Cambridge University Press).

KRISHNA, R. and G.S. RAYCHAUDHURY, 1982: 'Trends in Rural Saving and Capital Formation in India 1950–51 to 1973–74', *Economic Development and Cultural Change* 30 (Jan.): 2, pp. 271–98.

KUMAR, D. and J. KRISHNAMURTHY, 1980: 'Regional and Economic Disparities Since the Industrial Revolution: the Indian Experience', *in* BAIROCH (Ed.) 1980: pp. 361–72.

KUNDU, A. and M. RAZA, 1982: 'Indian Economy: The Regional Dimension', (Centre for the Study of Regional Development, Jawaharlal Nehru University, New Delhi.) (mimeo.).

MAHENDRA DEV, S., 1986: 'Growth of Labour Productivity in Indian Agriculture: Regional Dimensions', *Economic and Political Weekly* 21 (8 June): 25–6, pp. A65–74.

MISHRA, G.P. (Ed.), 1985: *Regional Structure of Development and Growth in India*, Vols. 1 and 2 (Delhi: Ashis).

MISRA, S.C., 1982: 'Commercialisation, Peasant Differentiation and Merchant Capital in the Late Nineteenth-Century Bombay and Punjab', *Journal of Peasant Studies* 10 (Oct.): 1, pp. 3–51.

MISRA, T., 1980: 'Assam: a Colonial Hinterland', *Economic and Political Weekly* 15 (9 Aug.): 32, pp. 1351–64.

MITRA, A., 1977: *Terms of Trade and Class Relations: An Essay in Political Economy* (London: Frank Cass).

MODY, A., 1983: 'Rural Resources Generation and Mobilisation', *Economic and Political Weekly* 18 (May): 19–21, pp. 789–824.

NADKARNI, M.V., 1987: *Farmers' Movement in India* (Delhi: Allied).

NATHAN, DEV, 1981: 'Movements For Agricultural Prices: A Comment', *Economic and Political Weekly* 16 (4 April): 14, pp. 635–6.

OMVEDT, G., 1981: 'Maharashtra: Rasta Roko, Kulaks and the Left', *Economic and Political Weekly* 16 (28 Nov.): 48, pp. 1937–41.

PARTHASARTHY, G., 1987: 'Changes in the Incidence of Rural Poverty and Recent Trends in Some Aspects of Agrarian Economy', *Indian Journal of Agricultural Economics* 42 (Jan.–March): 1, pp. 1–21.

PARTHASARTHY, G. and V. POTHANA, 1983: 'Capitalism in Agriculture and Peasant Farming: Case of Godavari Delta', *Economic and Political Weekly* 18 (24 Sept.): 39, pp. A93–97.

PATNAIK, P., 1984: 'The 1984–85 Budget: Gathering Fiscal Crisis', *Economic and Political Weekly* 19 (24 March): 12, pp. 501–4.

———, 1985: 'On the Political Economy of Economic "Liberalisation"', *Social Scientist* 13 (July–Aug.): 7–8, Nos. 146–7, pp. 3–17.

PATNAIK, U., 1975: 'Contribution to the Output and Marketable Surplus of Agricultural Products by Cultivating Groups in India 1960–61', *Economic and Political Weekly* 10 (27 Dec.): 52, pp. A90–100.

———, 1976: 'Class Differentiation Within the Peasantry: An Approach to the Analysis of Indian Agriculture', *Economic and Political Weekly* 11 (25 Sept.): 39, pp. A82–101.

———, 1986: 'The Agrarian Question and the Development of Capitalism in India', *Economic and Political Weekly* 21 (3 May): 18, pp. 781–93.

———, 1988: 'Some Aspects of Development in the Agrarian Sector in Independent India', *Social Scientist* 16 (Feb.): 2, No. 177, pp. 17–40.

PRASAD, P.H., 1988: 'Roots of Uneven Regional Growth in India', *Economic and Political Weekly* 23 (13 Aug.): 33, pp. 1689–92.

PURI, H.K., 1983: 'The Akali Agitation: An Analysis of Socio-Economic Bases of Protest', *Economic and Political Weekly* 18 (22 Jan.): 4, pp. 113–18.

RAO, C.H.H., 1986: 'Change in Rural Poverty in India: Implications for Agricultural Growth', *Mainstream* 24 (11 Jan.): 19, pp. 9–12.

RUTTEN, M., 1986: 'Social Profile of Agricultural Entrepreneurs: Economic Behaviour and Life Style of Middle–Large Farmers in Central Gujarat', *Economic and Political Weekly* 21 (29 March): 13, pp. A15–23.

SHARMA, H.P., 1973: 'The Green Revolution in India: Prelude to a Red One', *in* GOUGH and SHARMA (Eds.), 1973: pp. 77–102.

SHETTY, S.L., 1978: 'Structural Retrogression in the Indian Economy since the Mid-Sixties', *Economic and Political Weekly* 13 (Feb.): 6 & 7, pp. 185–244.

SRIVASTAVA, R.S., 1989: 'Tenancy Contracts during Transition: A Study Based on Fieldwork in Uttar Pradesh (India)', *Journal of Peasant Studies* 16 (April): 3, pp. 339–95.

———, 1994: 'Planning and Regional Disparities in India', in BYRES (Ed.), 1994, pp. 147–219.

———, (forthcoming): *Surplus Appropriation and Capital Accumulation by Rural Households in India* (Delhi: Oxford University Press).

THIMMAIAH, G., 1985: 'Inter-State Disparities in Financial Allocations', in MISHRA (Ed.), 1985: Vol. 2, pp. 68–105.

TYAGI, D.S., 1987: 'Domestic Terms of Trade and Their Effect on Supply and Demand of Agricultural Sector', *Economic and Political Weekly* 22 (28 March): 13, pp. A30–A36.

———, 1994: 'Planning and Regional Disparities in India, *in* Byres (Ed.), 1994, pp. 147–219.

VAIDYANATHAN, A., 1986: 'Labour Use in Indian Agriculture: A Study of Spatial and Temporal Variations', *Economic and Political Weekly* 21 (27 Dec.): 52, pp. A130–146.

VARSHNEY, A., 1989: 'From Urban to Rural Bias? Rising Peasantry, the State and Food Policy in India', unpublished Ph.D. Dissertation, Massachusetts Institute of Technology, Cambridge, MA.

VIRMANI, A., 1988: 'Terms of Trade Between Agriculture and Industry: Conceptual Problems and Alternatives', Economics Research Paper No. 1 (Planning Commission, Government of India, New Delhi) (mimeo.).

WALINSKY, L., (Ed.), 1979: *Agrarian Reform as Unfinished Business: Selected Papers of Wolf Ladejinsky* (Oxford: Oxford University Press).

WORLD BANK, 1989: *India: Poverty, Employment and Social Services* (Washington, DC: The World Bank).

Agrarian Capitalism and Political Processes in Punjab • 8

SUCHA SINGH GILL

Economic and political changes are two facets of socio-economic transformation. The broad dynamism of social change affects these changes and is itself affected by them. Economic change and political change also affect each other mutually and are dominant parts of the operation of the social system as a whole. Economic policy reflects a conscious effort on the part of a political regime to meet the short-term and long-term requirements of the system in order to consolidate its own position. Economic policy initiates economic change which has political implications. With the passage of time, economic change necessitates changes in economic policy in order to adapt to the changed situation. Political change leading to the formation of a new regime affects economic policy and the pace of economic development.

In the present world, considerable integration of the global economy has occurred. There are powerful institutions operating on a global level which exert economic and political influence on the societies of various countries. In spite of this, economic and political changes in different countries do not follow a uniform pattern. This is explained by the fact that every society has its own specificity and possesses its own internal dynamism of change. Though economic and political changes in various countries are influenced by global factors, yet their contents are largely determined by the internal dynamism of social change.

In this chapter, I consider the changes in political processes that have occurred in the wake of the development of capitalism in agriculture in India. Though the observations here are based on a concrete study of Punjab, yet they are largely applicable to other regions where agricultural capitalism has advanced with the spread of Green Revolution technology. In section 1 of this chapter contradictions created by agrarian capitalism in the State are identified; section 2 discusses the political changes that have taken place in the State; in the third section, the specificities of the situation prevailing in Punjab are summed up.

Contradictions of agrarian capitalism in Punjab

Economically, Punjab is the most advanced State of the Indian Union. In 1985–6, the per capita income of the State (at 1970–1 prices) was Rs 1600 compared to an all India average of Rs 779. Over a period of time, Punjab has not only maintained its lead but has also increased its advance over both backward as well as other developed States in the country. Being the most developed State, the contradictions of the Indian path of development are manifested here more fundamentally than elsewhere in the country, and are specific to Punjab. During the post-Independence period, Punjab has developed within the broad framework of controlled capitalist development in India. As such, it has developed its own specific model of transformation, based on rapid capitalist development of agriculture accompanied by the development of small-scale industries within the production structure of commodities. Economic transformation in Punjab is facilitated by a relatively more developed tertiary/service sector.

The Punjab model has generated contradictions at two levels: first, within the State itself, and, secondly, between the regional dominant class and the dominant class at the all-India level. Contradictions within Punjabi society arise from its mode of production as a predominantly agrarian economy. In 1980–1, agriculture and livestock contributed 49.04 per cent of the State Domestic Product (SDP) at 1970–1 prices (Government of Punjab, 1981) and employed 59.1 per cent of the total workforce in 1981 (Government of India, 1981).

Agriculture in Punjab has undergone capitalist development at a relatively rapid rate since 1965–6. A section of the peasantry exercising greater control over land and capital resources has been transformed into a class of capitalist farmers. In 1980–1, the most well off 7.19 per cent of farmers operating 10 hectares of land and more were in control of 19.17 per cent of the cultivated area. Farmers with 4 hectares or more constituted 33.39 per cent of the total number of farmers and operated 71.98 per cent of the total cultivated area (Government of Punjab, 1981). These farmers own capital assets such as tractors, threshers, pump-sets, etc. to an extent greater than the proportion of the total land held by them. They find their family labour insufficient to carry out farm operations and so depend principally on hired labour. A study of labour use pattern in Punjab agriculture shows that all farmers with different levels of technology hire labour power, but more mechanized farms owning tractors, pump-sets, threshers, and trolleys utilize hired labour to the extent of 76.27 per cent of their total requirements (Kumar, 1981: 44–5).

Along with capitalist farmers there has come into existence a class of agricultural labourers which owns neither land nor capital resources but only labour power which it sells to farmers. This class constitutes 38.6 per cent of the total labour force engaged in agriculture as against cultivators who constitute 61.4 per cent (Government of India, 1981). There is an irreconcilable difference between farmers and agricultural labourers on the issue of wages in the Punjab, and every year there are a number of cases of conflicts between them on this account. Sufficient documented evidence testifies to this (Singh, 1979; Singh, 1980; and Singh, 1987).

The agricultural labour class has emerged from four sources: first, some workers were already in the agricultural sector of the State on the eve of Independence. Second, with the introduction of land reform, tenants-at-will were ejected from the land and joined the ranks of agricultural labourers (Talib, 1979). With the introduction of new technology, pure tenants as a category ceased to exist when occupancy tenants became owners of land and tenants-at-will were ejected by the owners. (Gill, 1980). Third, with the modernization of rural society in the State, the traditional artisans became redundant, a majority of them joining the ranks of agricultural labourers (Gill, 1980a). Lastly, with the rapid increase in agricultural productivity and cropping intensity in Punjab, there has been a considerably increased flow of migratory labour from other regions such as Eastern U.P., Bihar, and M.P.

The capitalist production system in agriculture has created a distinct class which principally depends on the sale of labour power for its survival. Since wage labour is freely sold in the market and this class is devoid of ownership of the means of production, it can be termed the 'agricultural proletariat'. It has taken over most of the operations requiring hard physical labour. The rising component of hired labour in agriculture has given rise to the distinct identity of this class.

Available evidence suggests that, even before the Green Revolution, the proportion of hired labour over family labour had begun to rise. National Sample Survey (NSS) data suggest that, between 1956–7 and 1964–5, per worker employment increased by 12 days; in other words, hired labour increased by 20 days and family labour decreased by 8 days per worker (Bardhan, 1977). It is estimated that, between 1954–5 and 1967–8, human labour requirements per hectare increased from 51 days to 60 days and family labour decreased from 37 days to 31 days, while hired labour increased from 14 days to 19 days (Randhawa, 1974). This process continued further with the maturing of the Green Revolution. Between 1971–2 and 1981–2 the share of family labour in wheat cultivation decreased from 49.54 to 41.46 per cent, whilst that of hired labour increased from 50.46 to 58.54 per cent. The introduction of paddy cultivation has further strengthened this tendency as Punjab is not traditionally a paddy growing area.

Affluent farmers are withdrawing their family labour from farm operations and performing more or less exclusively managerial and supervisory functions. Indeed, their family labour is meagre compared to their requirements. Whereas capitalist farmers mainly depend on hired labour, even small and medium categories of cultivators also hire labour during the busy agricultural season. Therefore, on the wage issue there is a contradiction between the class of agricultural labourers, on the one hand, and, on the other, that of cultivators led by capitalist farmers. At the same time, differences in the caste composition of cultivators and agricultural labourers is superimposed on this contradiction. Whereas cultivators are Jat Sikhs, agricultural labourers are Scheduled Castes(SCs) and Backward Castes (BCs) known as Harijans (and Mazhabis).

Capitalist development in Punjab agriculture has had a differential effect on the different sections of the peasantry. Capitalist farmers have been the largest beneficiaries and are producing much more than is required for family consumption. They are producing surpluses which are invested in agriculture and in other income–generating activities. The small and marginal farmers experienced some gain in their income levels during the initial period which they are no longer able to sustain. Many small and marginal holdings have become unviable. According to a recent survey, in 1974 small farmers were incurring annually an average loss of Rs 125 per head, whereas farmers owning 5 to 10 acres of land were generating a per capita profit of Rs 50, and farmers with landholdings in excess of 20 acres were making an average profit of Rs 1200 per capita (RCPI, 1984). Consequently, 24 per cent of small farmers and 31 per cent of marginal farmers were living below the poverty line by 1979–80 (Bhalla and Chadha, 1982).

With agricultural development in crisis in more recent times, small and marginal farmers are finding it difficult to survive. Between 1970–1 and 1980–1, a large number of such holdings disappeared. In 1970–1, the total number of operational holdings in Punjab was 13,75,382. This figure subsequently declined to 10,27,127 (a decline of 25.3 per cent). This decline was completely due to the disappearance of marginal and small holdings. The marginal holdings declined by 61.9 per cent and small holdings by 23.3 per cent. This is revealed in the Agricultural Census data presented in Tables 8.1 and 8.2, and is also confirmed by NSS data. The estimated number of ownership holdings in the State declined by 2,33,500 from 24,40,700 in 1971–2 to 22,07,200 in 1982 (Government of India, 1972; Government of India, 1982).

As yet there is no evidence to suggest that the farmers giving up cultivation have joined the ranks of agricultural labour. Rather, there is a strong antipathy among Jat cultivators to joining the agricultural labour force, a majority of which is low caste Harijan in composition. Neither are the cultivators being released from agriculture absorbed in the industrial sector of the State. The rate of industrial growth in the State, on an average, has been higher than that of the Indian economy as a whole (Government of Punjab, 1986).[1] But its nature and structure is such that it absorbs largely migratory labour (Mehra, 1985: 51). In fact, wages are very low and working conditions are so unattractive that in small-scale and ancillary units, local labour with a peasant background has no wish to find employment.

At the same time, they do not have the requisite expertise and experience needed for jobs in the tertiary sector. For many years now, jobs in other States of India and abroad have been unobtainable. Thus, the cultivators released by capitalist development in agriculture in Punjab are not being absorbed outside agriculture and are consequently experiencing unemployment. This force is accompanied by a large body of educated unemployed youth from the rural areas. The youth in these areas has inflated aspirations and would like to become rich overnight, their reference group comprising in the main relatives who have migrated abroad.

Educated young people belonging to the poor peasant class find

[1] Since 1984–5, this tendency has been reversed. Punjab's rate of industrial growth has declined compared to the corresponding All-India figure (Government of Punjab, 1989).

their parents' holdings unviable and are therefore in dire need of employment outside agriculture. The same applies to ex-army personnel of low rank who retire around their mid- '30s or early '40s. Since jobs are scarce, a large number of youths and retired army personnel are unemployed and their number is increasing by the year. This constitutes the human elements available for various uses and manipulations, including violence unleashed by the political process in the State.

The well-off sections in the rural areas, who have been able to appropriate a major part of the prosperity generated by capitalist development, experience a different set of problems. They have not only gained from the rise in agricultural production and productivity, but also from the facilities extended by the government in the sphere of rural development. They are taking over the land of the poor peasantry through lease and mortgage or sale.

Big cultivators with land over 10 hectares constituted 5.01 per cent of the total and operated 26.86 per cent of the land (as at 1970–1). Their proportion increased to 7.19 per cent and they controlled 29.17 per cent of the land in 1980–1. These farmers have surpluses which they invest in land. But the return from this investment is falling, making it less and less profitable. The rate of return on wheat cultivation per quintal declined from 24.50 per cent in 1970–1 to 1.32 per cent in 1977–8 (Kaur, 1982: 275). As a consequence, the net income per hectare from wheat cultivation (at 1970–1 prices) declined from Rs 328 in 1971–2 to Rs 54 in 1981–2 (Swamy and Gulati, 1986).

Since agriculture is less rewarding for capitalist farmers, they are upset over this untoward development. Some of them invested their surpluses outside agriculture—particularly in transport, cinema, cold storage, etc. But these activities are also becoming less rewarding. It is very difficult for them to enter business and industry for a number of reasons. First, they do not have the necessary expertise and experience in these activities. Second, they have to face stiff competition from established business and industrialists. Only the traditional business community and artisans, particularly carpenters and blacksmiths (Ramgarhias), have been able to enter the emerging industrial sector of the State (Pandit, 1978 and Sahni, 1980). Thus there is a lack of integration of agricultural capitalists and urban-based business and industrial capitalists.[2] As a consequence, the former feel alienated and deprived.

[2] This is unlike coastal Andhra Pradesh where rich Kamma peasants are transforming themselves into industrialists by using agricultural surplus (*see* Ch. 3), and Kammaver Naidus of Kongunad (of Tamil Nadu) who similarly used agricultural surplus in order to develop cotton mills in Coimbatore during the '20s and '30s (Upadhyaya, 1988).

Such modest efforts to diversify in Punjab, as evidenced in the growth of the truck transport business, received a major setback following the November 1984 massacre of Sikhs in Delhi and other places outside Punjab.

Capitalist development has brought about fuller integration of agriculture with the market network. There is a greater degree of specialization in production. Farmers grow mainly wheat in the *rabi* season and paddy, cotton, and sugarcane in the *kharif* season. This has transformed farmers of different layers of society into commodity producers. They sell a major part of their output in the market. There is a considerable increase in the use of inputs such as HYV seeds, chemical fertilizers, insecticides, weedicides, pump-sets, electric motors, threshers, tractors, etc. Farmers sell a major part of their output to and buy all the modern inputs and a substantial percentage of their consumer goods from the market. Market operations thus have an important effect on the income and life of the farmers. Issues relating to price and smooth market operation are therefore of vital importance to the peasantry.

In the market, farmers come into direct contact with traders. Traders/commission agents are responsible for unloading and bagging the produce, preparing it for sale, providing display space, finding buyers and selling, stitching the bags, weighing them, collecting payments from buyers and making them available to farmers after making the necessary deductions for various intermediary charges and taxes. They also provide a considerable amount of credit to the farmers, both in cash and in the form of inputs on deferred payments. In the process, there is a predominance of commission agents who exploit the farmers through various malpractices and usurious rates of interest (Aulakh, 1983: 50, 69). Farmers buy various inputs and consumer goods from traders in the marketplaces of urban areas where they are overcharged. This gives rise to an acute contradiction between the farming community and the trading community. Farmers are suspicious of traders and regard them as cunning exploiters, whilst traders view farmers as rustic, rash, uneducated, and easy to cheat. Indeed, the peasantry nurtures a grudge against urban centres which harbour not only traders but also government officials who harass and oppress them.

Class contradiction is reinforced by communal division in this society. A study based on the 1971 Census data shows that 69.37 per cent of the Sikhs in Punjab are concentrated in the rural areas and are mainly cultivators, and 66.39 per cent of the Hindu population live in urban areas and specialize in non-agricultural occupations such as trade, industry, services, etc. In the urban-based occupations, Hindus are over-represented and Sikhs under-represented (D'Souza, 1982), and entry of Sikhs into these occupations on a significant scale creates communal polarization. This explains the contradiction between Hindu and Sikh traders and growing communalism among educated people in the urban areas. Due to the uneven numerical strength of

Table 8.1

Percentage variation in the distribution of number of operational holdings in Punjab over 1970–1 by major size groups

Size group	Individual holdings			Joint holdings			Total holdings		
	1970–1	1980–1	Percentage variation	1970–1	1980–1	Percentage variation	1970–1	1980–1	Percentage variation
Marginal (below 1 hectare)	395603	195060	–50.7	121965	2234	–98.2	517568	197322	–69.9
Small (1–2 hectares)	193016	196167	1.63	67067	3201	–95.2	260083	199368	–23.3
Medium (2–4 hectares)	203820	279422	37.1	77283	7981	–89.7	281093	287423	2.3
Large (4–10 hectares)	166961	253906	52.1	80744	15157	–81.2	247755	269072	8.6
Extra large (10 hectares and above)	37059	66256	78.8	31824	7649	–75.9	68883	73941	7.3
Total	996459	990811	– 0.6	378883	36222	–90.5	1375382	1027126	–25.3

SOURCE: *Agricultural Census of Punjab*, 1970–1 and 1980–1.

Table 8.2

Percentage variation in the distribution of operated area in Punjab in 1980–1 over 1970–1 by major size groups

Size group	Area under individual holdings			Area under joint holdings			Total area		
	1970–1	1980–1	Percentage variation	1970–1	1980–1	Percentage variation	1970–1	1980–1	Percentage variation
Marginal	171504.7	116680	–32.0	53750.9	1527	–97.2	225255.0	118230	–47.5
Small	275782.2	276467	0.22	96343.6	4483	–95.3	372130.8	280900	–24.5
Medium	574217.4	768686	33.9	220890.5	22118	–90.0	795107.9	790855	– 0.5
Large	1010325.5	1474348	45.9	503897.8	91290	–81.9	1514223.3	1565674	3.4
Extra large	531355.9	997842	87.8	523952.8	131822	–74.8	1067373.1	1434717	6.3
Total	2563160.9	3634023	41.8	1398835.6	251240	–82.0	3974090.1	4190376	– 2.1

1970–1			1980–1		
Total operated area	=	3974091.1 hectares	Total operated area	=	3890376 hectares
Total number of holdings	=	1375382	Total number of holdings	=	1027127
Area per holding	=	2.89 hectares	Area per holding	=	3.8 hectares

SOURCE: *Agricultural Census of Punjab*, 1970–1 and 1980–1.

Sikhs and Hindus in urban areas and strong social bonds, communal alienation has not been in the past and cannot in the present and the future be converted into communal clashes. As there is no tendency among Jats in rural areas to take up agricultural labour, and as that community exercises dominance in all spheres of life, this is reflected not so much in communal tension as in contradictions of class and caste.

There is a second set of contradictions between Punjab and the Centre. With capitalist development in agriculture, the class of capitalist farmers has come to dominate rural life and to wield decisive influence on the political processes and power in the State. The institutions of Panchayati Raj, the network of cooperative institutions, and the State Legislative Assembly are all controlled by people from this class. This class is using government machinery at State level to promote its interests, and in doing so comes into conflict with the class in control of government machinery at the central level. The former finds that the powers of administration at the State level are being continuously eroded by the central government and sees its growing assertion of its class interests thwarted. The growing economic and social role of the administration is being monopolized by the central government. Whether it be the sphere of economic planning or of the diversification of financial institutions or the radio–television communication system, policy choices are monopolized by the central administration. In such a situation, the role of promoting the interests of the ruling class at the State level through administration has extremely limited scope.

A major part of the agricultural surplus is marketed by capitalist farmers. The prices of major agricultural crops produced in the State such as wheat, paddy, sugarcane, and cotton are fixed by the central government on the recommendation of the Agricultural Prices and Costs Commission (APCC). There is another agency, the Bureau of Industrial Costs and Prices (BICP), which advises the central government on the prices of industrial goods. The central government intervenes in the pricing of various commodities under a regulated price system and has the capacity to change this system.

Over the past several years the price system has moved against agriculture and in favour of industry under pressure from the Indian monopoly bourgeoisie, which exercises a decisive influence over the central government. The terms of trade have moved against agriculture from 100 in 1970–1 to 81.8 in 1980–1.[3] This shift has been sharp and consistent from 1974–5 onwards. A major part of the surplus wheat, paddy, and cotton is purchased by central agencies such

[3] Calculated from Government of India, 1985.

as FCI and CCI, whilst inputs such as fertilizers, pesticides, insecticides, and electric motors are supplied by Indian (private and government owned) and foreign monopolies. Any shift in the terms of trade against agriculture results in a transfer of economic surplus from capitalist farmers to the industrial monopoly bourgeoisie. It is over the price issue and other market operations that a major agitation under the leadership of the Bharatiya Kisan Union (BKU), an organization of capitalist farmers, was launched in Punjab in 1983–4 (Gill and Singhal, 1984).

Monopoly capital's greater need for centralization has brought about greater encroachment of the central government in spheres of activity earlier demarcated for the States under the Constitution. This has led to a blurring of the distinction between Central and State subjects. The Centre's involvement in the State and in the Concurrent list is both direct and indirect. The massive involvement of the Centre is made possible by the enormous amount of financial resources at its disposal. Of the total budgetary resources of the Central and State governments, more than two-thirds accrue, in the first instance, to the former. Expenditure, borne by both the Centre and the States, is on a 50–50 basis. Thus, the States depend on the Centre for their resources to the extent of one-third of the total required (George and Gulati, 1985). This makes the position of the State administration precarious, particularly when it is ruled by parties in opposition to the party in power at the Centre.

Political changes

The development of agrarian capitalism has affected the political processes in the State. The emergence of a class of capitalist farmers/rich peasants with considerable command over resources has increased their economic, social, and political influence in Punjab in general and Sikh society in particular. Capitalist relations have freed them from arduous farm operations to pursue activities in other spheres. The entire gamut of the development process has given rise to much greater mobility (Oberai and Singh, 1980; Gill, 1990) leading to the activation of diverse segments of the population. Individuals with a rural background are active not only in rural areas but also in urban centres where they are employed in the tertiary sector. Agrarian capitalism has weakened traditional kin and caste ties and created a new set of relations governed by market conditions generally based on capitalist rules of the game. People's responses to various situations clearly reflects the impact of the changed situation. Apart from the economy, change is quite evident at the level of the political processes in the State. These changes can be observed at three levels:

1. changes in the support base of various political parties;
2. class base of the individuals holding elected offices at various levels; and,
3. the emerging pattern of class organizations in rural areas.

Changes in the support bases of political parties

The Shiromani *Akali Dal* and the Congress(I) are the two parties which have always competed for power in the State. Here the influence of the capitalist/rich farmers has been decisive, particularly at the level of the leadership of the parties. But, over a period of time, the Akali Dal has been able to emerge predominantly as the party of the rural bourgeoisie. Since the peasantry in the State is largely Sikh, the Akali Dal is able to exercise a wider appeal to various sections of cultivators. Being a regional party, it does not suffer from the handicaps of the Congress(I). It is thus able to promote exclusively the interests of the peasantry, and in particular the capitalist farmers. The Congress(I) competes with the Akali Dal in this sphere, yet the policies of the former (particularly when it is in power at the Centre) clearly operate more in favour of the industrial and commercial bourgeoisie than the rural bourgeoisie.

The Congress(I) has been consistently losing support among the Sikh peasantry. This is also evident from the declining percentage of Sikhs in the composition of the Congress Legislature Party from 1969 onwards. Table 8.3 shows that the percentage of Sikh MLAs belonging to the ruling Congress party stood at 68.4 per cent in the 1969 elections from which it declined to 59.1 in 1972, 47.0 in 1977, 44.0 in 1980, and 34.4 in 1985. Conversely, the percentage share of Hindu MLAs belonging to the same party consistently increased from 31.6 in 1969 to 39.4 in 1972, 53.0 in 1977, 54.0 in 1980, and 68.8 in 1985. This is on account of the fact that the Akali Dal has its support base largely in the rural areas where the Sikhs dominate numerically, whereas the Congress(I) support base is divided between rural as well as urban areas. The Congress(I) support base has been eroded in the rural areas over a period of time due to a decline in the peasant support base in urban centres. The Akali Dal has continuously replaced the Congress party in the rural areas.

Table 8.4 shows that between 1977 and 1985, the proportion of Akali Dal MLAs from rural areas was 96.6 per cent, 97.4 per cent, and 89.1 per cent in 1977, 1980, and 1985 respectively. Agriculturists

Table 8.3

Composition of the State Congress Legislature Party by religion in the Punjab Assembly

(percentage)

Religious community and percentage of population in 1981	1967	1969	1972	1977	1980	1985
Hindus (37.09)	33.03	31.6	39.4	53.0	54.0	68.8
Sikhs (60.75)	64.8	68.4	59.1	47.0	44.0	31.2
Muslim (1.00)	1.9	—	1.5	—	1.6	—
Unidentified	4.0	—	—	—	—	—
Total	100.00	100.00	100.00	100.00	100.00	100.00

SOURCE: Dhami (1988a: 389).

Table 8.4

Background of the Akali MLAs (1977–85)

(percentage)

Variables	1977	1980	1985
1. Rural/Urban			
Rural	96.6	97.4	89.1
Urban	3.4	2.6	10.9
2. Occupation			
Agriculture	70.0	76.3	76.7
Non-agriculture	29.3	23.7	23.3
3. Religion			
Sikhs	98.3	100.0	95.9
Hindus	—	—	2.7
Muslims	1.7	—	1.4
4. Caste			
Jat Sikhs	63.9	65.8	67.1
Scheduled Castes	27.6	26.3	24.7
Other Castes	8.5	7.9	8.2

SOURCE: Verma (1987: 281)

Table 8.5

Economic status of SGPC members

	Exit	Existing	Emergent	Total (per cent)
Upper: Farmers	16	9	23	48 (46.60)
Non-farmers	3	—	4	7 (6.80)
Middle: Farmers	6	3	11	20 (19.41)
Non-farmers	4	1	3	8 (7.77)
Lower: Farmers	6	2	—	8 (7.77)
Non-farmers	4	—	8	12 (11.65)
Total	39	15	49	103 (100)

Exit : members between 1965–79.
Existing : members elected in 1979 but were members earlier also.
Emergent : members elected in 1979 for the first time.

SOURCE: Singh, 1986: 150.

dominate this party. 70.7 per cent of the total number of MLAs in 1977, 76.3 per cent in 1980, and 76.7 per cent in 1985 were agriculturists by occupation. Their caste composition reveals that during the last three elections (as at 1991), Jat Sikhs cornered nearly two-thirds of the total number of seats won by the Akali Dal. Since Jat Sikhs numerically dominate the rural areas and a majority of capitalist farmers/rich peasants belong to this caste, the Akali Dal is obviously dominated by them (Verma, 1987).

Two other parties which have some support base among the peasantry are the two main Communist parties, namely the CPI and the CPI(M). These parties have also a share of their leadership from the middle and rich peasantry (Brar, 1988). These two parties exercise their influence through their Kisan Sabhas. Their followings are restricted to a few pockets in the State which are shrinking under the pressure of the current crisis there.

The support bases of various political parties in rural Punjab has been aptly described in a study based on fieldwork done in Sangrur district.

> In the rural areas, the main political force is Akali Dal which received predominant peasant support because of its image as a farmers' party. The Congress party's support in this region has been considerably eroded in recent years though the process has been under way since the mid-sixties when the Congress Party lost leaders like Partap Singh Kairon, Gian Singh Rarewala, and Giani Kartar Singh who mustered support for the Congress. The supporters of the latter two leaders now support the Akali Dal. Congress still gets some support from [the] peasantry primarily through patronage politics, especially when the party is in power. The more stable support for the Congress comes from Scheduled Castes, Caste Hindus and a majority support from servicing castes. The support of the two Communist parties is mainly confined to the peasantry. Though, of late, they have made some dent into Scheduled Castes support. The majority of Scheduled Castes still view these parties as the supporters of peasant interests which keeps them within the Congress orbit. [Dhami 1988, 228.]

The majority of SCs in rural Punjab are agricultural labourers. They have views on wage issues which contradict the views of other sections of the peasantry. Besides, there is also social oppression of the SCs by the peasantry as a whole on a caste basis. Therefore, the political party which is dominated by the peasantry, viz. the Akali Dal, hardly gets any support from the SCs/agricultural labourers. On the question of peasant support, the Congress and Communist parties are in competition with Akali Dal. The latter uses the ideology of Sikhism in order to gain advantage over secular parties. On the use of Sikh religion, the Congress did try to challenge the Akalis in the past with a view to building a support base among the peasantry. This was a major factor in the shape that the political crisis in the State has assumed during the last decade or so.

Influence on elected offices

There is a whole range of offices, beginning from membership of Panchayati Raj institutions such as village Panchayat, Block Samiti, Zilla Parishad to the Legislative Assembly and Parliament of the country, to which elections are held periodically on the basis of universal adult franchise. There is also a network of cooperative credit and marketing institutions open only to members, and to which elections are held periodically. Along with such a structure of secular electoral processes and elected offices, there is a religion-based election process for the maintenance of Sikh religious shrines popularly known as the Shiromani Gurudwara Prabandhak Committee (SGPC). With the emergence of agrarian capitalist production relations, the agrarian bourgeoisie has established its dominance over these bodies. The institutions of cooperative credit and marketing and Panchayati Raj such as village Panchayats, Block Samitis, and Zilla Parishads are the exclusive domain of the rural rich/agricultural bourgeoisie who belong to the Jat Sikh caste. Their influence on the State Legislative Assembly through the Akali Dal and the Congress(I) has been well established. As shown in Table 8.4, Akali MLAs mainly belong to this class, as do the Congress Party Sikh MLAs.

Capitalist farmers/rich peasants not only dominate the secular electoral process and structure of elected offices, but have also come to acquire effective control over the religion-based electoral process in the SGPC. A study of the SGPC elections conducted during the post-Green Revolution period clearly underlines this (Singh, 1986: 129–65). It is evident from Table 8.5 that, out of a sample of 103 SGPC members, 66.01 per cent were upper and middle farmers. The percentage of members with a farming background is 73.78. This largely applies to former members, existing members, and emergent members. It is further revealed that 68.93 per cent of the members studied belonged exclusively to the Jat Sikh caste.

A high percentage (nearly two-thirds) of the upper and middle Jat peasantry in the SGPC points to the dominance over it of individuals of this class. This is the reason why office-bearers and members of the executive of the SGPC bear the stamp of this caste and class.

In the wake of capitalism in agriculture, the class of capitalist farmers has increased its influence on elected offices and, through them, its influence over the administrative machinery at the State and local levels has become dominant. It is able to use the administrative structure at the State level in order to promote its class interests and to increase its area of influence. Its dominant influence at the level of the structure of the two major political parties, viz. the Akali and the Congress(I), helps them electorally. The relative changes in the support bases of these parties among the peasants reflect the dominant interests of this class. Recognizing the role of religion and the ideology of Sikhism in promoting the interests of the agricultural bourgeoisie, the capitalist farmers have come to dominate the SGPC. They have placed the issue of Sikh identity centre stage among the political issues facing Punjab.

The emergence of class organizations

The organization of agriculture along capitalist lines has led to the emergence of class organizations of peasantry as well as agricultural labour. These organizations exclusively fight for the demands of the respective classes; they also struggle against the government as well as against the contending classes in order to promote the interests of the class for which they were created.

Two types of organization represent farms. First, the organizations created by political parties as their Kisan wings. The two Communist parties have their Kisan Sabhas, the Congress party has its Kisan cell, and the Akali Dal has the Kisan Dal. There is a general decline in the influence of the front organizations of political parties among the peasants of the State. The second type of organization which has acquired prominence consists of non-party farmers' associations. At present two factions of the BKU are working in the State. Both are led by capitalist/rich farmers who are independent of the dominant political parties in the State. At the time of the parliamentary and Assembly elections, the BKU invariably reserves the right to support or oppose the different political parties with a view to promoting the interests of farmers in general and of rich/capitalist farmers in particular.

Though the Punjab unit of the BKU came into existence in 1980, efforts to create such an organization started much earlier. The Khetibari Zamindara Union (KZU), a forerunner of the BKU, came into existence in 1972 and was transformed into the BKU subsequently, in 1980. Initially, the Union had exercised much influence in Ludhiana district and in the adjoining parts of Patiala district. But as a result of various agitations, the BKU was also able to organize units in other districts of the State. By 1983, the BKU had fully functioning units in all the districts of Punjab except Kapurthala. The BKU has its units at four levels—the village, block, district, and State levels. The programme includes three types of demand (cf. Ch. 9).

First, it includes general rural demands such as the creation of more educational, health, and communication facilities, the reservation of jobs for rural people, and the industrialization of rural areas, etc. Second, demands of the peasantry are generally related to the supply of agricultural inputs at low prices and higher/remunerative prices for agricultural products, abolition/reduction of taxes on agriculture, cancellation of debt, etc. And third, general demands such as the ending of smuggling, corruption, licencing system, inefficiency, etc. also form part of the programme of the BKU. An analysis of various struggles launched by the BKU would reveal that these have been largely related to the cost and quality of agricultural inputs and the prices of farm output (Gill and Singhal, 1984).

In recent years, the BKU has raised three more demands: first, it has been demanding a share in the non-agricultural business—in particular, agencies to the farmers for the supply of tractors, insecticides, pesticides, weedicide, fertilizers, etc., and in the opening of commission agent shops by farmers. Second, it has been demanding that the management of cooperative processing units, such as sugar mills, should be transferred to it. The third demand is for a remission of farmers' loans.

In order to promote the interests of the peasantry, the BKU has been launching powerful struggles which have resulted in raising the level of mass-mobilization. These struggles have been largely independent of party politics. An equally powerful tool has been the opposition to or support of various political parties in the elections. The BKU was responsible for the victory of Congress (I) in the 1980 and its defeat in the 1985 State Assembly elections (and the corresponding victory of the Akali Dal). The victory of the Akali Dal (Mann) in the 1989 parliamentary election was due not a little to the support provided by the BKU. Agitational struggles and support/opposition during elections are tools used to exert pressure on the political parties and the government to promote rich peasant/capitalist farmers' class interests. The emergence of the BKU and the growth of its mobilizational capacity has increased the influence of the rich peasantry over the bourgeois/petit-bourgeois political parties, viz. the Akali Dal and Congress(I). The BKU in Punjab, like its units in other States, has also mobilized peasants with the aim of thwarting the struggles of agricultural labour (Gill and Singhal, 1984; Hasan, 1989).

Another Union with a militant posture has come into existence in the recent period. This is the Kirti Kisan Union (KKU). It is controlled by one of the Naxalite groups. Though most of its leadership belongs to the middle peasantry, the KKU has been taking up the same issues as those espoused by the BKU. The influence of this Union is at present concentrated, in patches, in 4 or 5 districts of the State.

With the emergence of agricultural labourers as a distinct class, an organized grouping has come into existence. The work of organizing this class started in the State in 1954, and by 1957 the Punjab Dihati Mazdoor (Rural Labour) Sabha had established working units in all the 12 districts of the present Punjab. In 1965, the Sabha was split into two, one controlled by the CPI and the other by the CPI(M). The CPI-controlled body, known (since 1971) as Punjab Khet Mazdoor (Agricultural Labour) Sabha, is affiliated to the Bharatiya Khet Mazdoor Union. The CPI(M)-controlled body is known as the Punjab Dihati Mazdoor Sabha and is affiliated to the All India Agricultural Workers Union (AIAWU). Both the Sabhas have their working units in all the 12 districts of the State with membership spread over a large number of villages.

These organizations have been raising various issues and demands of agricultural labourers, pertaining to their general welfare as well as taking up specific issues such as the wage rate and social oppression by the dominant peasantry in the villages. They have demanded intervention of the State machinery in order to protect the interests of their class through the implementation of various existing legislative measures and introducing new protective measures. At the village level, struggles have been organized against the economic and social exploitation of this class by the upper sections of the peasantry, particularly the rich peasantry (Gill, 1990).

In recent years, the issue of wage rates has assumed greater significance, leading to a contradiction between agricultural labour, on the one hand, and the peasantry as a whole, on the other, led by capitalist farmers. This is the reason why agricultural labourers who mostly belong to the lower castes do not support the political party (viz. Akali Dal) or organization dominated by the peasantry. In the light of this reality, the growing influence of the Bahujan Samaj Party (BSP) in the State can be understood.

The organized strength, mobilizing capacity, and influence on the State machinery is vastly greater in the case of peasantry led by capitalist farmers than in the case of agricultural labourers who are not mobilized and organized on a similar basis. But the existence of organized strength among both the sections has serious implications for political processes in the State.

The specificities of the political situation in Punjab

With the emergence of capitalism in agriculture, the influence of the agrarian capitalist class on the political parties, and on the elected offices in the State has grown. The emergence of non-party farmers' organizations has further increased the assertiveness of the peasantry. The growing pressure of socio-economic exploitation and the subsequent emerging awareness of the exploited classes has led to the formation of unions of agricultural labourers in the State. This fits in with the general pattern of changes in the political processes.

Punjab has its own specific characteristics which make this process distinct from the set pattern observed in other parts of the country in which Green Revolution has brought changes in agriculture along capitalist lines. The Punjab situation is marked by two special features. Most of the peasantry is Sikh and belongs to the Jat caste. At the same time, the non-agricultural occupations in urban centres are dominated by Hindus. Most of the agricultural labourers belong to the SCs and are non-Sikhs. The contradiction therefore between the peasantry and the traders (urban) on the one hand, and on the other, that between the peasantry and agricultural labour has communal implications.

The emergence of the issue of Sikh identity with the development of agrarian capitalism can be understood in this context. This has serious implications for the competition for (and the consolidation of) the support of this class by various political parties. The Akali Dal has used this as a stick with which to beat the secular Communist parties. Competition by the Congress(I) on this count has promoted a militant separatist movement which is threatening to make party-based political structures irrelevant. This factor has affected the whole process of mobilization of the peasantry through religious appeals from *gurdwaras* and appeals to Sikh symbols. The Punjab State unit of the BKU, unlike its counterparts elsewhere in the country, has also sometimes used Sikh *gurdwaras* at the local level and the recitation of the Sikh holy book (i.e. Guru Granth Sahib) to organize *dharnas*. This has given a distinct shape to political processes in the State, in a way that has not been observed elsewhere.

In the contemporary context, the issue of Sikh identity merits attention here. It is a question that arose with some force during the last two decades of the nineteenth century and the first two decades of this century, of whether or not Sikhs were a sub-set of Hindus. It was energetically debated between the Arya Samaj and the Singh Sabha Chief Khalsa Diwan, principally in the form of a war of pamphlets. This increased the awareness among the Sikhs of their separate identity. A fear psychosis gripped the Sikh élite, arising from a perceived threat of its reabsorption into the Hindu fold. This they sought to counter by asserting their separate identity.

Identity consciousness in the community received a big boost with the introduction of separate electorates for different religious communities (including the Sikhs) by the colonial authorities. This brought in its train communal competition among the English-educated members of different communities for political patronage and posts in the provincial legislative bodies (Kapoor, 1987). With the emergence of the Akali Dal and the Shiromani Gurudwara Prabandhak Committee (SGPC), the articulation of a separate Sikh identity as distinct from a Punjabi identity became a tool for the acquisition of political power. In the event, the colonial power treated the Sikhs, prior to Independence, as a separate negotiating party.

The sequence of events since Independence further consolidated the separate Sikh identity. The Arya Samaj, the Jan Sangh, and a section of the Congress Party joined together and manipulated the language issue by persuading Punjabi Hindus to repudiate their mother tongue in favour of Hindi. The idea of establishing a Punjabi-speaking State was vigorously opposed. After the reorganization of

States, the Akali leadership responded to the political onslaught of the Congress and the Communists by motivating its support base among the Sikhs to revive the issue of a Punjabi-speaking State. This issue was given emphasis in the Anandpur Sahib Resolution (ASR), first formulated in 1973 and finally adopted in 1978.

Competitive, opportunistic electoral politics was the game that the Congress chose to play with disastrous effect on the State. The Akali Morcha, accompanied by killings initially of Nirankaris and subsequently of innocent Hindus and Sikhs opposed to the process, resulted in Operation Blue Star in June 1984 and in the massacre of the Sikhs in Delhi and other places outside Punjab in November 1984. The signing of the Accord between Prime Minister Rajiv Gandhi on behalf of the Government of India and Sant Harchand Singh Longowal on behalf of Akali Dal (25 August 1985), and the subsequent failure of the Government of India to honour its commitments made under it, have contributed a great deal to the strengthening of the Sikh identity.

The current phase of long-drawn continuous struggle, spread over a period of over a decade during which the Sikh community has faced tremendous outside pressure and major psychological shocks, has provided an adequate ideological basis for the dissolution of numerous cleavages within the community as a whole and to merge a plurality of political and socio-economic identities into a single ethnic identity.[4]

In spite of a long historical background and the objective basis created by strong external pressure on the community in the past decade or so, several problems relating to the transformation of religious identity into a single political and socio-economic identity still remain. First, the working of the capitalist economic system is giving rise to contradictions within various classes among the Sikhs. The most important of these is the contradiction between cultivators (Jat Sikhs) and agricultural labourers (Harijan Sikhs), and the contradiction between traders and bureaucracy, on the one hand, and cultivators, on the other. Similarly, a large number of Sikhs have been settled as cultivators in Haryana, Rajasthan (Ganganagar), Uttar Pradesh (in the Terai belt), and other States, and also as a business community in major metropolitan centres and smaller cities. The economic interests of these Sikhs neither coincide with those of the Sikhs in Punjab, nor are they attracted to Punjab. On some of the issues, such as sharing the river water, the economic interests of the two clash.

[4] Bombwal (1987) believes that the emergence of a separate single Sikh identity is by now a *fait accompli*.

Second, the political process both within and outside the State provides alternative channels for promoting the interest of differentiated groups/classes among the community. At present, the Akali Dal, Congress(I), Communists, Bahujan Samaj Party (BSP), and Janata Dal (JD) contend with each other in order to build a support base for themselves amongst the Sikhs. In spite of exaggerated claims, the Akali Dal has not become the embodiment of the aspirations of all sections/classes of Sikhs. After the trauma of 1984, when the election to the State Legislative Assembly was held in 1985, even under most favourable political conditions the Akali Dal could secure only 38.5 per cent of the total votes compared to 37.8 per cent secured by the Congress(I).

It is well known that the Sikhs constitute 60.75 per cent of the population of the State and an equal proportion of the electorate of Punjab. Even if we assume that only Sikhs voted for the Akali Dal, 35.8 per cent of Sikhs did not. It is well known that in the 1985 elections, some non-Sikhs (Hindus and Muslims) won on Akali Dal tickets and consequently a section of voters of these communities voted for the Akali Dal. This means that the proportion of Sikhs who did not vote for the Akali Dal in 1985 was higher than 35.8 per cent.

With the passing of the first deadline (26 January 1986) to implement some of the clauses of the 1985 Punjab Accord, the influence of the Akali Dal began to decline and the role of various militant groups expanded at its expense. At present, militant groups and their front organizations are divided into three major segments around three Panthic Committees. As a consequence, there is no single militant group which can claim to have a dominant support base and thus become the embodiment of the aspirations of Sikh masses. This has further eroded considerably the Akali Dal's claim to be the sole representive party of the Sikhs.

Third, on a theoretical level there are several problems of articulation of the Sikh identity in a particular region in the face of the universalism of Sikh religion. The interpretation of various religious praxis and theoretical issues as to what constitutes Sikhism and who is a Sikh have emerged as important questions which call for and need clear answers (McLeod, 1989) in view of the violence perpetrated by the various militant groups in the State. Currently, the debate on these issues is suppressed due to the fear psychosis plaguing intolerant militant groups. Confusion on these issues is in itself an indication of a deepening crisis of identity.

Violence on the part of militants and counter-violence on the part of the State has caused much bloodshed (to an increasing extent since the arbitrary dismissal of the Barnala government in 1987), and until the assumption of power by the Beant Singh government in 1992, completely disrupting the social and political life of Punjab. Indeed, the democratic process was completely paralysed and the State was under central rule for five years (1987–92). The process of violence was unleashed by the militants in 1981, and has since grown in scale over the last decade. Now, it is directed against both Hindus and Sikhs in equal measure.[5]

Fear of violence has overawed politicians, press, intellectuals, and the people at large. Public debate on sensitive issues such as religion, culture, language, and the future of Punjabi society has been virtually paralysed. A brief comment about the socio-economic background of the militants is therefore appropriate.

This is based on information relating to 80 militants from the districts of Amritsar, Kapurthala, and Ropar (Dang, 1988; Singh, 1990). 70.80 per cent of the militants belong to the families of peasants. Middle peasant families account for the largest proportion of militants, followed by an equal proportion from amongst rich and poor peasants. This runs counter to the fact that, out of all the cultivators in the State, 38.62 per cent, 27.9 per cent and 33.40 per cent respectively were from the poor (small and marginal) middle and rich peasantry. The remaining 29.2 per cent of the militants were drawn from the urban middle class and other sections of the State.

Agricultural labourers (Mazhabi Sikhs) are, by and large, bypassed by this movement. Rather, there is evidence of opposition on the part of this class to attempts at its suppression by other segments of the peasantry which are backed by militants. We found, too, that 85 per cent of the militants were less than 30 years of age. Further, highly educated (graduates and post-graduates) comprised 31.2 per cent and the less well-educated (below matriculation) consisted of 65 per cent of the total number of militants. Evidence suggests that the educated youth coming from the rich peasantry and urban middle classes provide leadership for the militant movement in the State. The background of the militants reveals that the recruiting base and backbone of the movement is provided by the Jat Sikh peasantry.

The second specificity of the State stems from the fact that capitalist relations in agriculture are more advanced in Punjab than in other parts of India. The contradictions of this path are therefore more clearly evident here, as are the inflated aspirations of the peasantry.

[5] After a lapse of nearly five years during which the Punjab State was under President's rule, the Centre held elections to the Punjab State Legislative Assembly and to the Lok Sabha (for the Punjab constituencies) on 19 February 1992. They were boycotted by the Akali Dal parties. With a turnout of 28 per cent of the electorate, a thoroughly unrepresentative State Legislative Assembly (reminiscent of the 1983 election to the Assam State Legislative Assembly) was ushered into existence. All but one of the parliamentary constituencies returned Congress(I) candidates to the Lok Sabha.

This is why the peasantry is much more concerned with the issue of its share in the development process of the economy and hence about the policies of the Centre, greatly emphasizing the issue of Centre–State relations.

At the same time, the peasantry—and particularly its leadership—is demanding an active partnership in trade, business, and industry. Here, its behaviour follows the pattern of the sugar lobby in Maharashtra and, to some extent, the aspirations of coastal farmers in Andhra Pradesh. It is interesting to note that the arrival of the multinational Pepsi company in agro-processing in the State resulted from a campaign launched by the State unit of the BKU which was subsequently espoused by various factions of the Akali Dal.

REFERENCES

ANON,* 1987: 'Punjab: The Real Resistance—I', *In Defence of Democratic Rights* 1 (May–June): 6.

*an unsigned report; (*see also* Anon., 'Other Side of [the] Punjab Crisis: Agricultural Workers' Struggle', *Economic and Political Weekly* 22 (15 Aug. 1987): 33, pp. 1386–7.

AULAKH, H.S., 1983: *Changing Foodgrains Market Structure in India* (New Delhi: B. R. Publishing).

BARDHAN, K., 1977: 'Rural Employment Wages and Labour Market—Survey of Research II', *Economic and Political Weekly* 12 (2 July): 27, pp. 1062–74.

BHALLA, G.S. and G.K. CHADHA, 1982: 'Green Revolution and Small Peasant: A Study of Income Distribution', *Economic and Political Weekly* 17 (22 May): 21, pp. 870–7.

BOMBWAL, K.R., 1985: 'Ethno-Nationalism', *in* SAMIUDDIN (Ed.), 1985, pp. 3–11.

———, 1987: 'Sikh Identity and Federal Polity—A Critique of Akali Position' *in* SINGH (Ed.), pp. 156–66.

BRAR, J.S. 1988: 'Emerging Pattern of CPI Leadership in the Punjab', *in* WALLACE and CHOPRA (Eds.), 1988, pp. 356-76.

DANG, S., 1988: *Genesis of Terrorism—An Analytical Study of Punjab Terrorists* (New Delhi: Patriot).

DHAMI, M.S., 1988: 'Caste, Class and Politics in the Rural Punjab: A Study of Two Villages in Sangrur Districs', *in* WALLACE and CHOPRA (Eds.), 1988, pp. 218–48.

———, 1988a: 'Religious-Political Mobilisation and Shifts in the Party Support Base in the 1985 Punjab Assembly Elections', *in* WALLACE and CHOPRA (Eds.), 1988, pp. 377–98.

DHILLON, B.S., 1986. An Economic Analysis of the Small Scale Bicycle Industry–A Case Study of Ludhiana, unpublished M. Sc. (Hons) Dissertation, Guru Nanak Dev University, Amritsar.

D'Souza, V. S., 1982: 'Economy, Caste, Religion and Population Distribution: An Analysis of Communal Tension in Punjab', *Economic and Political Weekly* 17 (8 May): 19, pp. 783–9.

George, K.K. and I.S. Gulati, 1985: 'Centre–State Resource Transfers 1951–84—An Appraisal', *Economic and Political Weekly* 20 (16 Feb.): 7, pp. 287–95.

Gill, S. S., 1980: 'Land Reforms and Peasant Movements in the Punjab', *Mainstream*, 18 (14 June): 42, pp. 23–8.

——————, 1980a: 'Impact of Economic Development on Rural Artisans in the Punjab', *Indian Journal of Labour Economics* 23 (Oct.): 3, pp. 154–65.

——————, 1990: 'Organisation of Rural Labour in Northern India with Special Reference to the Punjab', (Department of Economics, Punjabi University, Patiala) (mimeo.).

——————, 1990a: 'Migratory Labour in Rural Punjab' (Department of Economics, Punjabi University, Patiala) (mimeo.).

Gill, S..S. and K. C. Singhal, 1984: 'Farmers' Agitation—Response to Development Crisis of Agriculture', *Economic and Political Weekly* 19 (6 Oct.): 40, pp. 1728–32.

Government of India, 1972: *National Sample Survey 26th Round* (1971–2) (New Delhi: Government of India).

——————, 1981: *Provisional Population Tables, Census of India, Series 17, Punjab 1981* (New Delhi: Census Commissioner, Government of India).

——————, 1982: *National Sample Survey 37th Round* (1982) (New Delhi: Government of India).

——————, 1985: *Economic Survey, 1983–1984* (New Delhi: Government of India).

Government of Punjab, 1981: *Agricultural Census of Punjab 1980–81* (Chandigarh: Economic Adviser to the Government of Punjab).

——————, 1984: *Statistical Abstract of Punjab 1983* (Chandigarh: Economic Adviser to the Government of Punjab).

——————, 1986: *Statistical Abstract of Punjab 1985* (Chandigarh: Economic Adviser to the Government of Punjab).

——————, 1989: *Statistical Abstract of Punjab 1989* (Chandigarh: Economic Adviser to the Government of Punjab).

Hasan, Z., 1989: 'Self-serving guardians: formation and strategy of Bhartiya Kisan Union', *Economic and Political Weekly* 24 (2 Dec.): 48, pp. 2663–70.

Kapoor, R., 1987: *Sikh Separatism: The Politics of Faith* (New Delhi: Vikas).

Kaur, R., 1982: Agricultural Price Policy in Developing Countries with Special Reference to India, (unpublished Ph.D. thesis, Punjabi University, Patiala).

Kumar, V., 1981: Impact of Technological Change on Farm Employment—A Case Study of the Punjab, (unpublished M.Phil. dissertation, Punjabi University, Patiala).

McLeod, M.H., 1989: *Who is a Sikh?—The Problem of Sikh Identity* (Oxford: Clarendon Press).

MEHRA, R.K., 1985: Some Aspects of Small Scale Chemical Industry in Amritsar, unpublished M.Sc.(Hons) Dissertation, Guru Nanak Dev University, Amritsar.

OBERAI, A.S. and H.K.M. SINGH, 1980: 'Migration Flows in Punjab's Green Revolution Belt', *Economic and Political Weekly* 25 (29 March): 13, pp. A2–12.

PANDIT, M. L., 1978: 'Some Less Known Factors Behind Recent Industrial Change in Punjab and Haryana', *Economic and Political Weekly* 13 (25 Nov.): 47, pp. 1935–40.

RANDHAWA, M. S., 1974: *Green Revolution: A Case Study of the Punjab* (New Delhi: Vikas).

Revolutionary Communist Party of India,* (13 March) 1984: *Punjab Da Kisani Masla* (in Punjabi) (Jalandhar: RCPI).

*RCPI

SAHNI, S., 1980: Entrepreneurial Structure and Performance (A Case Study of Woollen Goods Industry in Amritsar), (unpublished M.Sc. dissertation, Guru Nanak Dev University, Amritsar).

SAMIUDDIN, A. (Ed.), 1985: *Punjab Crisis: Challenge and Response* (New Delhi: Mittal).

SINGH, A., 1979: 'Farm Workers *vs*. Rich Farmers: Class and Caste in a Punjab Village', *Economic and Political Weekly* 14 (27 Oct.): 42–3, pp. 1753–4.

SINGH, G., 1986: *Religion and Politics in Punjab* (New Delhi: Deep & Deep).

———, (Ed.), 1987: *Punjab Today* (New Delhi: Intellectual Publishing House).

SINGH, Master H., 1980: *Agricultural Workers' Struggle in the Punjab* (New Delhi: People's Publishing House).

SINGH, M., 1990: On the Character of Militancy in Punjab (Patiala), unpublished note.

SWAMY, D.S. and A. GULATI, 1986: 'Farm Prosperity to Retrogression : Indian Cultivators during the 1970s', *Economic and Political Weekly* 21 (21–8 June): 25–6, pp. A57–64.

TALIB, B.D., 1979: 'Agrarian Tensions and Peasant Movements in Punjab' (Centre for Rural Development, Jawarharlal Nehru University, New Delhi) (mimeo.).

UPADHYAYA, C.B., 1988: 'The Farm Capitalists of Coastal Andhra Pradesh', *Economic and Political Weekly* 23 (9 July): 27, pp. 1376–82.

VERMA, P.S., 1987: 'The Akali Dal: History, Electoral Performance and Leadership Profile', *in* G. Singh (Ed.), pp. 257–85.

WALLACE, P. and S. CHOPRA (Eds.), 1988: *Political Dynamics and Crisis in the Punjab* (Amritsar: Guru Nanak Dev University).

Aspects of the Farmers' Movement in Uttar Pradesh in the Context of Uneven Capitalist Development in Indian Agriculture

UTSA PATNAIK AND ZOYA HASAN*

Introduction

From the second half of the '70s, farmers' movements emerged in a number of States in India (Tamil Nadu, Karnataka, and Maharastra) which were noticeably different in character in comparison to earlier peasant movements in the period immediately prior to and after Independence. The earlier movements had a mass character and had in the main raised anti-landlord demands, while also confronting the colonial state; most saw themselves as an integral part of the freedom struggle (Desai, 1979; Dhanagare, 1983). The post-Independence

* In the presentation of jointly authored papers, the convention followed here is not to name the authors in alphabetical sequence but according to the author to whom the editor in the first instance extended an invitation to contribute.

farmers' movements of the '70s onwards, on the other hand, have been led invariably by well-to-do surplus producers and have focused mainly on the issue of prices, i.e. demanding higher prices of final output, on the one hand, and lower prices of production inputs, on the other (Nadkarni, 1987). These movements represent in a very clear form the articulation of the demands of the emerging classes of rural capitalist producers which have been in the process of formation during the past three decades. At the same time, it has been widely observed that these movements have drawn into their ambit also the middle peasants and, in some cases, small peasants, although their appeal is limited (for obvious reasons) as far as the really poor peasants and rural labour are concerned.

While the farmers' movements in southern and peninsular India were on the upsurge in the '70s, they have become noticeably quiescent after the mid-'80s in particular; whereas, by contrast, the farmers' movements in northern India have been fairly active during the last decade. We should make it clear, at the outset, that by 'farmers' movements' we refer to organizations, whether unitary or of a more loosely structured type, which are concerned specifically with articulating and agitating for the demands of the rural producers. Of course, the landed interest has been represented traditionally within organized political parties as well, such as the Congress, and might even represent the dominating interest in some political parties; but the basic difference between the latter and farmers' movements is that organized political parties in principle have something programmatic to say on most aspects of the economy and society, whereas farmers' movements have arisen from and consciously articulate the interests of specific classes, narrowly confined to the rural sector.

Our concern here will be to explore the overall politico-economic trends, within which the rise of this farmers' movement in northern India may be located, and to a certain extent, made explicable. We will focus in particular on Uttar Pradesh (U.P.) and the emergence of the Bharatiya Kisan Union (BKU) in that State.

This chapter is divided into two parts. In the first, we look at northern India and within it at U.P. in relation to the India-wide trends of agricultural production during the last three decades. This section argues that the growth of capitalist production in Indian agriculture, involving investment embodying technical change and growth in productivity has been concentrated in northern India to a far greater degree than is usually appreciated. The growing dependence of the urban food economy on this region has increased considerably the importance and bargaining strength of the rural capitalist producers of northern India. In the second part of the chapter, we look at the spatial concentration of capitalist investment in the western part of Uttar Pradesh and relate this to the phenomenon of the emergence of the BKU.

REGIONAL CONCENTRATION, THE FOOD ECONOMY AND THE POSITION OF UTTAR PRADESH

Capitalist production and the new technology

In a seminal study of regional trends in agricultural production during the period 1897 to 1947, Blyn (1966) found that foodgrains production per head of population actually *declined* in every region of British India after the Great War, the largest decline (38 per cent) taking place in eastern India, where absolute total output levels had also declined, the smallest decline taking place in 'greater Punjab' where total foodgrains output was growing but not fast enough to keep pace with population growth. By contrast, non-foodgrain crops production expanded at a respectable rate, some ten times the rate at which foodgrains was growing: not very surprising in a colonized economy catering to the industrial raw materials needs of metropolitan countries. On the eve of Independence, the typical Indian faced a 20 per cent lower availability of basic foodgrains than was the case at the turn of the century: in one part of the country hyperinflation, induced by reckless War financing, had led to at least three million deaths in the great Bengal famine (Sen, 1981), the memory of which was still fresh at Independence.

It is not surprising that the overriding concern of policy-makers in independent India was to achieve rapid foodgrains output growth and maintain stable prices. The necessity for land reforms was seen not so much in terms of distributive justice, as in terms of freeing production from the incubus of a class of parasitic *rentiers* politically nurtured by British rule. The initial rhetoric of egalitarianism and the slogan 'land to the tiller' received short shrift as, in State after State, the legislatures, substantially influenced by the landed interest, passed laws permitting intermediaries to retain large areas as home farms (*khud-kasht*, *sir*) or permitting landlords to evict tenants provided they claimed 'personal cultivation' of hitherto tenant-cultivated land.

'Personal cultivation' as a concept was divorced from the exclusive idea of working directly in the fields, and defined to include cultivation through the hiring of labour (paid in cash or kind, but not in share of crop). The thrust of the reforms was not radical land redistribution or effective 'levelling' of inequalities at all, but the attempt to push and cajole *rentiers* to turn themselves into capitalists, while giving an optional right of purchase of land vesting in the state to tenants—a provision of which the well-to-do minority of tenants in every area availed themselves with alacrity. The majority of poorer tenants were evicted and either joined the ranks of landless labour or became insecure tenants on unrecorded, oral leases.[1]

The characterization of the social and economic impact of land reform measures has been put forward elsewhere by one of the authors as the promotion of 'landlord capitalism' to which emerging peasant capitalism is a subordinate trend (Patnaik, 1986). This characterization seeks to emphasize two related and important features of the evolving agrarian structure. Firstly, the socially conservative character of the reforms, which have preserved the dominance of the traditional upper caste social groups controlling landed property provided they adapted minimally to changing times by taking an interest in production. Those who could not so adapt themselves found over time that their economic and social importance tended to decline and that they were supplanted by upwardly mobile sections of the rich peasantry. This latter process, the emergence of a peasant–capitalist element, is the second important feature, which has a bearing on the spatial concentration of investment to which we turn below: for, on the whole, there would appear to be a positive association between the importance of the peasant–capitalist element relative to the landlord–capitalist one, and the extent to which a socially broad-based investment trend emerges.

It is arguable that land reform measures by themselves simply create a suitable institutional framework for the growth of capitalist production, which in the last resort requires the stimulus of profitability. In the event, the implementation of a strategy of planned investment and other expenditures by the Indian state, on a large scale from the mid-'50s onwards, created an expanding domestic market for necessities, particularly foodgrains, and raised the profitability of foodgrains production substantially. The ratio of prices of foodgrains relative to manufactured goods (for consumption and productive use) rose by nearly 50 per cent from the late '50s to the mid-'70s (Mitra, 1977). We need not enter in detail into the controversy over the statistical estimation of terms of trade (Kahlon and Tyagi, 1983; Thamarajakshi, 1990) which tends to become mired in the minutiae of choice of base-year, weights, and the like. Suffice it to note that, despite the debates, the results of all estimates are rather similar. There is an improvement for agriculture in the decade 1963–74 and a deterioration thereafter, and the improvement is

[1] For the evaluation of specific land reform and tenancy laws, carried out under the auspices of the Programme Evaluation Organization of the Planning Commission in different States, *see* Basu and Bhattacharya, 1963; Singh and Misra, 1964; Parthasarathy and Rao, 1969; Khusro, 1958; Dool Singh, 1964; Nanekar, 1968.

substantially larger than the subsequent deterioration, giving an overall rising trend in the terms of trade for agriculture. Similar conclusions are reached by Rao (1983) and Nadkarni (1987) on the basis of use of the implicit price deflators for agriculture and non-agriculture calculated from national income series available in current and constant prices at the sectoral level. Nadkarni (1987: 169–170) finds a sharp upward deviation from the trend in the ratio of agricultural to non-agricultural prices during the 1964–75 period, with a deterioration which leads to a reversion of the trend thereafter.

Rising final output prices, especially of foodgrains, compared to input and manufactured consumption goods prices, have diametrically opposite effects on the rural well-to-do compared to the rural poor. At one pole of the rural class–caste structure, the emerging capitalist producers employing hired labour and contributing the bulk of all sales, not only gain by virtue of selling output to the largest extent, but also gain by virtue of being employers, since the real wage of labour falls with rapid food price rise. This 'profit-inflation' undoubtedly provided a strong stimulus for the switch from rent-appropriation to direct capitalist investment embodying new 'Green Revolution' techniques for landlords, and similarly stimulated the expansion of the activities of the rich peasantry. During the same period that such profit-inflation was taking place, benefiting the emerging rural capitalists, the *Rural Labour Enquiries* of 1963–5 and 1974–5 showed a drastic fall in the real earnings of rural labour and a rise in indebtedness: annual real earnings of male workers declined by about 33 per cent and of females by nearly 50 per cent (Patnaik, 1983).

The dissemination of manufactured modern inputs at subsidized rates as part of the 'new agricultural strategy' from the early '60s onwards came, thus, at a time of rising profitability of foodgrains production in particular, and also demand–pull inflation in certain commodities such as sugar, which increased the profitability of sugarcane cultivation. The new technology 'package' of HYV seed –fertiliser–pesticides, applicable only under conditions of assured irrigation, was expensive. It required considerable additional investment over and above the normal level of investment (required at the pre-existing level of technique to keep reproduction going). It is hardly surprising that only the top minority of well-to-do cultivators of landlord and rich peasant origins were in a position to switch their investible surpluses, hitherto put to other uses, into direct agricultural investment embodying these new techniques.

We have elsewhere estimated on the basis of some field data that, as far as the landlords were concerned, a profit rate on direct investment of at least 50 per cent on capital was required to make it pay for them to switch from leasing out land to direct capitalist cultivation. This is a very high profit rate requirement, dictated by the fact that the landlords obtained a rent earlier without tying up their money in the land, and had that money free to earn profit elsewhere (trade, moneylending, etc.) in addition to rent. The rich peasantry, on the other hand, habituated to work themselves on the land, faced no such stringent profitability requirement. In those regions where there existed a substantial stratum of well-to-do peasants in addition to the landlords (as in the case of Punjab, subsequently partitioned into Punjab and Haryana), a strong peasant–capitalist trend could emerge. The social–class basis of investment could be much wider than elsewhere. Conditions similar to Haryana can be identified to some extent in the contiguous western districts of Uttar Pradesh, where the proportion of farmers cultivating above four hectares is larger than in any other region of this enormous State (except the arid and low-productivity Bundelkhand rgion). Punjab and Haryana, with high levels of irrigation dating from colonial times, enjoyed a higher average farm size and a much higher proportion of total farms above four hectares in size than any other region of India in the early '60s (excepting, again, desert and arid regions such as Rajasthan).

Two trends were observable in north India: with the rising profitability of direct cultivation embodying new technology, the landlords evicted their petty tenants on a massive scale, while the well-to-do minority of peasants set about enlarging their holdings through purchase and appropriated common land wherever possible. The number of recorded tenants in Punjab dropped from 5,83,000 in 1955 to 80,000 by 1963; further cessations of 1,80,000 pure tenancies are observed on comparing the agricultural census of 1980–1 with that of 1970–1, raising the number of operated holdings above four hectares, whilst at the same time there occurred a drastic fall in the number of holdings below two hectares (Sandhu and Grewal, 1987).

In Uttar Pradesh there was a long three-year lag between the placing before the U.P. Legislature of the *U.P. Zamindari Abolition and Land Reforms* (ZALR) bill on 7 July 1949 and the enforcement of the *ZALR Act* with effect from 1 July 1952 (Singh, 1986). This gave ample time to landlords to claim larger areas as *khud-kasht* than was actually the case by putting pressure on the poorer tenants to surrender, and to engross waste and common lands. The fact that the *U.P. Land Reforms Act* had no formal provision permitting landlords to 'resume' land for 'personal cultivation', as in other States, appears not to have inhibited *de facto* evictions of the poorer tenants or Adivasis (as they were to be designated under the *ZALR*). In the event, large numbers were evicted even before they could be recorded as Adivasis. The *khud-kasht*, *sir* and grove lands claimed by the landlords automatically became their *bhumidari*, carrying full rights of mortgage and transfer, while they received compensation in cash and bonds for that part of their former estates which was taken over by the government. The tenants of this part became *sirdar*s, obliged to pay the same amount as revenue to the state as they were paying earlier to the *zamindar* as rent, and if they wished to acquire *bhumidari* right they had to pay for it at the rate of ten times the standard rent; subsequently they would continue to pay revenue but at half the former rate (Singh, 1986).

Thus, the *U.P. ZALR Act* forced the tenants to pay the capitalized value of feudal rent—capitalized at ten per cent interest rate—for ownership right, while it conferred ownership free to the *zamindar*s. Further, the cash raised from the tenants was used to finance compensation to the *zamindar*s. The U.P. land reform in many ways was far more conservative than even the Meiji Reforms in Japan after 1868 had been, for that had divorced the intermediaries completely from landed property whereas in India, generally (within which U.P. is a specific case), the former intermediaries continued their predominant control over land. Singh and Misra (1964), in their evaluation based on sample studies before and after enforcement of the *ZALR Act*, found that 52.8 per cent of all cultivators had no *bhumidari* right over any part of their farms, and over 70 per cent cited lack of cash as the reason.[2] Of the farms above 50 acres in size, all had *bhumidari* rights, while of the farms below three acres, only one-third claimed *bhumidari* over some part. Clearly, only a minority of tenants comprising the well-to-do section could afford to purchase *bhumidari*, which remained the substantial monopoly of landlords.

It is hardly surprising that the former *zamindar*s of substantial standing were in the best position to take advantage of the improving profitability from the late '50s to the mid-'70s; and large numbers used their considerable funds (augmented by compensation paid by their former tenants) to go in for more intensive cultivation on their slightly curtailed but still substantial farmlands. Investment in pumpset–tubewell irrigation and in new HYV seed–fertilizer technology, along with mechanization, was to be observed on the part of former *rentiers* turning capitalist. At the same time, the well-to-do section of the Jat tenants also improved its position through purchase of *bhumidari*, contributing the peasant element to the trend of growth of capitalist production.

The western part of U.P. comprising 19 districts has had a more dynamic growth profile than any other part of the State. This region, covering Meerut, Agra, Bareilly, and Moradabad divisions, may be

[2] It must be borne in mind that between 1952 and 1953–6 all prices were falling so that the real burden of rent payment was rising for tenants, making it much more difficult to raise ready cash.

considered to be contiguous with Haryana and Punjab in constituting that area of the country with the most wide-based growth of rural capitalist investment in the 'Green Revolution' as well as, of late, in processing and small-scale rural industries. Before discussing some specific indicators of the lead enjoyed by western U.P., let us look at the place occupied by U.P. as a whole, as part of the northern region of India, in the food economy of the country.

Regional per capita food output trends and the PDS

We saw that in the half century preceding Independence, foodgrains output per capita had declined in every region of British India, as had availability. It is naturally of interest to see what the regional trends have been since Independence. It is not possible to recapture the regionalization in Blyn (1966) as that related to British India alone. Nonetheless, a broad order of comparison is possible at a different level of aggregation, in which four large regions are defined: *East India* comprising Assam, Bihar, Orissa, and West Bengal; *South India* comprising Andhra Pradesh, Karnataka, Kerala, and Tamil Nadu; *West–central India* comprising Gujarat, Madhya Pradesh, Maharashtra, and Rajasthan; and finally, *North and North–west India* comprising Haryana, Himachal Pradesh, Jammu & Kashmir, Punjab, and Uttar Pradesh.

Table 9.1 reveals a remarkable picture of the impact on the new agricultural strategy nearly three decades after its inception with regard to foodgrains output per capita. The time-series employed is of physical output (in million tonnes) from 1960–1 to 1986–7, and three triennia have been chosen, centred on 1961–2, 1972–3, and 1985–6, respectively. The small overall all-India rise in foodgrains output per head of 7.4 per cent is the outcome of *diametrically opposite trends* in North India compared to the rest of India. In the north (plus north–west) region, there has been a large rise of nearly 65 per cent, comparing the last triennium with the first; but, in every other region, the pre-Independence trend of decline has persisted, even into the era of 'Green Revolution' being considered here. In the Eastern region, the decline is small enough at 5.7 per cent for 'stagnation' to be an appropriate description, and this small order of decline is mainly on account of the modestly rising trend in West Bengal almost offsetting the falling trends in the other three States.

In the southern region, the decline has been substantial at 12.1 per cent, and reflects the decline in every State making up the region.

Most dismal, however, has been the situation in the West–central region where the decline has been as much as 19.3 per cent, reflecting varying orders of decline in all the four component States.

Table 9.1
Foodgrains output per head per annum by regions in India

Region	*Triennial Average, in kilogrammes* 1960–1962 (1)	1972–1974 (2)	1984–1986 (3)	*Change (3) over (1) per cent*
North and North-west	204.6	234.7	337.2	+64.8
East	162.2	152.9	152.9	– 5.7
South	152.3	150.4	133.9	–12.1
West–central	198.6	158.3	160.3	–19.3
All-India	178.9	172.0	160.3	–19.3

NOTE: *North*—Punjab, Haryana, and Uttar Pradesh; *North-West*—Jammu & Kashmir, Himachal Pradesh;
East—Assam, Bihar, Orissa, and West Bengal; *South*—Andhra Pradesh, Karnataka, Kerala, and Tamil Nadu;
West-central—Gujarat, Madhya Pradesh, Maharashrta, and Rajasthan

SOURCE: Calculated from basic data in Government of India, 1987 *(passim)*. These results are discussed in greater detail in Patnaik (1994). The authors wish to record their gratitude to S. Batra for computational assistance.

North India has emerged as virtually the sole supply area of genuinely surplus food production for feeding the non-rural areas. By 'genuinely surplus' we refer to a situation where marketed supply to non-rural areas need not be at the expense of depressing the rural population's own consumption. Within this region, Punjab–Haryana has registered a phenomenal 134.4 per cent rise in per capita foodgrains output, whilst Uttar Pradesh follows more modestly with a creditable 31.6 per cent rise[3].

The share of the dynamic northern region in the country's total

[3] Himachal Pradesh and Jammu & Kashmir together have a very low weight in total food production and their exclusion would make little difference to the averages.

food production has risen from around one quarter to nearly two-fifths, but more dramatic is its contribution to the urban food economy. We have not investigated the data at present on the results of *private* trade in channelling foodgrains to urban areas; but *government* procurement of foodgrains for the public distribution system (PDS) has become increasingly dependent on this region. In the triennium ending 1988, of the total procurement of wheat, 99 per cent was from the northern region, while even in the case of rice as much as 67 per cent of total procurement was from this region (with Andhra Pradesh in the South trailing with 11 per cent and the balance being divided between the remaining rice-producing States).

The bulk of foodgrains procured by the government goes to feed the urban and semi-urban population, as over 60 per cent of the fair-price shops (selling foodgrains and some manufactured necessities such as sugar and fuel at low subsidized prices) are located in cities and towns or on their rural periphery, serving a quarter of the total population. Government procurement operations are a very important part of total sales of cereals: during the period 1981–5, for example, procurement of wheat and rice amounted to an annual average of 16.73 million tons which made up 16.5 per cent of average annual production of these crops. Assuming that about 40 per cent of food crop output is sold for all purposes, including transactions within rural areas, government procurement from farmers amounted to as high as 43.75 per cent of all sales by them. It would be a much larger share of all sales to urban consumers alone.

The public distribution system has played a politically crucial role in providing a buffer against inflation to the volatile and vocal urban population. In effect, the surplus production from the dynamic, growing region of North India has been used by government, not in making good declining availability for rural consumers in the areas of maximum decline, but in purchasing urban peace. The targeting of the benefits of the PDS may not be optimal in that it is not only the urban poor but also the urban middle classes and well-to-do who have access to it: but some automatic screening out of the really well-to-do takes place owing to the quality factor. With all its drawbacks, the operation of the PDS has been largely responsible for preventing the immiserization of the urban poor under a regime of inflation. If the system ceases to operate or is curtailed, urban discontent and unrest on a mass scale may be confidently anticipated, much as occurred during the period of rapid inflation in the early '70s which had given rise to the price-rise resistance movement culminating in the imposition of internal political Emergency.

The overwhelming bias of the PDS towards urban areas is not true of Kerala, which has long been a foodgrains deficit State; in the '80s, the governments of the three southern States as well as, more recently, of Gujarat, have partly extended the scope of the fair-price shops to outside urban areas. But, for the rural poor in these areas of declining availability per head, food security is far from assured since physical supply is only one part of the picture; for effective offtake purchasing power is required, which appears to be stagnant or declining in real terms for the poorest rural classes (Dandekar, 1986; Patnaik, 1994).

Under the new 'economic liberalization' policies followed by the government sporadically from 1985 and with great vigour from mid-1991, the rate of inflation has accelerated, particularly since mid-1991 after the devaluation of the rupee. The potential importance of the PDS has grown commensurately. Part of the conditionality required to be fulfilled by a country taking a large loan from the IMF, as India has done in 1991, is that subsidies be cut. The cutting of food subsidy would constitute a suicidal policy for any government in power; the compromise has been a cut in fertilizer subsidy in August 1991 which has raised the price of this crucial input by 30 per cent at a stroke. The adverse effect on production has been sought to be offset by a rise in procurement prices for foodgrains, thus adding to the inflationary trend.

The importance of the surplus-producing minority of capitalist farmers in the northern region of the country has thus grown greatly owing to a combination of diverse macroeconomic trends. On the one hand, owing to successful investment in Green Revolution technology, this region has emerged as the exclusive supply area for wheat, and the largest supply area for rice going to the PDS. The political importance of the PDS in minimizing urban discontent under an inflationary regime, on the other hand, has itself grown as the inflationary trend has been fuelled further by new economic liberalization measures politically entailed by heavy external borrowing. Against this background, the increasing objective importance for the government of the behaviour of the surplus-producing farmers of this region should be evident. These farmers' bargaining position *vis-à-vis* the state is a great deal stronger than is that of farmers in other parts of the country; the quiescence of farmers' movements elsewhere, compared to their activity in northern India, is probably largely to be traced to these differing conditions.

Sub-regional concentration: western Uttar Pradesh

The rise of the farmers' agitation in northern India in the fastest growing areas of this region reflects its close nexus with the extent of growth of capitalist production. Within Uttar Pradesh, which is the largest State in the country, the western sub-region has shown markedly more rapid growth and is the base for the emergence of the BKU.

The western region of U.P. comprises 19 districts in all:[4] all the 17 districts of the Agra, Meerut, Moradabad, and Bareilly divisions plus two districts of the Allahabad division. By the early '80s, this region was substantially ahead of all others owing to the more rapid process of capitalist investment. Although accounting for less than three-tenths of all holdings in U.P. and 27.3 per cent of the total rural area, the western region had as high as 74 per cent of all Persian wheels for irrigation, 48.4 per cent of all pump-sets as well as 43.7 per cent of all private tubewells. As might be expected, it had the highest proportion of gross sown area under irrigation in the State at 76 per cent, compared to the next highest figure of 45 per cent in the central region and a State average of 49 per cent. However, the cropping intensity at 150 was rather lower than might be expected. This region's share of total livestock was also rather low, but this is to be explained by the substitution by machinery, of bullock labour for tillage, irrigation, and traction to a much greater extent. This is evident from the fact that as high as 54.5 per cent of all tractors, 48.3 per cent of all improved threshing and chopping machines, and 50.1 per cent of all oil-engines and electric pumps were located in this region, although it had only 27.3 per cent of the total area and 28.8 per cent of all farms (Government of Uttar Pradesh, 1984). On average, the number of these modern machines per unit of area was about double the level in the eastern region, the other very large region in the State with a slightly larger total area and population than the western region.

In sugarcane production too the western region maintained the lead even though important concentrations of cane production exist in the more populous eastern region: of the total tonnage of sugar-cane crushed, as high as 57 per cent was in the western region while it accounted for 57.5 per cent of sugar produced—owing to a marginally better juice recovery rate. The region lagged surprisingly in the number of electrified villages, however, with only 30 per cent of villages having access to electricity, close to the State average. Evidently, machinery is to a large extent powered by other fuel sources. The western region had brickmaking as an important activity, accounting for 47.7 per cent of all bricks produced in the State.

An important indicator of the capitalist character of the investment taking place is the concentration of not only physical productive asset formation, but eventually increasing class-concentration of land itself.

[4] According to official classification.

The landlords and rich peasants, increasingly turning into rural capitalists, first respond to good profitability conditions by investing more intensively on the land they already have; when this reaches a plateau, they look for more land to augment their farms. This is particularly the case with the rich peasant Jat cultivators who started with a much smaller land base than the ex-*zamindar*s. We would expect a more rapid process of land transfers from small farms to the well-to-do to be taking place in the faster-growing areas. This is indeed what was revealed by a recent study of land transfers spanning the period between 1952–3 and 1982–3 (Kripa Shankar, 1990), based on a 0.2 per cent sample of *nyaya panchayat*s in the State.

A significant finding of the study was that on average it was the smallest landholders with less than 2.5 acres who had sold land to the greater relative extent: 7.3 per cent of the land owned by them compared to around 3 per cent or less by other size-groups of farmers. In western U.P., the extent of land transfer was higher than the State average, and the holders with less than 2.5 acres had transferred 10 per cent of their land. Most of the transfers had taken place during the last five years (1978–9 to 1982–3) of the three-decade period covered. The households which became totally landless after transferring land accounted for nearly 42 per cent of the total land transferred (Kripa Shankar, 1990).

On the other hand, of the total land that was purchased, some 60 per cent went to owners of more than 5 acres, of which 28 per cent went to those with over 10 acres. There was thus a net loss of land for those owning below 2.5 acres and a net gain for all others with the largest gains recorded by owners of 10 acres and more. The study found that more land transfers had taken place in the post-Green Revolution period than earlier; the trends of net transfer from poorer cultivators to the well-to-do were confirmed by the income and asset-wise analysis, which showed that 56 per cent of total area sold came from farmers with annual income below Rs 10,000 while, conversely, 61 per cent of all land purchased was by farmers with over Rs 10,000 income. As usual, the Green Revolution area, the western region of U.P., showed a more marked picture of concentration, with 75 per cent of area purchased going to the well-to-do.

Regarding the caste and community status of the cultivators, the net losers of land were found to belong to the category of high-caste Hindus, on the one hand, and Sikhs, on the other; while the net gainers were mainly from the 'backward castes', accounting for 84 per cent of the net area transferred, with farmers from the Muslim and Sikh communities making up the remainder. This confirms an observation made by many on the rapid rise of the 'middle castes' in the rural areas to well-to-do peasant status.

Why do middle and small peasants join the farmers' movements led by the well-to-do?

When we look at the nature of the demands articulated by the farmers' movements, whether in southern and peninsular India or in northern India, we find that they represent the interests of the new rural capitalist producers clearly and consistently. The demands relate to raising final output prices to 'remunerative' levels, reducing or at least checking the rise in prices of modern inputs such as fertilizers, pesticides, and electricity, and against cesses and levies on irrigation provided through state investment. Nevertheless, even though it is the minority of capitalist producers (monopolizing the use of modern techniques and total market sales) who are the obvious potential beneficiaries, medium- and even small-scale producers join the movements enthusiastically. In social terms, while it is the dominant upper castes (who traditionally control land property), plus some members of the upwardly mobile sections of the backward castes who provide the leadership to these movements, the participants include not only the economically worse-off members of the same castes, but also cultivators holding relatively little land belonging to the 'backward' and 'other backward' castes as well.

Caste–class solidarity can thus only partly account for the surprisingly broad support enjoyed by these movements. The major explanation is to be sought in the economic imperatives facing these cultivators under a regime of deteriorating terms of trade for their output, such as has prevailed since the mid-'70s. The economic imperatives can be spelt out in terms of a simple example which illustrates the fact that the smaller and poorer the cultivator, the greater the adverse impact of a decline in the terms under which he sells his output.

In Table 9.2 we assume that there are two families, identical in size (5 members) and composition (2 workers and 3 dependents), the first with a 1 hectare farm which leaves family labour underemployed, and the second with a 10 hectare farm. Assuming an identical yield of 25 quintals/hectare, the first produces a net output (gross output *less* replacement of all non-labour costs, assumed at 30 per cent of gross output) of 17.5 quintals and the second produces a net output which is ten times higher. Assuming a minimum level of grain consumption of 500 gms. per head daily, we obtain an annual minimum consumption per family of roughly 9 quintals. Assuming basic grain consumption is half of total expenditure, we obtain a total consumption of 18 quintals in wheat equivalent.

Table 9.2

Small peasant and rich peasant families compared

	Small Peasant (1 hectare farm) *Quintals of Wheat*	*Rich Peasant* (10 hectare farm) *Quintals of Wheat*
1. Gross output	25.0	250.0
2. Net output (@ 0.7 of 1.)	17.5	175.0
3. Minimum grain consumption of family	9.0	9.0
4. Wheat equivalent of total consumption	18.0	18.0
5. Wages	0.0	29.0
6. Surplus/deficit (2.–(4. + 5.))	0.5	128.0

This minimum consumption leaves the 1-hectare farmer in marginal *deficit* and the well-to-do farmer with a grain surplus of 157 quintals. Suppose the well-to-do farmer employs hired labour to a greater extent than family labour and pays out 29 quintals of grain as wages: he will then have a 128 quintals surplus. The consumption level of the rich farmer can be assumed to be double that of the 1 hectare farmer; the former, it is evident, will still have a large surplus of over 100 quintals of wheat.

If the final output price changes adversely relative to prices of the inputs into production and of manufactured consumer goods purchased by both types of farmer, the impact will be much greater on the small farmer. Suppose production costs go up from 30 to 34 per cent, this will leave the 1 hectare farmer with 16.5 quintals of net output instead of 17.5 quintals; and a mere 10 per cent rise in consumption goods prices will entail selling 9.9 quintals of grain instead of 9 quintals, as earlier. The small farmer's deficit will widen to 2.4 quintals of grain equivalent—or, in short, he will be unable to meet minimum consumption needs and to purchase basic manufactured necessities. For the rich farmer, on the other hand, this order of adverse movement of the terms of trade will merely reduce his surplus slightly by 10.9 quintals to 117 quintals and will thus have no impact on his consumption level.

To sum up, adverse output price–cost and output price–consumer goods movements affect the very subsistence of the small farmer, and in cases of large movements even his survival; whereas they merely eat into the profits that large farmers make. It is hardly surprising that small farmers enthusiastically support agitations for reduction in the rate of rise in input costs or to raise final output prices. There can be no doubt that the greatest beneficiaries are the well-to-do farmers, but owing to the precarious economic position of the small farmers, they must perforce support any agitation—whoever might lead it—in order to have some hope of preventing their immiserization. Those poor peasants and labourers who are substantial net purchasers of foodgrains are the only classes who lose out from these types of agitation and are therefore entirely outside their support base.

Capitalist investment in western U.P. and its relationship to the emergence of the BKU[5]

In U.P., the farmers' movement today is virtually a monopoly of the BKU. The BKU emerged as a political force in the late '80s with its main sphere of activity in the western region of the State which, as we have earlier noted, has had a more dynamic trajectory of capitalist investment than any other region. However, the BKU is not limited to U.P.

It was formed in 1978 and units were established in Delhi and Haryana, as well as in western U.P. In 1980, the Punjab Khetibari Zamindara Union (PKZU) was converted into the Punjab unit of the BKU. This gave a big fillip to the farmers' movement in north--west India.

A U.P. branch of the BKU was set up in 1978, but it failed at that time to show much activity. In fact, the BKU came to public notice only in 1987, when a concerted effort was made to resurrect the organization in order to fill the political vacuum created by the death of Charan Singh in May that year. Farmers' politics at this juncture were at the crossroads, largely because of the Lok Dal's failure to extract any major gains from the government on the price issue. This was compounded by the fissures in the Lok Dal itself in which two leading factions emerged—one headed by Ajit Singh (son of Charan Singh), and the other by H. N. Bahuguna, a maverick ex-Congressman who had been Chief Minister of U.P. during the early '70s. Both were experiencing difficulties in mobilizing farmers who were being steadily alienated from the Lok Dal (Dhanagare, 1988). The revival of the BKU was a manifestation of this trend, an expression of the desire to create a space for the articulation and assertion of the growing demands of farmers.

[5] The factual and narrative material in this section relating to the BKU is drawn from English medium daily newspaper sources (mainly national) and the *Frontline* (a fortnightly) from 1987 to the time of writing (February 1992).

Initially, the leadership of the BKU was in the hands of Chaudhri Sukhvir Singh, who led the largest *khap* (caste *panchayat*) of western U.P. Jats. But, by 1987, Mahendra Singh Tikait (whose daughter was married to Sukhvir Singh's son) displaced the latter as leader of the BKU by organizing a successful agitation in 1988 in Shamli in Muzaffarnagar district. The BKU raised demands against a rise in power tariff, and protested against the erratic power supply to the irrigation pumps which are crucial to the growth of the capital-intensive agriculture in the Meerut and Agra divisions.[6] This initiative, which led the government to reduce the power tariff by one-sixth, bolstered the image of Tikait and established his hold over the farmers' movement in north India. In the wake of the Shamli agitation, thousands of farmers joined the BKU. Its support, which until 1986 was limited to Meerut and Muzaffarnagar, now extended to Aligarh, Bulandshahr, Bijnor, Moradabad, Pilibhit, Ghaziabad, and Saharanpur (Hasan, 1989).

In 1988, the BKU organized two major agitations: the *dharna*s in Meerut and in Delhi. Both agitations were of a militant nature, with frequent resort to both *gherao* and *dharna*. Both the agitations received widespread and enthusiastic support. According to one estimate, more than 2,00,000 farmers joined in the Meerut *dharna* which lasted for 25 days. Equally impressive was the solidarity expressed by the farmers and their tenacity in sustaining the *dharna*, which was largely peaceful—apart from sporadic altercations with the police, leading to some casualties.

The BKU's main demands centred on electricity, remunerative prices, low import costs, and the inclusion of BKU representatives on committees appointed to fix prices. These demands were similar to those put forward by other farmers' movements (Nadkarni, 1987; Omvedt, 1988). Nevertheless, in recent years no leader of the farmers' movement has managed to focus attention on their demands as dramatically as Tikait of the BKU. Both the Meerut and Delhi *dharna*s gave BKU the kind of national attention that was perhaps lacking in movements led by Narayanswamy Naidu in Tamil Nadu, M. D. Nanjundaswamy in Kanataka, and even the nationally well-known Sharad Joshi in Maharashtra.

Tikait succeeded in gaining considerable attention by dramatically spurning the efforts of politicians to take part in the BKU agitation. He had already repulsed the efforts of the Lok Dal to participate in the BKU-led farmers' agitation. Although Charan Singh had been the chief inspiration behind the formation of the BKU, Tikait did not allow even Gayatri Devi, Charan Singh's widow, and Ajit Singh, his

[6] The two divisions accounted for over 20,000 tubewells.

son, to share the BKU platform. Charan Singh himself, whom Tikait acknowledged as his mentor, had been deeply involved in party politics. He pursued power within and throughout the existing party system, with the object of bringing about a shift in the balance of power away from cities in the direction of rural areas. Charan Singh had been the undisputed leader of the peasantry of this region for nearly half a century. Through his writings and his involvement in political struggles (such as abolition of *zamindari* holdings and resistance to land taxation) he had skilfully promoted the interests of the peasantry (Byres, 1988). Furthermore, Charan Singh challenged the supremacy of the Congress in western U.P. by forging an alliance of middle and backward castes. Eventually, the Congress party was marginalized in the area. At no stage did he separate the activities of the BKU, once it was formed, from those of the Lok Dal; rather, his aim was to shape the BKU into an instrument of the Lok Dal. However, his efforts in this direction were unsuccessful because the BKU itself remained a rather weak organization during his lifetime.

After Charan Singh's death, and particularly since 1988, Tikait's distrust of political parties has received wide coverage. The BKU echoed the widespread feeling in the countryside that the legitimate interests of *kisan*s had been neglected by urban-based politicians, political parties, bureaucracy, and government. Tikait believed that in order to be an effective vehicle of agitation fighting against the government's urban bias and against the deterioration in the terms of trade, BKU should organize itself as a non-party force, exert pressure on an insensitive government, and extract concessions through farmers' struggles. The BKU justified its self-conscious autonomy from political parties on the grounds that 'all parties are parties of India, not Bharat' (Omvedt, 1988: 42). Yet this autonomy, which sometimes borders on hostility to political parties , does not mean that BKU has placed itself beyond the pale of politics. On the contrary, the BKU would appear to entertain hegemonic aspirations. This should be evident from the fact that it is not content merely to remain outside established party politics, but keen to articulate a populist, specifically *kisan* ideology. In this, BKU's aims are similar to those of other farmers' movements in India. Sharad Joshi has provided the clearest exposition to date of the new ideology which is based on the concept of a fundamental dichotomy between rural Bharat and urban India.

The BKU's version of this *kisan* ideology is inspired by Charan Singh's ideas. The BKU speaks a language that evokes elements of Charan Singh's discourse on agrarian problems. Its insistence upon the primacy of agriculture sometimes borders on the notion of an urban bias in development. Charan Singh deprecated the urban bias of planning and heavy industrialization, and held it responsible for the diversion of resources from agriculture (Byres, 1988: 179). The BKU, in a somewhat similar vein, blames the city-based *sarkar* for the problems faced by the peasantry and argues that it is imperative that the entire peasantry should be mobilized to oppose the privileged position of the urban industrial sector.

The emphasis on the promotion of rural economy and agriculture creates a political space in which prosperity in the countryside can be promoted by stressing the common interests of all rural families. Thus the BKU underscores the vital interest of all cultivators in the overall development of agriculture. In this framework, the well-being of farmers is entirely linked to higher prices for farm produce because deteriorating terms of trade affect the entire peasantry. All producers, then, have a shared interest in demanding the principle of parity with industry.

The BKU's interventions are not confined to the ideological sphere. Although its stance against parties prevents the organization from entering into direct dealings with political parties, this has not stopped it from actively supporting parties and politicians who are regarded as pro-farmer. Tikait maintained close relations with Devi Lal whose party was in power in Haryana during the period 1987–91, because he believed that Lok Dal(B)'s position came closest to that of the BKU. On the other hand, the BKU was categorical in its opposition to the Congress because the latter was viewed as being indifferent, if not hostile, to farmers' problems. BKU activists enthusiastically supported the Janata Dal in the 1989 elections; but they played a key role in the defeat of the Janata Dal in 1991 in western U.P.

In the 1993 U.P. State Assembly elections the mainstream (BJP and Congress-I parties were defeated by the Samajwadi Janata Dal with the active support of the BKU. A government, made up of a new balance of castes, is in power. The BKU is visibly active in politics in U.P. Similarly, the Shetkari Sangathana in Maharashtra and the Karnataka Ryota Sangha have also adopted an overtly political stance, supporting opposition candidates in elections. The Karnataka Ryota Sangha played a major part in the defeat of the Congress in Karnataka, whilst the Shetkari Sangathana was widely believed to have played a critical role in reducing the margin by which a number of Congress candidates won their seats in the 1984 elections to the Lok Sabha seats from Maharashtra and to the Maharashtra State Legislative Assembly (Lenneberg, 1988: 450).

The BKU supporters were quite clear that while it was admissible for them to support or oppose a political party, the BKU as such should keep 'out of politics' (Gupta, 1988: 2694). In other words, the BKU's attitude towards parties was pragmatic; in fact, its supporters saw considerable merit, in a strategic sense, in keeping the BKU independent of political parties. The BKU leaders calculated that their effectiveness would be maximized if they functioned as a pressure group outside the established party system.

The hegemonic thrust of farmers' organizations can also be seen from the numerous attempts, though as yet unsuccessful, to forge a united farmers' movement. An Inter-State Coordinating Committee under the leadership of Sharad Joshi has been established. However, the growth of a united farmers' movement was prevented by a rift between Tikait and Joshi on the question of leadership. Mutual recriminations were triggered off by the deep mutual suspicion the two leaders nurtured. Tikait was vehemently opposed to Joshi's support for V. P. Singh in the 1989 elections, whilst Joshi was disdainful of Tikait's unrealistic attitude towards political parties.

Nonetheless, the prospect of a potentially unified farmers' movement enhanced the political importance of surplus-producing farmers as political parties competed with each other to extend support to the BKU. The Congress(I) was particularly concerned about the rural appeal of the BKU because it was apprehensive that a united farmers' movement was likely to support the opposition, which had a generally pro-rural image unlike the pro-industrial and pro-urban image of the former. Consequently, systematic efforts were made by the Congress(I) to solicit endorsement of organizations which would mobilize farmers and shape rural consciousness and politics.

Inevitably, farmers' movements have been drawn into the contest of party politics, especially in States such as U.P. where, in recent years, centrist parties have faced a stiff challenge from right wing opposition forces. Thus, Tikait has been courted by important leaders of the Congress, Janata Dal, and (since 1990) the Bharatiya Janata Party (BJP). The political ambivalence of BKU leaders and their shifting political loyalties laid them open to overtures from all the national parties.

In the May–June 1991 Lok Sabha election, Tikait switched support from the Janata Dal to the BJP, which reportedly allowed BKU complete freedom to decide its party candidates for a number of constituencies in western U.P. Senior Janata Dal leaders such as V. P. Singh and Ajit Singh went to considerable trouble, during both the 1989 and 1991 elections, to canvass support from Tikait. They sought to convince the farmers' constituency that they were committed to a reorientation of policy and to a reorientation of investment patterns that would reflect the primacy of agriculture.

Likewise, the Congress(I) too made concerted efforts to penetrate the *kisan* movement. Rajiv Gandhi declared that the highest priority would be accorded to the development of agriculture in the Eighth Five Year Plan. Repeatedly emphasizing the vital role played by agriculture in the country's development, he announced plans to initiate parliamentary legislation aimed at treating agriculture on a par with industry. This policy was intended to help farmers who suffered a loss of crops due to floods or drought, in much the same way as sick industry was compensated. Soon after assuming power, P. V. Narasimha Rao, the Prime Minister of the minority Congress(I) government, has affirmed[7] his government's commitment to restore the primacy of agriculture and rural development.

The real gain made by the BKU in conjunction with similar movements in other States was to draw national attention to farmers' demands. In a sense, all political parties responded to the electoral and mobilizational imperative by addressing—in word if not in deed—the demands of the agrarian sector of the economy (both with respect to general demands such as supply of potable water, adequate roads, etc. as well as specific demands related to the issue of prices). Farmers' power has thus acquired such importance that 'no State and no political party today can dare to either ignore them or openly oppose their cause' (Nadkarni, 1987: 1).

We have stressed the overall economic context of the emergence of the BKU in U.P. politics because, in our view, an assessment of the movement cannot be undertaken only in terms of the success of the specific agitations, however spectacular, that it has shown itself capable of organizing. Rather, the movement should be located within the general prevailing political and economic trends in north India which have led to the growing importance of the surplus-producing market-oriented farmers.

This point takes on an added significance in the case of the U.P. farmers' agitation, precisely because the BKU managed to sustain its agitation even when farmers' movements elsewhere were on a decline, and often despite a lack of immediate concrete results. In actual fact, even the concessions extracted by the BKU in the course of the Meerut and Delhi *dharnas* were not substantial (Hasan, 1989: 2667). The judicial inquiry into the police firings in which six farmers in Aligarh district were killed, the waiver of electricity bills in 1987–8, and the establishment of a statutory committee on agriculture including its representatives, were among the features of an agreement in June 1989 between the BKU and the U.P. government in the midst of much fanfare and publicity. But this was not sufficient to ensure that the BKU would cease to engage in political agitation in the future.

Indeed, an opportunity for agitation presented itself in August 1989 and BKU seized the initiative. Naiyma, a Muslim girl from

[7] At a meeting of Chief Ministers (National Development Council) in December 1991.

Bhopa village, was abducted and murdered by a gang allegedly associated with Shuja-ur-Rehman, the *sarpanch* of Sikri village, who happened to be a close relative of the Home Minister of U.P. The incident rapidly developed into a major agitation as Tikait marched with thousands of farmers to the banks of the Ganga Canal to demand that the government should recover Naiyma. Following the recovery of her body, thousands of farmers gathered near the canal to protest against her murder. This uprising resulted in a series of concessions made by the U.P. government to the BKU, which included an inquiry into complaints by the BKU against police officials connected with the Bhopa incident, an inquiry by the Central Bureau of Investigation (CBI) into Naiyma's murder, withdrawal of cases against agitating farmers except those relating to arson and murder, new power connections, the appointment of a State level committee to decide whether government could waive dues accumulated by farmers in the course of their two-year agitation, and, finally, regular meetings between the President of the BKU and the Chief Minister of U.P. to resolve farmers' problems. This agreement shored up BKU's support among farmers and its determination to continue the fight for more substantial gains.

A large section of western U.P. farmers—the principal agents and beneficiaries of capitalist investment in agriculture, who have, obviously, a vested interest in its preservation and expansion—are dissatisfied that farm prices have not kept pace with costs of production and the cost of living since the mid-'70s. The BKU demands not merely higher procurement prices from the government, but prices based on the 'principle of parity', which would compensate adequately for the rising prices paid for inputs. The discontent over prices is located within a more generalized resentment over the undoubtedly real and large disparity in urban and rural standards of living, especially in respect of public goods and services such as drinking water, energy, roads, communications, health, and education. Finally, the BKU representatives made their strong case by contrasting the rate of return on their investment in agriculture and that achieved by those engaged in urban occupations in general and government service, trade, and professions in particular.

The aspirations of farmers were heightened by the critical importance of surplus food production in this region in the country's food economy. As we have observed in the previous section, north India has become the main supply area of surplus food production needed to feed the urban areas. This has given the farmers of north India a bargaining strength which their counterparts in western and southern India do not have at the national level. This bargaining power has been augmented by the fact that the demands put forward by surplus producers in these regions were increasingly supported by small peasants who, for reasons of survival, are left with no option but to support these agitations. We find that farmers' agitations have been particularly successful in regions in which small peasants have been drawn into the support base. This was the case in Maharashtra where commercial crops of onion and tobacco are grown by small peasants as well, and they were fairly enthusiastic supporters of the Shetkari Sangathana (Nadkarni, 1987).

Organization and support base of the BKU

The organization and consolidation of the BKU was helped by several factors, including the caste and community configuration of western U.P. The most important caste groups in the region are the Jats, Ahirs, Yadavs, and Kurmis. Jats comprise nearly 40 per cent of the population in Meerut, Muzaffarnagar, Saharanpur, and Bijnor districts of western U.P. Because of their numerical preponderance as well as their control over a sizeable proportion of land, Jats dominate both the politics and economy of the region. In the districts where BKU is particularly strong the legacy of the *bhaichara* system still prevails among the Jats who are the archetypal peasant proprietors in northern India, drawing upon a genuine peasant tradition marked by a willingness to work the soil themselves (Byres, 1988: 146). The other peasant castes, for example Rajputs and Tyagis,[8] also cultivate their own land and consider themselves to be the Jats' equals in status. Many of these peasants are linked by ties of caste and kinship which have helped to establish a rough correspondence between caste and class in western U.P. The BKU derives a major share of its support from these peasant castes. Its support is not only concentrated in western U.P. where these castes predominate but is even more visible in the Jat-dominated districts of western U.P.

The specific structure of caste support is not unique to the BKU. The support of most such movements is located in the landholding dominant castes. But what distinguishes the BKU from other farmers' organizations is the organizational structure of the Union at the district level. The BKU functions on the basis of some kind of collective leadership provided by the local *khap* councils. Tikait may be the strong central figure in the BKU, but his leadership is under constant observation by *khap* heads and his followers who regard themselves as members of 'a community of near equals' (Gupta, 1988: 2694). This consciousness is inspired by the coparcenary traditions of the *bhaichara* system.

Most decisions are made at a *kisan panchayat* in Sisauli in Muzaffarnagar district which meets on the seventeenth of every month. The

[8] The Tyagis happen to be the *only* Brahmins who work the land with their own hands.

Sisauli *panchayat*, which is the fulcrum of decision-making, is in turn an extension of the Jat *khap* councils, a form of caste *panchayat* which has existed for as long as five hundred years as an administrative apparatus running the affairs of the Jat community (Pradhan, 1966).

A *khap* consists of 50 to 100 villages, each *khap* has its own *panchayat* and a hereditary *chaudhary*, and *khap* members claim a common ancestry and belong to a single *bhaichara* community. The *chaudhar*ies of the seven *khaps* that are still active provided the secondary leadership of the BKU after they were designated as heads of the local units following the BKU's reorganization in 1986 (Kant, 1990). The new quasi-political role assigned to *khap* leaders offered them an opportunity to participate in the decision-making process which proved to be extremely beneficial in building and sustaining support for the BKU. Tikait's own rise to leadership was facilitated by the status he enjoys as *chaudhary* of Baliyan *khap*, claiming a history of more than five centuries of authority and domination over 84 villages. The reorganization of the BKU on caste lines assisted Tikait in maintaining his leadership by restoring a balance among different *khap*s, implicit in the recognition of the *chaudhari*es as local unit heads.

The caste–clan solidarity, however, is not sufficient to explain the broad-based support which the BKU has enjoyed, which goes well beyond the Jats of western U.P. For example, it has mobilized the support of Gujars and Muslims as well. It can be argued that this is necessary because the percentage of Muslims in western U.P. is higher than in U.P. as a whole, and the proportion of Muslims in districts such as Muzaffarnagar, Meerut, Moradabad, and Bijnor is particularly high. The demographic composition of western U.P. might explain why the BKU made a special effort in Meerut to restore communal harmony which was destroyed by the brutal killing of Muslims in May 1987. The fact that a sizeable part of the BKU's support structure consists of Muslims makes it difficult to understand the motives underlying Tikait's decision to support the BJP in the 1991 elections to the Lok Sabha and the U.P. State Legislative Assembly.

Equally significant is the wide class base of the movement which again cannot be explained solely in terms of caste–community ties and sentiments. A modest survey conducted in Meerut and Moradabad points to the conclusion that the bulk of BKU support came from rich and middle level rural producers, regardless of caste and community (Hasan, 1989: 2669). This is confirmed by a more recent study, which found not only that membership and active participation in the BKU was dominated by the rich and middle peasants, but also, more importantly, that every rich peasant household in the five villages surveyed had taken part in BKU agitations (Kant, 1990). This study also points to the considerable participation of small peasants in BKU-led agitations.

The available evidence suggests a measure of antipathy towards the

BKU among landless peasant tenants and agricultural labourers. Agricultural labourers, mostly belonging to the Scheduled Castes (SCs), dread the additional power the dominant caste peasants would derive from the farmers' movement (Hasan, 1989; Kant, 1990).

Class differentiation is often concealed by the widespread existence of *bhaichara* practices and peasant proprietorship in western U.P. The social base of the BKU is concentrated in districts dominated by Jat peasant proprietors, but the Jat peasantry is by no means homogeneous in character. The self-cultivating peasantry is a distinctly differentiated one (Byres, 1988: 146). A recent study of Meerut points out that capitalist producers and rich peasants, who constituted 7.69 per cent of the households, controlled the largest proportion of area (i.e. 47.53 per cent), whilst the middle peasants, who constituted 20.44 per cent of the households controlled 39.3 per cent, leaving only 13.17 per cent of area with the remaining 71.87 per cent of small and poor peasants (Singh, 1988: 100).

The BKU has clearly articulated the interests of the minority of well-to-do and medium-scale producers (accounting for the bulk of market sales) through highly specific and concrete demands; while the interests of the small-scale producers are represented, if at all, only in terms of broader and vaguer demands of rural–urban parity in facilities. The BKU does not concern itself with the problems faced by the rural poor. Minimum wages for rural labour have thus never figured among the demands raised by the BKU. Nor has the BKU ever sought to link the demand for higher procurement prices for foodgrains and other crops with a demand for provision of fair-price shops in rural areas, or with a demand for the enforcement of minimum wage laws already in existence. On the contrary, the interests of the rural poor have been perceived as antithetical to the interests sought to be articulated by the BKU.

REFERENCES

Basu, S.K., and S.K. Bhattacharya, 1963: *Land Reforms in West Bengal—A Study in Implementation* (New Delhi: Oxford Book and Stationary Co.).

Blyn, G., 1966: *Agricultural Trends in India 1891–1947: Output, Area and Productivity* (Pennsylvania: University of Philadelphia Press).

Byres, T.J., 1988: 'Charan Singh 1902–1987: An Assessment', *Journal of Peasant Studies* 15 (Jan.): 2, pp.139–89.

————, (Ed.), 1994: *The State and Development Planning in India* (Delhi: Oxford University Press).

Dandekar, V.M., 1986: 'Agriculture, Employment and Poverty', *Economic and Political Weekly* 21 (20–7 Sept.): 38–9, pp.A90–100.

DESAI, A.R., 1979: *Peasant Struggles in India* (Bombay: Oxford University Press).

DHANAGARE, D.N., 1983: *Peasant Movements in India: 1920–1950* (Delhi: Oxford University Press).

————, 1988: 'An Apoliticist Populism', *Seminar*, Dec., No. 352, pp.24–31.

GILL, S.S. and K.G. SINGHAL, 1984: 'Farmers' Agitation: Response to Development Crisis of Agriculture', *Economic and Political Weekly* 19 (6 Oct.): 40, pp.1728–32.

GOVERNMENT OF INDIA, 1987: *Area and Production of Principal Crops 1981/84* (Delhi: Controller of Publications).

GOVERNMENT OF UTTAR PRADESH, 1984; *Statistical Abstract* 1983–4 (Lucknow: Economic and Statistical Division, State Planning Institute, Government of Uttar Pradesh).

GUPTA, D., 1988: 'Country–Town Nexus and Agrarian Mobilisation: Bharatiya Kisan Union', *Economic and Political Weekly* 23 (17 Dec.): 51, 2688–96.

HASAN, Z., 1989: 'Self-serving Guardians: Formation and Strategy of the Bharatiya Kisan Union', *Economic and Political Weekly* 24 (2 Dec.): 48, pp.2663–70.

KAHLON, A.S. and D.S. TYAGI, 1983: *Agricultural Price Policy in India* (New Delhi: Allied).

KANT, A., 1990; Agrarian Mobilisation in Western Uttar Pradesh: A Case Study of the Bharatiya Kisan Union, unpublished M.Phil Dissertation, School of Social Sciences, Jawaharlal Nehru University, New Delhi.

KHUSRO, A.M., 1958: *Economic and Special Effects of Jagirdari Abolition and Land Reforms in Hyderabad* (Hyderabad: Osmania).

KRIPA SHANKAR, 1990: *Land Transfers: A Case Study* (New Delhi: Gian Publishing House).

LENNEBERG, C., 1988: 'Sharad Joshi and the Farmers: the Middle Peasant Lives!', *Pacific Affairs* 61 (Fall): 3, pp.446–63.

MITRA, A., 1977: *Terms of Trade and Class Relations* (London: F. Cass).

NADKARNI, M.V., 1987: *Farmers' Movements in India* (New Delhi: Allied).

NANEKAR, K.R., 1968: *Land Reforms in Vidarbha* (Calcutta: IBH Publishing Co.).

OMVEDT, G., 1988: 'New Movements', *Seminar* (Dec.): No. 352, pp. 39–44.

PARTHASARATHY, G. and B.P. RAO, 1969: *Implementation of Land Reforms in Andhra Pradesh* (Calcutta: Scientific Book Agency).

PATNAIK, U. and M. DINGWANEY, (Eds.), 1985: *Chains of Servitude: Bondage and Slavery in India* (New Delhi: Sangam).

PATNAIK, U., 1983: 'Classical Theory and its Application to India: Some Preliminary Propositions, with some thoughts on Sharecropping', *Journal of Peasant Studies* 10 (Jan.–April): 2–3, pp.71–87.

————, 1986: *The Agrarian Question and the Development of Capitalism in India* (New Delhi: Oxford University Press).

————, 1991:'Food Availability and Famine: A Longer View', *Journal of Peasant Studies* 19 (Oct.): 1, pp. 1–25.

————, 1994: 'India's Agriculture Development in the Light of Historical Experience', in Byres (Ed.), 1994, pp. 265–90

PRADHAN, M.C., 1966: *The Political System of the Jats of Northern India* (Bombay: Oxford University Press).

RAO, V.K.R.V., 1983: *India's National Income, 1950–1980* (New Delhi: Sage).

SANDHU, P.S., and S.S. GREWAL, 1987: 'The Changing Land Holdings Structure in Punjab', *Indian Journal of Agricultural Economics* (July–Sept.): 3 pp. 294–300. (Special Issue/Conference Number).

SEN, A., 1981: *Poverty and Famines* (Delhi: Oxford University Press).

SINGH, B. and S. MISRA, 1964: *A Study of Land Reforms in Uttar Pradesh* (Calcutta: Oxford Book and Stationary Co.).

SINGH, C., 1986: *Land Reforms in U.P. and the Kulaks** (New Delhi: Vikas).

* As told to an anonymous author by Charan Singh.

SINGH, D., 1964: *Land Reforms in Rajasthan* (New Delhi: Research Programmes Committee, Planning Commission, Government of India).

SINGH, J., 1988: Changing Agrarian Relations and Politics in Western Uttar Pradesh: A Study of Meerut District 1952–1985, unpublished PhD Thesis, School of Social Sciences, Jawaharlal Nehru University, New Delhi.

THAMARAJAKSHI, R., 1990: 'Intersectoral Terms of Trade Revisited', *Economic and Political Weekly* 25 (31 March): 13, pp.A48–52.

Reflections on Land Reforms in India Since Independence • 1(

D BANDYOPADHYAY

The pre-Independence background

Land reform, in some sense, has been on the agenda of political action as an integral part of the freedom movement in India. On his return from South Africa, the first major action initiated by Gandhi was in favour of the oppressed tenants of Champaran in 1917. He went to Champaran in April 1917 to enquire into the conditions of the peasants who were being oppressed by the indigo planters. He was served with a prohibitory order by the local administration. It electrified the nation when Gandhi pleaded guilty and declared before the magistrate

> that I have disregarded the order served upon me not for want of respect for lawful authority, but in obedience to the higher law of our being, the voice of conscience [Tendulkar, 1951: 204].

Champaran was followed soon after by a peasant movement in Kheda (1918).

> A condition approaching famine had arisen in the Kheda district owing to a widespread failure of crops [Tendulkar, 1951: 221].

The tenantry demanded suspension of revenue assessment which was admissible under the revenue regulations. A callous government did not respond. After enquiring into the actual conditions of the tenantry, as he had done in Champaran a year previously, Gandhi launched a 'no tax campaign'. After some time the dispute was amicably settled. These events took place before the non-cooperation movement of 1921.

In Nehru's *Autobiography* (1984) one encounters numerous references to agrarian problems in the then United Provinces (U.P.) and to political action taken in respect of them under the leadership of the United Provinces Provincial Congress Committee (UPPCC). Under the aegis of the Congress, the peasantry in U.P. organized a movement for the fulfilment of the just demands of peasants and for the removal of oppressive levies by the landowners. According to Nehru (1984: 58):

> the taluqdars and the big zamindars, the lords of the land, 'natural leaders of the people', as they are proud of calling themselves, had been the spoilt children of the British Government.

It is evident from Nehru's account that peasants participated both in the Civil Disobedience movement of 1921 and in the non-cooperation movement of 1931 in large numbers, and suffered imprisonment and other disadvantages in the hope that the political freedom of the country would be accompanied by their emancipation from the oppression and bondage of the landowning class.

In 1936, the Indian National Congress (INC) set up a front organization known as the All India Kisan Congress (AIKC) with the aim of organizing the peasantry to fight against the abuses of landlordism. The initial thrust of the AIKC (subsequently re-named All India Kisan Sabha (AIKS)) was directed against the *zamindari* system. Swami Sahajanand, the first President of AIKS, asserted (in 1936) that

> no compromise was possible between the peasants and the landlords except dispossession of zamindars of their lands [Rasul, 1974: 5].

The landowners—*zamindars*, *taluqdars*, etc.—generally identified themselves with British rule. An anti-imperialist mass movement, therefore, required a multi-interest alliance directed against not only the British rulers but also those institutions and classes which had an abiding interest in the continuation of the 'Raj'.

In the initial stages, not only was the leadership of the AIKS exercised by the rural 'gentry', but its membership also largely belonged to the various categories of secured tenants including the rich and middle peasantry. The rich and middle peasantry had suffered under the *zamindari* system. Moreover, irrespective of their actual economic position, the landlords enjoyed great social prestige and political power, and exercised patronage. Thus, in the initial phase of its existence, the AIKS had under its banner various conflicting interest groups loosely knit together against the common enemy, namely the *Raj* and the class of landowning proprietors which was its protégé. Though there were tensions within the organization of the AIKS, they managed to maintain a common front until 1942. A sharp divergence of opinion between the Congress leadership and the Kisan Sabha (by then dominated by the communists) on the issue of India's support to the war effort and on the 'Quit India' movement within the Congress fold led to the disintegration of the movement. But there was no change in the anti-*zamindari* stance of both the INC and the Kisan Sabha (KS). So it was on the cards that, if and when Congress came to power, the *zamindari* system would be abolished.

After Independence, this was precisely what happened. Right from the First Five Year Plan (1951–6) onwards, the main elements of the land reform policy have been fivefold:

1. abolition of intermediary tenures;
2. tenancy reform comprising regulation of rent, security of tenure, and conferment of ownership right on tenants;
3. ceiling on landholding and distribution of surplus land;
4. consolidation of holdings; and
5. compilation and updating of land records.

There is general agreement that out of these five items, the abolition of intermediary tenures has been carried out successfully throughout India, barring a few insignificant and isolated exceptions. There is equally strong general agreement that in all other respects very little has been accomplished and much remains to be done.

But this generally dismal scenario must be qualified by a few brilliant localized exceptions. In West Bengal, both on detection of land in surplus of the ceiling and on conferment of non-evictable rights to sharecroppers, a fair measure of success was achieved.

This chapter is divided into two parts. The first part presents a synoptic view of the land reform experiment in West Bengal: a story of a fairly successful attempt. The second part contains some general comments on the land reform issue in an all-India context.

Land reform in West Bengal

Since I was personally involved in the land reform programme in West Bengal through legal process supported by mass action both in the United Front (UF) Governments of 1967 and 1969, and (years later) the Left Front (LF) government during the period 1978–81, I shall discuss the subject from the point of view of an operational person rather than that of an ideologue or a simulated academic. In the first UF government of West Bengal, Harekrishna Konar served as the Minister for Land Revenue. With the decline in land revenue as a proportion of the total revenue of the State, the Revenue Department of the government of West Bengal (as elsewhere) had ceased to enjoy bureaucratic 'prestige'. It therefore came as a surprise to many, including myself, that a charismatic mass leader of Mr Konar's stature was placed in sole charge of the Land Revenue department.[1]

In my innocence, I could not help asking Konar himself why he was placed in sole charge of such an unimportant department. His reply was simple. He said that the activities of this Department directly touched the masses of the rural poor, and without active mass support the basic objective of redistributive land reform through the legal channel could not be accomplished. That explained why he was where he was. By 1967, the *West Bengal Estate Acquisition Act 1953* had been in operation for over 12 years. The total area of land vested under this *Act* until then was little over 350,000 acres. These lands had been mostly surrendered by big landowners (including big *zamindars*). But it was common knowledge in the countryside that good available land belonging to these landholders was being kept by them under various ruses in contravention of the provision of the law. No serious attempt had been made by the government thus far to get these lands vested.

The countryside was seething with discontent, and the first arrow of the well-known militant Naxalbari movement was shot in the village of that name in northern West Bengal soon after the United Front Government assumed power in February 1967. Konar's analysis of the situation was that the rural masses would not support a movement which was overtly 'illegal'. Such a movement would at best be spasmodic and was bound to die down soon. For a sustained movement to take root, it would be necessary to make it clear that its objective was not only legitimate but also legal. That was why Konar laid a great deal of emphasis on the vesting of the clandestinely held surplus land of the big landowners through the legal process. He issued a directive calling for a number of case studies on the actual evasion of ceiling on landholding, in an effort to identify the various methods used to evade the *1953 Act*. This was to be an essential preliminary step before taking prompt corrective action in order to ensure that all surplus land over and above the ceiling was vested without unnecessary delay. This ministerial directive resulted in the preparation and publication of a brochure entitled *A Note on Evasion of Ceiling Law in West Bengal* (Bandyopadhyay, 1967), giving details of the methods employed by different families in different parts of West Bengal to evade the

[1] Another charismatic mass leader, Gowriamma, was given the land Revenue portfolio in the CPL led Government of Kerala (1957–9) and in the United Front Government (1967–9).

ceiling law. This publication was withdrawn from circulation when the State was brought under President's Rule in 1968. Hardly any copy is available now of this extremely useful document. The return to power of the United Front government in 1969 gave a tremendous boost to the legal process of vesting surplus land.

An analysis of the case studies on evasion of the ceiling law showed that a number of steps were required to detect surplus land over and above the ceiling and eventually vest them with the State. Thereafter, these lands would have to be redistributed. The following stages would be involved in any all out effort to vest land in excess of the ceiling:

1. identification of families possessing or suspected of possessing lands above the ceiling;
2. locating and identifying all plots of land in the effective and real possession of every such family (both according to the record of rights or title deeds), and the tracing of *benami* lands (i.e. lands held clandestinely or in somebody else's name, real or imaginary) through fictitious or collusive transactions;
3. initiating the quasi-judicial process of the vesting of all surplus land (including land kept through collusive transfers and *benami*) by gathering adequate evidence which would stand the scrutiny of review/appeal at higher levels of administrative tribunals or courts of the judiciary;
4. taking over possession of vested land after completing quasi-judicial and other administrative processes;
5. assigning such land according to legally/administratively prescribed priority among the landless and land poor peasantry;
6. providing a mechanism to prevent illegal physical eviction from the assigned land;
7. some provision for consumption and production credit in order to enable the resourceless new allottee to start cultivation without getting into the debt trap of the former patron with the real possibility of alienating the land.

If we analyse some of these stages (say the first and the second) more closely, it will be evident that much of the crucial information and intelligence could be obtained only from the agricultural workers, tenants, or sharecroppers working on the land in the possession of such families. Even though the record or the document might show that a particular family had only a landholding within the ceiling limit, agricultural workers could give evidence that they were employed and paid by the said family for the cultivation of land nominally held in someone else's name, that all the management decisions regarding crops to be sown, fertilizer to be applied, produce to be stored or marketed, were taken by the said family, thus clearly establishing that this was the family in real and effective possession of the land and enjoying its usufruct to the exclusion of others. Effective possession and enjoyment of usufruct of land proved that the land really belonged to them, even though the documents might show otherwise. Similarly, the sharecroppers could identify those to whom they delivered the share of the crop and by whom they were engaged as sharecroppers in the first place. If such evidence, though oral in nature, could stand the test of cross-examination, it could be used to overturn the documentary evidence to the contrary and would be acceptable even in courts of appeal.

In the past, even a conscientious officer attempting to act positively in such matters was faced with the difficulty that, against the documentary evidence of the landowners, he would be in no position to obtain any acceptable and reliable data to contradict them, even though he might be convinced that the documents were collusive or fraudulent. Now, however, if the sharecroppers, agricultural workers, casual cart drivers, or petty middlemen gave reliable facts and were able, even under severe cross-examination, to establish the facts, the documents could be disbelieved and credence could be given to overwhelming oral evidence to the contrary. But, ordinarily, agricultural workers or sharecroppers would not dare come forward and tender evidence against their employer–landowners for fear of severe reprisal which might consist of anything from loss of employment or sharecropping rights and social ostracization to physical elimination. Under ordinary circumstances, it was well nigh impossible to persuade these valuable and reliable witnesses to depose against their landowners. Hence no information worthy of recognition by the quasi-judicial processes would be forthcoming.

Notwithstanding the stringency of the legislation, the landowning class therefore had nothing to fear from a half-hearted administrative system which had no back-up support from the potential beneficiaries of the land reform laws. Moreover, normal bureaucratic prudence would suggest that it was better to steer clear of any controversy since the landowning class wielded enormous political clout. Hence, the law retained its pristine lustre untouched by human hand in the statute book. Landowners remained happy after voluntarily declaring and surrendering whatever was not worth keeping, the bureaucracy continued to be 'overworked' with the processing of surrender documents and preparation of compensation statements, the potential beneficiaries of land reform watched in fearful silence the farcical pantomime being performed ostensibly for their benefit but without an iota of their own involvement in it, and the government could afford to be smug and complacent that it had done all that could be done for the abolition of intermediary rights and redistributive land reform. That was the situation obtaining prior to the UF government's being returned to power in West Bengal in 1967.

Though the UF came to power with tremendous electoral support, it had to function strictly within the rigid parameters of the Indian Constitution, the established basic laws, judicial review of executive action, set legal and administrative procedures and practices. Any threat to any of the established parameters would have led to a summary dismissal of the by now not-so-friendly government at the Centre. The political genius of Harekrishna Konar lay in his ability to play his own ball game with the same set of rules which were apparently set against it.

The Constitution of India guarantees the right to form associations and unions and to assemble peaceably. The *Indian Evidence Act* permits the disbelieving of documentary evidence by overwhelmingly reliable oral evidence. *The Criminal Procedure Code* (under Section 110, commonly known as the bad livelihood section) allows for some sort of public participation for gathering evidence against a person allegedly engaged in bad livelihood in order to bind him down for good behaviour. Nowhere is it stated that the restraining powers under the *Criminal Procedure Code* should always and invariably be used against the peasantry and workers. If the agricultural workers and sharecroppers assembled peacefully in order to espouse a cause, if the public order was threatened by the machinations of landowners, they could be restrained under the *Criminal Procedure Code* in the interest of the maintenance of public order. Combine the essence of these constitutional and legal rights and procedures and you have the 'Konar recipe' of legal land reform with popular participation. It was so simple, so audacious, so innovative.

Peasants' organizations of the Left parties, particularly those led by the Communist Party of India (Marxist) (CPI(M)) and the Communist Party of India (CPI) had so far been engaged in agitations against various injustices suffered by the peasantry with varying degrees of success. The administrative machinery, including the police, being generally hostile to such agitation and group action, it was difficult to ensure willing and large-scale participation of the poor masses for any length of time except in cases of extreme injustice and oppression. Basically, these agitations consituted protest actions which were inevitably followed by some harassment against the poor in the nature of arrest, court action, etc. However sympathetic they might be, large sections of poorer peasantry preferred to stay away from such trouble. After all, at the end of it all, they had to go back to the same landlords or their relations for jobs and grain loans. And they would have to face, apart from insults, much harsher and adverse terms of employment and even more usurious loans than in normal times. They had no protective mechanism to fall back upon. They would suffer silently for a long time after a short and noisy outburst.

The situation changed dramatically with a sympathetic political executive at the helm and an administrative machinery that was undoubtedly less hostile at the cutting edges even if it might not be in full sympathy with the programme of the government. When the administrative system sought the cooperation of the mass organization for gathering information and intelligence in order to identify families which clandestinely or illegally held considerable amounts of land, a strange metamorphosis beset the mass psyche. After the initial identification of *benami* land, the administration was faced with the more difficult and arduous task of gathering evidence from the witnesses who had been so long under the absolute control of the landowners. A state of disbelief prevailed on both sides of the barrier. The witnesses could not imagine that the authorities would believe their oral testimony denouncing such well-crafted documents. The landowners could not believe that their own agricultural workers and sharecroppers would have the impudence and temerity to tell the truth against them in front of public officials. The frustrated wrath on their demented faces often sent shivers down the spine of the witnesses. But many held on. Many did not buckle under fierce cross-examination. Indeed, in many cases, through cross-examination, many a startling revelation was made which had been initially withheld by witnesses, perhaps out of fear, even perhaps out of respect for old relationships.

When, after such a gruelling process, clandestinely held lands were identified and vested in the state, it had an electrifying effect on the poor peasantry. What started off as a trickle induced from without, soon turned into a voluntary deluge of evidence, coming from the organized or even not-so-organized peasants and peasant groups. Evidence of a totally unassailable character was brought out into the open. Landowners were often frightened by the vehemence with which these facts were asserted. The process of eliciting information also had a salutary impact on the bureaucracy itself. Officialdom was now totally exposed to the glare of peasant action groups. The only protection that it had was strict neutrality and proper appreciation of evidence. A proper appreciation of evidence was essential to save officials from severe strictures handed down by the High Court which landowners moved *en masse*. Neutrality was equally essential to assure the peasant action groups that there was no inherent bureaucratic bias in favour of the landlords and against the peasants. Any indication of partisanship resulted in peasant action groups virtually laying siege to the office premises, but in most cases when the points at issue were properly explained, they left peacefully.

An element of direct public accountability, hitherto totally absent in the functioning of the bureaucracy, now came into existence. It did exercise a chastening effect on the system. That an officer of the State would have to explain his/her conduct to a group of ill-clad, unshaven peasants speaking in a language which was not quite the correct *officialese* was an unpleasant and novel experience for many. A few resented this 'ignominy'. There were a few cases of excesses and, in a couple of incidents, it subsequently became clear that landowners had indeed organized such nasty demonstrations with the aim of breaking down the morale of honest and committed officers. The struggles waged by the peasantry under such conditions resulted in the vesting of approximately 900,000 acres during a period of 42 months (1967–70). A large proportion of the land thus vested was good and cultivable—land which landlords had kept to themselves by evading the law.

The vesting of land surplus over the ceiling had three major effects. In the first place, it reduced to some extent the concentration of landholding in a small number of hands and, to a certain extent, led to a break in the feudal practices associated with large holdings. An environment suitable for capitalist agriculture was thus created. Shorn of excess land, landowners with resources tried to make the best out of their shrunken holdings by utilizing new high yield variety (HYV) seeds of paddy and improved biochemical techniques. A segment of the parasitic landowning class was transformed into active agents taking part in improved agricultural production.

The second effect was that, after experiencing the buffeting between a newly mobilized poor peasantry and an entrenched landlord class, the State bureaucracy became a slightly better instrument for carrying out a programme of social interventionist action. The third effect had a major political implication. Through the vesting of land, and even more so through the occupation of such vested land as a consequence of peasant initiative, the Left parties (and in particular, the CPI(M)) were able to penetrate deep into rural West Bengal. The CPI(M), which was a party based on organized workers in the industrial sector and lower middle and middle classes in the urban services sectors, came to be known as the friend of the rural poor. It opened up the possibility of major political action through an alliance of workers and peasants, even though electoral compulsion eventually led to a different course of action.

An ugly facet of the land reform programme during the 1967–9 period in West Bengal was the fierce internecine fight among the UF partners for the occupation of vested land. Konar did not approve of the occupation of private property by peasants, even though such land had in many cases been held in *benami*. He believed that occupants of such land would be vulnerable to feelings of guilt which could well deprive them of the courage needed to fight against new threats of eviction. The peasant psyche regarded private property as inviolable. And if one encouraged such expropriation for the sake of political expediency, one would be compelled to follow the same logic to the extent of supporting banditry and brigandage. Konar therefore favoured the legal process of vesting land in the State. Once the land became the property of the State, what happened to it automatically became a matter of State policy and no individual's proprietary right would be involved in it.

On the assignment of land, Konar did not favour a dominant role for the bureaucracy. Deprived peasants, who had lost their rights over land through usury, rack-renting, or sheer physical violence, would be allowed direct participation through their own organizations to occupy such vested land in order to satisfy their land hunger and also as a measure of historical revenge for the involuntary alienation of their land in the past. Moreover, in the event of a change of political environment, new occupants would be in a position to defend their possession through collective action. Furthermore, resourceless new occupants would also feel an urge to engage in collective action for productive purposes. Once they felt the need for mutual help, they would be well placed to lay a good foundation for the development of service cooperatives. The whole process would result in significant political education for the masses.

So far so good. But a problem arose when each partner of the UF thought that the easiest way to extend its political influence would be by letting loose its genuine or temporary or fake supporters to occupy the vested land. Moreover, landowners who had lost their land thought that if they could have their land occupied by their own hoodlums and goons under a party banner, no one would be able to dispossess them of it. Some party flags were available to the highest bidders. Opportunistic politicking reached its nadir in some parts of rural West Bengal. Intense dissension at the lowest levels of political parties had an adverse effect on unity at the top. The lower echelons of many partners of the UF were penetrated by local vested interests. The Second UF Government could barely survive in power for 18 months.

Looking back on the period of the UF governments, one wonders why Konar took two different lines: one on the vesting of lands and another on the distribution of vested lands. Vesting was carried out in an orderly manner through a strict legal process, even though collective action of the peasantry was allowed in order to gather and present incontrovertible evidence against fictitious transfers, fraudulent tenancies, mis-classification of land, *benami* holding, etc. Many cases, which subsequently went up even to the Supreme Court on final appeal, were upheld. But on the distribution of vested land, Konar took a completely different line by refusing to allow the official machinery even to carry out a routine filtration of the applicants.

Perhaps he took a deliberate risk in this area because he aimed at rapidly politicizing the poor peasantry of West Bengal. Konar was well aware of what was happening in the State. In the Autumn number (1969) of *Desh Hitaishee*, he wrote:

> No doubt the consciousness of the peasants has increased during the time under discussion, but we cannot say, even now, that it has completely come out of the sphere of economism. Militant land struggle does not mean political struggle for land. It is not free from the influence of economism, when it aims at some immediate gains. A militant land struggle, lacking in revolutionary political task and a clear conception about enemies and allies at different stages of struggle may even be misdirected and enmeshed into revisionism and sectarianism. [Konar, 1977: 69–70.]

He was also fully aware of the inter-party clashes (within the UF) over land, and their political implications. In the same article, Konar wrote that

> It is not easy for *jotedars* to launch direct attacks everywhere, because of the existence of the United Front Government in West Bengal. So, they are resorting to other devices. Owing to the rapid disintegration in Congress, some of these *jotedar*s have been conspiring to plot disunity and dissension under the cover of different partners of the United Front. The impact of it has been felt behind the clashes between partners. Of course jealousy, the petit-bourgeois mentality for inflating a party and the sectarian attitude in the peasant movement have also contributed to these clashes ... The militant peasant movement will have to take its guard against this danger of disruptive efforts. All these disruptive machinations will have to be defeated by making the people conscious about it and by raising the slogan of a united movement of all the toiling peasants without displaying any sectarian outlook. [Konar, 1977: 71–2.]

Konar knew what was happening, but he was confident that through these struggles—with their successes and failures—he and his comrades would be able to raise the consciousness of the poor peasantry, and to forge it into a mighty weapon for the revolutionary struggle ahead. But he hardly had time. A major shift in the political scenario, and his untimely death soon thereafter, destroyed the possibility of any such movement.

While it is true that the prospect of immediate gain might have been a major motivating factor in the mobilization of poor peasantry in order to gather evidence against the landlords, such a process also clearly showed that collective action did produce results and provide some measure of protection. During the ensuing backlash (1971–7), many lost possession of the lands which were occupied without any official authorization, but some were able to hold on to their land through collective or group action. But when large-scale eviction of sharecroppers took place, those evicted were given no legal protection because most of their leases were oral and they did not have a scrap of paper to support their case. They suffered a great deal because the relevant land revenue authorities under the UF governments attached great weight to the evidence that they gave (unlike that given by agricultural workers) because of their closer and more continuous relationship with the landowners, unlike the much more casual relationship which agricultural workers had enjoyed with the latter. Landowners took particular care to evict and to socially ostracize those sharecoppers who had played a prominent role in the peasant mobilization during the period 1967–70.

Lack of administrative support, the natural hostility of the law and order machinery towards mass action of any kind, the readiness of the government to suppress any threat of rural violence in the name of containing the Naxalite movement of the 'ultra left', the resurgence of political power of the landowning class and the imposition of an internal Emergency in 1975 by the central government had an influence, at once demoralizing and debilitating, on the peasant movement and land reform programme. While nothing much happened, gimmickry in the name of land reform reached new heights. A ceiling on family landholding was introduced. The law was made more favourable to sharecroppers.

The government claimed that it had distributed the entire area of 1.3 million acres of vested land to genuine and deserving poor peasant families by evicting the occupants who had taken the land by force. But the government's claim was largely false. Nothing much happened on the ground because no one in the higher echelons was really interested in getting anything concrete done, except to tell the rural masses how badly the previous regimes (i.e. the UF governments) had mismanaged the land reform programme. Unfortunately for them, the more they criticized the UF governments, the more the rural people disbelieved their propaganda, because many a poor peasant and agricultural worker had participated in the programme and a number of them had received some benefit. In the 1977 general election, the left parties were returned to power in West Bengal. A Left Front (LF) government was formed.

The new LF government was keen to pick up the thread of the programme left unfinished by the second United Front in 1970.

Benoy Krishna Chowdhury, the new Land Reforms minister, declared the twin objectives of the land reform programme of the LF government to be:

1. vesting of surplus land after the imposition of a ceiling on family landholding; and
2. recording the names of sharecroppers in the record of rights in order to give them the protection and benefit to which they were entitled under the law as amended by its predecessor government.

Strangely enough, in spite of a great deal of enthusiasm on both these counts at the political level, the response of the rural masses was somewhat muted. The upsurge of spontaneous enthusiasm which greeted the earlier programme was now conspicuous by its absence. Instead, a clear disinclination on the part of many to come forward to be associated with either of the programmes was clearly evident.

Indeed, the progress of vesting land was disquietingly slow. Instead of potential witnesses besieging the revenue offices, even official requests failed to meet with any willing response in the initial stages. It quickly emerged that the sharecroppers who, during the 1969–70 period, had provided unassailable evidence of collusive transfer, fictitious recording, and *benami* holding were now quite unwilling to do so. The unpleasant experience, so recently, of eviction and subsequent compromise on harsher terms was still fresh in their minds. They had no legal document to prove that they were sharecroppers. As a consequence, there was no way a sharecropper could muster evidence against a landowner disclaiming any relationship with him. It was the landowner's word against the sharecropper's. Sharecroppers became silent sufferers. As there was no guarantee that the LF government would survive its full term, sharecroppers were loath to take the risk of disturbing the existing arrangement.

It was, however, obvious that no major dent could be made on the land question as long as the defeatist mentality of potential witnesses prevailed and they were not confident that their interests would be protected. That apart, the registration of sharecroppers (*bargadars*), by itself, constituted a major programme which had been given no special attention by either of the UF governments. Hence, recording the names of sharecroppers and giving them certificates of sharecropping became priority number one on the agenda of land reform in West Bengal under the LF government.

Recording the names of sharecroppers was by no means a novel element in the land reform administration of the State. Revenue offices were opened in different parts of rural areas at which names could be recorded on the basis of application, followed by routine verification. When instructions were sent to intensify this already existing element of the programme, strangely enough they evoked only a lukewarm response.

It was ironic that the LF government which had been returned to power with the massive support of the rural poor was unable to generate enthusiasm for policies designed to benefit the poor on a scale comparable to that witnessed during the tenure of the UF governments during the late '60s. The usual political clichés or jargon of the Left could not explain the public's indifference to reform. A deeper understanding of the peasant psyche was necessary.

I was recalled from the central government in April 1978 and given charge of the land reforms programme of the State. My first task was to make an assessment of the situation on the ground. It was essential to gain an insight into the real constraints on a quick recording of sharecroppers and to devise a mechanism for direct participation by the beneficiaries. One could legitimately raise a question on the latter point.

Participation through mobilization of the beneficiaries is a political process, so why should the bureaucracy become involved in it? After the two experiments with the United Front there was a major change in the tactical line of the CPI(M), the dominant partner of the LF. It was probable that in order to deny the central government an excuse to intervene, the LF government eschewed the line of mass mobilization which it had consciously followed on an earlier occasion. That apart, the inter-party clashes over the occupation of vested land in 1969–70 which led to the rapid disintegration of the second United Front government could best be avoided by handing over the initiative to the administrative machinery, and providing it with support from below, whilst at the same time applying pressure from above. An important change from the late '60s and early '70s consisted of the removal from the scene of Hare Krishna Konar, whose political determination had contributed to the viability of that line during the previous decade.

The methodology of rural workers' camps developed by the National Labour Institute (New Delhi) during the internal Emergency (1975–7), for the implementation of rural mass-oriented projects of the first Twenty Point Programme, appeared to be the best way of delving into the peasant psyche of West Bengal in mid-1978. Such camps—where a number of selected agricultural workers, bonded labourers, tenants-at-will, and the like lived together with academics, administrators, and social workers, sharing the same environment, same food and same facilities in a school building or a *panchayat 'ghar'*—set into motion a magnificent two-way learning process without any artificial distinction between the teachers and the taught. Each participant, official, or non-official, academic or illiterate, learnt something new through this direct face-to-face sharing of raw experience and information. It created a new awareness of the realities of life.

When this methodology was proposed in West Bengal in 1978, it created ripples in political circles. Was it a new strategem aimed at undercutting the base of the *kisan sabhas* (peasant organizations)? Some viewed it as constituting an open encroachment on the closed preserves of political organizations. The association of Paulo Freire's name with this methodology of conscientization of the rural poor raised dark suspicions of interference by a 'foreign hand'. After a great deal of persuasion at higher political levels, it was agreed that this experiment should be tried out in the presence of the Minister for Land Reforms.

The first camp in Medinipur turned out to be a disaster. The selection of the participants went wrong. The second camp at Halushai (in Hooghly district) was a great success. The main point to come to light was related to the overwhelming sense of fear among the poor peasantry that the landowners had the power to take reprisals against them. The LF government, however sympathetic to them, was situated too far away to do anything to mitigate their suffering. The painful and ugly experience of the political backlash of the 1971–7 period was too raw and fresh in their minds to be easily forgotten. Moreover, the method of recording so far followed was faulty from the sharecroppers' point of view. The whole methodology of Operation Barga, the hallmark of the LF government's land reform programme, was evolved at Halushai camp on the basis of an exposition by Shambhu Tudu, an *Adivasi* activist sharecropper:

> His points were very simple. He said that recording of sharecroppers was done during the usual office hours, say between 11 am and 4 pm. This was the period when the sharecropper was out in the field and the time did not suit him. Secondly, the camps or offices in rural areas were invariably situated in *pucca* buildings. Such buildings were owned by affluent persons in the village. In such camps or offices, when recording took place, one would find the landowners getting their property rights recorded or mutated with various documents and papers. Outside the building, there would be touts and stamp vendors, busy preparing the types of documents that might be necessary to substantiate the claims of landowners. According to Shambhu, it was very difficult, if not impossible, for a lone sharecropper to penetrate this ring of landowners and their henchmen to establish his claim before the Revenue Officer. The location and the method of recording were totally against even those sharecroppers who were intrepid enough to risk the landowner's anger. Thirdly, unless the *bargadars* were told what real benefits would flow from such recording and were provided with some forum for collective action and a mutual support system, it would not be possible

> for them to overcome their fear. They suffered from fear of economic sanction, physical violence and various types of social oppression. [Bandyopadhyay 1981.]

Hence, when Operation Barga was launched towards the end of 1978 in West Bengal, emphasis was placed on developing group action among potential beneficiaries in order to create a mutual support system that would enable them to overcome fear. An important feature of the operation consisted of group meetings each evening with the potential beneficiaries in each locality at the end of a day's work. At such meetings, sharecroppers were encouraged to express their grievances and to publicly relate their experiences of various social and economic injustices.

Teams of officials who were put in charge of this operation would camp in neutral premises such as the *dak* bungalow or school building or *panchayat* house, scrupulously avoiding the large mansions of the rural rich. They would attend the evening meetings and explain the benefits to which the *bargadar*s were entitled under the law and the advantages that would accrue to them through recording their names, namely protection against eviction, obtaining a higher share of the produce according to the law, the right to bequeath the rights of sharecropping to a member of the family in the event of death, and the possibility of obtaining crop loans from institutional sources.

The evening meetings were followed by officials drawing up tentative lists of claimants. On the following morning, these lists were publicly verified in the field in the presence of the claimants, landowners, and other villagers. After the field verification, a provisional list would be displayed in all important public places in the cluster of villages where the operation was in progress. Landowners were given the opportunity to file objections on the following day. These would be heard and verified again in the field. Thereafter, the lists would be finalized and a certificate of sharecropping would be distributed before the team moved out to another cluster of villages. (Bandyopadhyay, 1980.)

This methodology, which depended heavily on the group action of sharecroppers, was qualitatively different from the traditional revenue court approach in which the poor were at a disadvantage because they, unlike the rich, were in no position to engage lawyers and to produce documentary evidence (which they did not have). Sharecropping leases being oral in character, sharecroppers have no documentary proof to support their claims. At the same time, sharecropping is a question of verifiable fact and not a matter of law, and therefore permits verification in public in the field.

Operation Barga thus enabled rural workers' organizations to participate effectively in the programme of registration by organizing sharecroppers in groups. Claims were also put forward in groups, each of which was verified in the presence of the group concerned. At the same time, collective action enabled the claimants to produce sufficient evidence in the field in support of their claims, a procedure which could not be followed in revenue courts. The quality of verification in the fields was of a very high order. One could hardly get away with a lie in public and cases were disposed of at considerable speed. The new procedure also enabled sharecroppers to overcome their fear and significantly reduced the possibilities of severe reprisal by landowners because they knew that they would have to face organized resistance against such attempts.

The procedure followed in the implementation of Operation Barga was safe, fast, and transparent. It resulted in the recording of over 1.2 million sharecroppers within a period of three years (1978–81). The land reform policy of the LF government had the effect of rekindling confidence and enthusiasm among the rural poor throughout the State. By 1982, however, the original thrust of Operation Barga had begun to wane. It was difficult to sustain an administrative delivery system at such a high pitch over a long period of time. With the recording of 60 to 65 per cent of the sharecroppers in three years, Operation Barga substantially achieved its objective.

It would be wrong to create the impression that it was a cakewalk all the way. Unlike the late '60s, when there was a coalescence of interests of the marginal, small, middle, and upper middle peasantry as well as landless labour against the big landlords, during the late '70s there was a conflict of interest between the sharecroppers, on the one hand, and the medium and even small, absentee landowners, on the other. At the same time, the large urban following which the Left parties had among the white collar workers and in the service sector was threatened. They feared that once sharecroppers were recorded, their land would be alienated.

Social contradictions developed. In the meantime, the *panchayat* system, which had been introduced in 1978 with considerable devolution of financial and developmental authority, gradually became a focus of power to be reckoned with. Even though, *prima facie*, the new structure of local government supported Operation Barga, contradictions similar to those that had arisen amongst white collar workers surfaced in the rural areas too where *panchayat*s were dominated by middle/upper peasantry and rural professionals. Lastly, *ex parte* injunction orders issued by courts and the failure of lawyers acting on behalf of the State to fight their legal battles in the courts with due diligence and commitment created an environment unfavourable to the aim of Operation Barga. That the affected landowners would move the court *en masse* had been anticipated, but the lawyers responded to their challenge totally inadequately, displaying no more than nameplate allegiance to the cause for which they had been engaged by the State. Their conduct was not only shocking but also set back the implementation of the land reform programme.[2]

In this section, I have described two fairly successful attempts at land reform in West Bengal in two different time settings, with the aim of drawing some lessons that would be applicable to India as a whole. The positive factors underlying these efforts can be described as follows:

1. On both these occasions, political parties of the Left with a genuine commitment to the rural poor formed the government. There was a manifest and visible 'political will' to set in motion land reform measures.

2. The government, being quite aware of the limitations of the legal administrative and judicial system, allowed mass mobilization in one case and created conditions for the participation of rural workers' organizations in a significantly complementary and supportive role to the efforts of bureaucrats in the other case. In both cases, the governments concerned displayed a commitment to people's organizations playing an activist role in bringing about social change.

3. The bureaucracy showed itself to be capable of providing appropriate and adequate leadership from within, which would enable the system to adapt itself to the needs and requirements of the task assigned to it. It was able to motivate a large army of people in the system to perform their task as something worth doing.

[2] On the occasion of the third anniversary of the Left Front Government (1980), I was officially requested to write a brochure on land reform. A booklet entitled *Land Reforms in West Bengal* was published by the West Bengal Government. It contains some remarks about the inadequate performance of State lawyers.

> What appeared to be very surprising was the role of the legal profession. Whether because of constant espousal of the cause of the propertied classes, or because of the impulse to be on the right side of the Establishment to protect their business interests, or, because of the lack of inner conviction, by and large, the profession in the State failed to rise to the occasion in the legal battles in the law courts. Perhaps, their class interests often transcended their professional ethics with the result that the programme was being stalled through innumerable court injunction orders. Land reform continues to be a major victim of the private property cult of the judicial system even after the abolition of the Right to Property from our Constitution. [Bandyopadhyay, 1980: 7.]

This passage created a furore among the lawyers and particularly among those who enjoyed the patronage of the State government for displaying a Left nameplate on their chests. They demanded that I should be removed from my post. The government half conceded by officially denouncing the booklet and withdrawing it from circulation. This episode illustrates the extent to which vested interests had penetrated even a government which was genuinely pro-poor in character.

This last point would not be easily accepted in political circles, which have a tendency to appropriate all successes to themselves and to attribute all failures to the easily identifiable scapegoat, the bureaucracy. In a non-revolutionary situation, any attempt at altering, within a strictly legal framework, the existing asset-owning pattern or, *inter se*, the rights of different groups with conflicting interests, would require cooperation between the political and bureaucratic wings of executive power in the State; at the very least, the two should not be at cross-purposes. Mutual understanding between the two sides of each other's role is essential; much would depend on the sagacity and astuteness of the leadership on both sides of the divide. The limited success of the land reform programme in West Bengal (and Kerala) and its not-so-satisfactory performance elsewhere in India clearly points to the absence of such positive factors in the latter.

Reform: The all-India picture

India's Five Year Plan documents make dismal reading as far as land reform is concerned. In the litany of failures recorded, great importance is given, albeit in veiled terms, to the hiatus between the political leadership at the Centre and the States, on the one hand, and their implementing agencies, on the other. Thus, for example, let us consider the following passage from the Sixth Five Year Plan (1980–5):

> If the progress of land reforms has been less than satisfactory, it has not been due to a flaw in policy but to indifferent implementation. Often, the necessary determination has been lacking to effectively undertake action, particularly in the matter of implementation of ceiling laws, consolidation of holdings and in not vigorously pursuing concealed tenancies and having them vested with tenancy/occupancy rights as enjoined under the law. [Government of India, 1980: 115.]

A plain reading of this statement would convey the idea that whilst the political leadership in charge of the government was wholly committed to the programme of land reform, the main problem was posed by the administrative apparatus which failed to implement the policy guidelines. One marvels at the naïvete of the statement. I do not deny that lack of appropriate administrative structures may, in some cases, frustrate the social interventionist policies of the State. But this plea is not applicable to the Revenue Departments of State governments in India.

The revenue apparatus of the State government has at its disposal a very strong organizational network right down to the village level, which is known to be capable of executing difficult tasks quite satisfactorily even under trying circumstances. Failure of implementation would not result from a basic weakness of the administration, or unwillingness of the system to carry out the necessary tasks; rather, it would result from a tacit understanding between the executing agencies and the policy-makers to the effect that the powers-that-be would overlook bureaucratic negligence. It does not stand to reason that a machinery, efficient in conducting elections or resettling/rehabilitating vast masses of displaced people (e.g. as a result of submersion in large irrigation projects) or coping with natural calamities or the collection of land revenue and other taxes, could suddenly become inefficient and lackadaisical in the implementation of land reform.

The attitude of the bureaucracy is certainly a major factor in the quality of implementation of any programme—but near total non-performance could only be explained on the ground of the bureaucracy being convinced that such 'failures' would not only be condoned but appreciated by the political bosses. It is not the perfidy of a few individuals or of a strategically placed group of individuals which is behind such a state of affairs. In no sphere of governance in India is the divergence between rhetoric and reality so large and so palpably obvious as in the case of land reform. A cynic may even propound a new law of inverse proportion between promise and performance in respect of land reform.

*Zamindar*s and other rent-receiving intermediaries, created by the British colonial power, always stood by the imperial power during the national movement for independence. In his autobiography, Nehru (1984: 58) wrote that

> as a class, they [*zamindar*s and *taluqdar*s] are physically and intellectually degenerate and have outlived their day; they will continue only so long as an external power like the British government props them up.

The INC was serious about abolishing the *zamindari* system after Independence. Despite the efforts of the *zamindar* class to ingratiate itself with the new political order, the long-standing commitment of the Congress Party, among other factors, made sweeping legislation aimed at abolishing the system unavoidable.

Agrarian legislation during the 150 years preceding Independence invariably consisted of the colonial government's response to intense agrarian unrest and peasant insurrection. This broad feature of colonial policy has largely continued since 1947. Even during the British days, the 'Blue' mutiny against the indigo planters led to the promulgation of the *Rent Act* in the Bengal Presidency in the mid-nineteenth century (Guha, 1974). A series of peasant revolts resulted in the promulgation of the *Bengal Tenancy Act* during its last decade. The aim was to give some protection to the tenants against rack-renting and indiscriminate eviction.

At the time of Independence, the country was witnessing two major peasant movements—the Tebhaga movement in United Bengal and later on in West Bengal, and the Telangana movement. These two militant and widespread peasant upsurges imparted urgency to the whole agrarian situation. The crisis faced by landless labourers, sharecroppers, and other categories of poor and informal tenants was brought into sharp focus by these two movements before the public and the new policy-makers. Two consequences followed:

1. A number of legislative measures concerning the abolition of intermediary rights, protection of tenants, and imposition of ceiling on landholdings; i.e. the first round of land reform legislation, were enacted;

2. The Bhoodan (gift of land) movement, aimed at rural reconciliation, was undertaken by Vinoba Bhave.

It is necessary to digress on Bhoodan in order to understand the implication of the agrarian question for the rural political scene. Even though the powers-that-be appreciated the inherent injustice of the situation, they were anxious to avoid, at all costs, class conflict in rural India which could well spill over into other sectors of the economy with serious implications for the newly established political equations in independent India. They believed that a major programme of reconciliation was essential in order to avoid class conflict. Rather than the landless wresting land from the landowners, thus destabilizing the political economy, a much safer procedure would involve persuading the rich to relinquish some of their land voluntarily to the land–poor. Such a gesture would help the haves to retain the initiative in their own hands.

Vinoba Bhave's philosophy consisted of an extension of the Gandhian concept of trusteeship. During the early '50s, the Bhoodan movement was launched in the then recently disturbed areas of Telangana. Bhave carried the message of peace which urged landowners to part with some of their land in order that the landless could be given a means of livelihood, and peace could be assured in the countryside. Through *padayatra* (pilgrimage on foot), Bhave and his followers covered the whole of India—a remarkable feat for an old and ailing person. Though his campaign was not formally associated with the government, it had the backing of the political leadership at the Centre as well as in the States. It was anticipated that through the Bhoodan movement enough goodwill would be generated to heal the wounds of Tebhaga and Telangana.

Unfortunately, however, Bhoodan did not result in any such upsurge of emotion or goodwill. It was claimed that about 4.2 million acres of land were donated for redistribution. But this was principally uncultivable wasteland involved in litigation. Not infrequently, land not belonging to putative 'donors' had been given away! Though the Bhoodan movement did not succeed in overcoming class conflict by encouraging a change of heart on the part of the landowners, it constituted a proof of the awareness on the part of the ruling party of the beneficial impact of redistributive land reform.

There was also some governmental response to the agrarian unrest which had started before Independence, and continued for a few years thereafter. Land being a State subject under the Constitution of India, almost all States passed legislation abolishing intermediary rights, guaranteeing protection of tenants, and imposing ceilings on landholdings. The INC appointed a Committee on land reform under the chairpersonship of J. C. Kumarappa. States ruled by the Congress party followed its recommendations to varying degrees.

On the abolition of intermediary rights, a majority of tenants succeeded in obtaining the land they cultivated. They were thus brought into direct relationship with the state. But this benefit was enjoyed largely by those who, under the existing tenancy laws, were eligible for occupancy rights through continued possession of the same holding over a prescribed number of years. But there were other grades of under-tenants, sub-tenants, sharecroppers, and tenants-at-will who were denied this benefit. The incidence of tenancy in terms of operated area registered a sharp decline from 35.7 per cent in 1950–1 to 10.7 per cent in 1960–1, and to 8.8 per cent in 1970–1. This sharp decline is explained largely by the conferment of ownership title (either on payment or otherwise) on erstwhile tenants belonging to various intermediate categories.

Although there is no denying the fact that a large number of occupancy or settled tenants received substantial benefits, a doubt still persists. It is common knowledge that a fairly large number of tenants was evicted during this period in order that land might be brought under personal cultivation. The exact figure for this number is not known. But intelligent estimates may be made from some other relevant figures. The phenomenal rise in the figure of agricultural labour from 28.26 per cent in 1951 to 37.79 per cent in 1971 could be explained on the strength of eviction on a large scale of those belonging to this category.

In absolute figures, their number rose from 27.05 million to 47.49 million during the same period. Another ominous pointer in the same direction is the sharp decline in the number of cultivators from 99.51 million in 1961 to 78.18 million in 1971. There are, of course, some definitional problems relating to the different census documents. But even after ironing out such issues, a significant fall in the number of cultivators accompanied by an equally significant rise in the number of agricultural labour is clearly noticeable (United Nations/ESCAP, 1985: 21). It tends to reinforce the doubt that tenants of different grades were evicted during this period in large numbers.

The task of vesting surplus lands, following the first series of legislation in the early '50s, was not seriously undertaken. Land ceilings were fixed at very high levels, and were invariably based on an individual ceiling and not on a family basis. Moreover, in most States, ceilings were fixed on the basis of the quality of land. The variation of the ceilings fixed by different States was immense, the lowest being 7.5 and the highest 131 hectares. Even these half-hearted measures were further attenuated by numerous exemptions granted in respect of orchards, plantations, tank-fisheries, mechanical farming, tubewell irrigated land, etc. Landowners were allowed to give away land to relatives out of 'love and affection'. Existing laws were applied to ensure that ceiling surplus land was not in fact formally acquired by the State. Thus, with the exception of Jammu & Kashmir in the early '50s and West Bengal in the late '60s, not much happened on the land reform front in India as a whole. Most States made a mockery of it.

Conditions of drought and food scarcity during the mid-'60s, the urgency of attaining self-sufficiency in food production, and the euphoria of the miracle HYV wheat seed constituting the core of the Green Revolution in the latter half of the '60s contributed to a state of official amnesia so far as land reform was concerned.

Renewed agrarian unrest which took the form of the Naxalbari movement compelled the Centre to renew its interest in land reform. Though the movement started in West Bengal, it soon spread to other parts of the country. Concerned about the intensity of the movement and its capacity to spread quickly, Prime Minister Indira Gandhi organized a conference of Chief Ministers in 1971 at which the Home Minister declared that the central government was committed to preventing the Green Revolution turning into a Red Revolution by legislating genuine land reform. At a subsequent conference of Chief Ministers (July 1972) national guidelines on land reform were formulated. The principal features of the government's land policy were: a lower level of ceiling for a family of five, fewer exceptions, payment of compensation to landowners almost at an expropriatory rate, and retrospective application of laws for annulling fraudulent transfers made with a view to evading the ceiling. All the States (with the exception of the North–Eastern States in which land is held communally) enacted laws along these broad guidelines. These laws were more serious in character than those made during the early '50s. Even so, a major loophole still remained unplugged. It related to *benami* holdings.

In quantitative terms, approximately 3 million hectares of land exceeding the ceiling—consituting less than 2 per cent of the cultivable land in India—had vested in the States through the operation of both the original and revised sets of laws. More interestingly, against the above figure, different estimates based on different rounds of agricultural census and National Sample Surveys (NSSs) indicated that, with all the existing constraints, between 5.95 and 12.10 million hectares should have been vested (Bandyopadhyay, 1986). Through distribution of land exceeding the ceiling and through the normal process of devolution, the number and percentage of area held by small and marginal holders (i.e. those holding less than 2 hectares) registered an increase between 1970–1 and 1980–1. Even so, the concentration of land in large holdings (i.e. holdings of 10 hectares and above) is still very high, with 2.4 per cent of the total number of landholders operating 22.8 per cent of the total cultivated area.

These all-India figures do not indicate the large unevenness in performance between different States. There is however little to be gained by labouring on variations between different States, excepting that where the State authorities did not seek (or deliberately shunned) the cooperation of organizations of the rural poor, the results were uniformly dismal. But even despite the built-in constraints and obstacles to an equitable distribution of land, those State governments which sought and depended on the support of organizations of the rural poor obtained encouraging results. Thus, as in the case of West Bengal, so too during the late '60s in the State of Kerala, a massive programme of conferment of titles to land on hutment dwellers and tenants was undertaken with the help of rural workers' organizations. The result was indeed commendable. In a short time, most of the eligible hutment dwellers and tenants obtained their legal documents. Whenever the State authorities resposed trust in organizations of the rural poor, they were handsomely rewarded with better performance in the sphere of implementation of land reform.

Land reform basically implies a restructuring of the productive asset-holding pattern in rural areas. It involves sharing and, in a sense, some transfer of economic (and, with it, political) power from the established landowning class to the rural poor, namely marginal farmers, agricultural workers, and small unsecured tenants of various types. At its core, such a transformation is a political process. Therefore, unless there is at least a willing acquiescence and, at best, a conscious political commitment to this process of empowering the poor through a redistribution of productive rural assets, it cannot be expected to progress smoothly and successfully. The all-India scene viewed over time, would point to the grudging way in which the Indian state has responded to the interests of the rural poor. Its main aim has been not to empower the poor but to prevent the spread of rural unrest.

The landowning class, including the rich and upper peasantry, has consolidated its position within the power structure. Obviously, concessions to the poor are given under pressure. The moment there is a let up in such pressure, there is a tendency to relapse to the original situation.

The gap between policy and implementation is directly attributable to the ascendancy of the landowning classes. If reforms have to be conceded under duress, whatever might be conceded in law can be undermined in implementation. First, laws are apt to be made in such a way that loopholes are left which would permit evasion and deception. Second, whatever may be the form of the law, executing agencies are susceptible to pressure from the rural rich to bypass the law or stall its implementation.

The attitude of the bureaucracy and the role of organizations of the rural poor are vital to the success of the land reform process in a constitutional set-up. At Independence, India inherited a strong and well-organized bureaucracy. The administrative structures have been retained without much change. To maintain *Pax Britannica*, the bureaucracy, including the police, developed a tradition of maintaining order with or without law. This tradition, unfortunately, continues even after four decades of Independence.

It is well known that a significant segment of the bureaucracy (particularly at the State level) is characterized by class bias. It is characterized by direct links with the rural élite. But, strangely, even where the bulk of the bureaucracy comes from the urban middle classes with no links with land, its behaviour is no different from that of those with landed interests. This can be explained by the tendency of any bureaucracy to imbibe the dominant ethos of the state.

In a market economy, where the propertied classes are in control of the state apparatus, the basic laws and administrative and judicial traditions would be naturally weighted in favour of private property. The propertied class and the existing social order would enjoy special protection of its interest. In such a situation, the bureaucracy would automatically develop a bias against any movement, or concerted popular action, or even against an isolated law aimed at bringing about an alteration of the existing social order based on pre-existing economic arrangements. Any organizational effort requiring changes in the existing system would be looked upon as a threat to the established order and retaliatory action would be set in motion as a reflex action.

Thus, a concerted move by the sharecroppers or tenants to withhold delivery of produce in excess of the legal dues to the landlords, or to insist on cultivating land which they are illegally prevented from doing by landowners, or to occupy vested land, is often treated as a breach of peace and they are dealt with accordingly under the preventive and criminal laws of the country.

Though of late there has been some recognition of the need for the development of beneficiaries' organizations to enhance the efficiency of the delivery system to the rural masses, there are difficulties inherent in the formation and development of genuine organizations of the rural poor which obstruct the espousal, promotion, and protection of the interests of different sections of the rural poor. The fact that such organizations have not developed to a significant extent should suggest that there are numerous factors—some of which are external and some internal—which inhibit their growth.

Among the factors external to the system are the coercive power of the propertied class, the economic dependence of the poor, the built-in bias of the law and order machinery in favour of the *status quo*, the lack of supportive legislation and non-implementation of the existing laws in favour of the poor (Bandyopadhyay, 1979). Internal factors would include the precarious economic existence of the vast mass of the rural poor which renders them easy victims of manipulation by the landed class. Often the seeds of dissension are skilfully and deliberately planted among them in order to incite the poor to fight the poor. The disability of destitution is so cleverly played upon that the poor tend to look upon their exploiters for support and deliverance, rather than on their own organizational resources through a path of unity.

In India, it is a moot question whether multi-interest umbrella organizations such as *kisan sabhas* can properly espouse the cause of one segment stridently when it affects the interests of another within the same organization. The lukewarm support of the middle peasantry belonging to the *kisan sabha* espousing the cause of Operation Barga in West Bengal in a number of places pointed to the existence of non-antagonistic contradictions among different interest groups within the same organization.

There is a refreshingly candid paragraph in the Seventh Five Year Plan (1985–90):

> Land reforms have been recognized to constitute a vital element both in terms of the anti-poverty strategy and for modernization and increased productivity in agriculture. Redistribution of land could provide a permanent asset base for a large number of rural landless poor for taking up land-based and other supplementary activities. Similarly, consolidation of holdings, tenancy regulation and updating of land records, would widen the access of small and marginal landholders to improved technology and inputs and thereby directly lead to increase in agricultural production. [Government of India, Planning Commission, *c*. 1985: 62.]

Empowering the powerless—the rural poor—which is the aim of land reform, cannot be brought about without bringing about a more than corresponding diminution of power of the rural rich. This basic political process can hardly be successful without a vibrant mobilization of the masses supplemented by equally responsive and vigorous administrative action. The 'Konar model' of the late '60s in West Bengal, perhaps, provides the only viable route to progress in the context of rural India.

REFERENCES

BANDYOPADHYAY, D., 1967: *A Note on Evasion of Ceiling Law in West Bengal* (Calcutta: Directorate of Land Records & Surveys, Department of Land and Land Revenue, Government of West Bengal).

——————, 1979: *Rural Development Administration: Some Emerging Power Issues* (Bangkok: Economic and Social Council for Asia and the Pacific (ESCAP)).

——————, 1980: *Land Reforms in West Bengal* (Calcutta: Department of Information and Cultural Affairs, Government of West Bengal).

——————, 1981: 'Rural Workers' Camps Show The Way', *Mainstream* 19 (4 July): 44.

——————, 1986: 'Land Reforms in India—An Analysis', *Economic and Political Weekly* 21(8 June): 25–6, pp. A50–6.

Government of India, *c.* 1980: *Sixth Five Year Plan (1980–1985)* (New Delhi: Planning Commission, Government of India).

——————, *c.* 1985: *Seventh Five Year Plan (1985–1990)* (New Delhi: Planning Commission, Government of India).

GUHA, R., 1974: 'Neel-Darpan: The Image of a Peasant Revolt in a Liberal Mirror', *Journal of Peasant Studies* 2 (Oct.): 1, pp.1–46.

KONAR, H., 1977: *Selected Works* (Calcutta: Gour Saha).

——————, 1977a: Agrarian Problem of India (Calcutta: Gour Saha).

NEHRU, J., 1984:* *An Autobiography* (New Delhi: Oxford University Press).

*Third Impression

RASUL, M.A., 1974: *A History of the All-India Kisan Sabha* (Calcutta: National Book Agency).

TENDULKAR, D.G., 1960*: *Mahatma, Life of Mohandas Karamchand Gandhi, Volume 1, 1869–1920* (New Delhi: Government of India).

*Revised Edition; orig. published in 1951.

United Nations Economic and Social Commission for Asia and The Pacific,* 1985: *Role of Participatroy Organisations, in Agrarian Reform* (Bangkok: ESCAP).

*ESCAP.

GLOSSARY

bargadar	Sharecropper.
benami	Literally, without name. It refers to the system of holding land or property nominally in the name of another person while the actual owner retains the right to usufruct.
bhoodan	Land gift.
Operation Barga	The LF government's massive drive (1978) to record the names of sharecroppers.
pucca	Solidly built.
taluqdar	Owner of a revenue estate—*taluq*.
zamindar	Tax farming proprietor of land who paid a fixed revenue to the state.

The Political Economy of the Evolution of Anti-Poverty Programmes · 11

A VAIDYANATHAN

Raising the standard of living of the poor?: The first four Plans

The notion that the state should intervene on behalf of the underprivileged is embedded in the directive *Principles of State Policy* and in the mandatory provisions for positive discrimination in favour of Scheduled Castes (SCs) and Scheduled Tribes (STs) in the Indian Constitution. While the former was no more than a declaration of intent, the concerns it emodies have been invoked in the formulation of state policy and are reflected in the Five Year Plans. In the case of SCs and STs, as well as a reservation of quotas for admission to educational institutions and government jobs, the Five Year Plans include special programmes to provide financial assistance for education, water supply, housing, and other welfare programmes. The Five Year Plans have consistently addressed the problems of poverty and inequality by attempting to achieve sustained expansion and diversification of economic activity.

The first Five Year Plan stressed that:

> The elimination of poverty cannot, obviously, be achieved merely by redistributing existing wealth! Nor can a programme aimed only at increasing production remove existing inequalities. The two need to be considered together.... The problem is not one of merely rechannelizing economic activity within the existing socio-economic framework. That framework has itself to be remoulded so as to enable it to accommodate progressively those fundamental urges which express themselves in the right to work, the right to adequate incomes, the right to education and a measure of insurance against sickness, old age and other disabilities. [Government of India 1953: 8.]

Nor is full employment (with rising wages) possible without growth:

> A programme for full employment with assurance of at least the current level of real wages to the newly employed can get into swing only as the capital formation in the community goes up [Government of India, 1953: 8].

Socialism, and later 'socialistic' patterns of society, were, of course, highlighted in the political rhetoric of the Indian National Congress. But, apart from land reform, no serious effort to alter the distribution of existing wealth was ever contemplated in any radical way. The land reform programme itself was not really pursued, then or subsequently, with conspicuous vigour or success. The expansion of public ownership through fresh investment, public sector control of 'the commanding heights', and regulation of private sector activity were seen as the means of containing the concentration of wealth, income, and power in the private sector.

This broad approach informed Indian planning well into the '60s. The Mahalanobis strategy was explicitly designed to rapidly expand the capacity of the Indian economy to achieve and sustain a high rate of growth on a self-reliant basis. Apart from reiterating the role of land reform, cooperatives and the public sector, the conflicts between growth and employment were seen as essentially transitional and containable by encouraging labour-intensive techniques in construction and light industry (especially in the production of mass consumer goods). Specific programmes—both protective and positive—were designed to help traditional and small industry. But in terms of allocation, these were relatively small, adding up to a mere 5 per cent of total public sector plan outlay.

Even as the overall development strategy was crystallizing, the '50s witnessed significant new initiatives for promoting rural development with the involvement of the people. The early years after Independence saw a great deal of interest in reviving the village *panchayats* and *gram sabhas* as a means of encouraging participation in local development. The first Plan referred to the need to establish village *panchayats* which would, amongst other things, be concerned with civic and economic activities in the community. Some States passed legislation for this purpose. Meanwhile, the central government sponsored a nationwide community development progrmme to encourage the rural population and enlist its cooperation in development.

Under the community development programme, a complement of staff was appointed specifically to carry out comprehensive development works in each block—the basic administrative unit for development. A substantial amount of state funds was placed at the disposal of the block development staff. The use of these funds was to be decided in consultation with local leaders through nominated committees. But the committees were rarely 'representative' enough of different sections of the rural population to generate the expected broad-based popular enthusiasm and participation. Also, the block level programmes throughout the country had to conform to a standard, schematic budget specified by the Centre, leaving little room for adapting the content of programmes to specific local conditions.

Dissatisfaction with the way the community development programme was evolving led to the appointment of the Balwantrai Mehta Study team in 1957 to review the working of the programme and suggest ways in which it could be made more effective.[1] The team came to the conclusion that administrative decentralization was essential for the effective implementation of the development programme and that elected bodies should control a decentralized administrative system. They recommended the early establishment of statutory elected local bodies and devolution to them of the necessary resources, power, and authority.

The recommendations had Nehru's support. The National Development Council (NDC) passed a resolution accepting them. The late '50s and early '60s saw a spate of Panchayati Raj legislation. By the mid-'60s, Panchayati Raj was in place in most States, but mostly in name only. From the beginning, it was clear that the members of the State legislatures, the State-level leaders and the bureaucracy were reluctant to devolve power and resources upon the elected local bodies. The latter were not enthusiastic about mobilizing resources on their own initiative and consequently depended almost wholly on the State and central governments for their resources. Once the intensive

[1] This review of the Balwantrai Mehta team's recommendations and its implementation is largely based on the report of the Asoka Mehta Committee (Government of India, 1978a).

phase of community development (lasting five years) was over, the quantum of resources available for use by the *panchayat*s declined significantly.

In some cases—e.g. Maharashtra and Gujarat—the initial enthusiasm for Panchayati Raj was dampened once the State-level politicians and the bureaucracy realized that their 'influence' would be greatly reduced if local bodies were given substantial resources *and* the freedom to allocate them. So, the role of *panchayat*s was curtailed by making them responsible only for the implementation of projects, with an advisory body comprising ministers, MLAs, MPs, and officials actually making decisions. The launching of the Intensive Agricultural Development Programme in 1960 and, subsequently, the Small Farmer Development Programme, the Drought Prone Area Programme, and others outside of the Community Development projects, further eroded the fledgling experiment with Panchayati Raj and strengthened the hold of the line departments on development at district level.

Criticisms of the development strategy, voiced even at the time of the Second Plan, began to find stronger political articulation in the early '60s. The draft Third Plan was criticized for not paying sufficient attention to the unemployment problem on the ground that, despite a decade of planning, the majority of Indians did not earn even as much as 6 annas per head per day, and that, if anything, the rich were getting richer and the poor poorer. The immediate response of the government was to assuage such criticism by incorporating a special employment programme as part of the Plan; appointing an expert Committee to review trends in the distribution of income and wealth; and ascertaining the extent to which the operation of the economic system may have resulted in concentration of wealth and means of production.

The Plan document stated that until the economy was strengthened, it would be difficult to provide work at an adequate level of remuneration to the entire labour force. In the meantime, the government would follow a strategy of maximizing the employment potential of the Plan programmes; speeding up and even enlarging programmes under the Plan with large employment potential 'especially in areas with heavy pressure of population and in which there is considerable underemployment'; and introducing a special rural work programme (RWP) to provide 100 days of employment per year for 2.5 million people. Programmes for village water supply and local development works were accelerated and later developed into the 'minimum needs programme'.

At the same time, a section of the Planning Commission put forward its view that accelerated growth is the best means of resolving the problem of poverty and employment. This viewpoint was articulated in a Perspective Planning Division (PPD) paper which outlined a long-term plan whose central aim was to assure a specified minimum standard of living within 15 years to everyone (Government of India, 1964). The paper ruled out as impractical any drastic redistribution of existing wealth and focused instead on using progressive taxation and public expenditure policies, along with an expanding and efficent public sector—all in a context of rapid overall growth—to bring about better distribution of incremental income and wealth.

The PPD view did not survive the droughts of the mid-'60s and the Indo–Pakistan conflict of 1965. The emphasis had shifted significantly by the time the final version of the Third Plan was completed. The latter part of the '60s saw a period of low investment, high inflation, and persistent balance of payment problems. Several studies highlighted the failure to check increased concentration of industrial activity among big business houses, and the skewed-off benefits from various rural development programmes.

Strategies such as the Integrated Agricultural District Programmes (IADP) and the use of High Yield Variety (HYV) seeds were allegedly aggravating inequalities (Das Gupta, 1977; Frankel, 1971). Confidence in growth-centred strategies (among economists and politicians) weakened. Interventions aimed more directly at particularly 'vulnerable' segments came to the fore. The Rural Credit Survey of 1960–1 had argued for special programmes to help small and marginal farmers increase their asset base and productivity (Reserve Bank of India, 1969). An incipient RWP was launched in the Third Plan, which also recognized regional disparities as an important problem calling for, among other things, special programmes for backward regions.

The planners continued to maintain that no significant reduction in disparities could be effected by redistributive fiscal, pricing and other policies; that surpluses derived from the rich were needed for investment; and that socio-economic goals had to be pursued through overall growth. But they were also forced to respond to the growing perception that the pace of alleviation of mass poverty and unemployment was unacceptably slow and needed to be addressed more directly.

At one stage, the requirements of rapid growth and those of speedy redressal of mass poverty and unemployment were thought to be reconciliable. The Planning Commission paper setting out the approach to the Fourth Plan (Government of India, 1972; Minhas, 1974) outlined a strategy consisting of two key elements. One was the proposal to launch a basic minimum needs programme which would earmark part of the Plan funds to ensure that all parts of the country had a specified minimum level of education, public health, transportation, and water supply facilities. The other was an integrated local works programme centred around improved land and water resource utilization, which would generate additional employment opportunities in rural areas so as to augment the production potential of agriculture and related activities.

The idea of 'basic minimum needs' was incorporated in the Fourth Five Year Plan and also figures in all subsequent Plans. It was not, however, backed up by adequate resources. Such resources as were made available were, however, not distributed (either to the States or eventually to the districts) in accordance with local needs.

If the Minimum Needs Programme existed only in name, the concept of an integrated land and water development programme for employment does not even figure in the Five Year Plans. Both would have called for major changes in the way rural development programmes were planned, funded, and implemented. Evidently no one was eager to confront the thorny issues which this would have raised; they were side-stepped in favour of a variety of programmes to benefit particular regions and classes of people who were poor, backward, or underprivileged.

Political pressures impinging on poverty alleviation programmes

Developments in the political arena gave a powerful impetus to the movement towards targeted poverty alleviation programmes. The split in the INC, and the subsequent internal struggle for control over the party, saw Indira Gandhi using a 'radical' platform in order to seek popular support. Bank nationalization and abolition of privy purses were the initial moves in this strategy, but they proved to be of limited usefulness and, prior to the 1972 general elections, Mrs Gandhi had adopted *Garibi hatao* as the main plank of her manifesto. The persistence of unemployment, the slow pace of improvement in living standards, and the fact that poor and unemployed people comprised a large part of the population compelled the politicians to pay attention to their problems.

It has been pointed out that the last 30 years have witnessed a major transformation in Indian politics. Large numbers of illiterate and impoverished people have become active participants in the political arena. They are aware of their political rights and, increasingly, have become more capable of wielding political power. Even conservative parties, if they are to remain committed to a competitive electoral political system, will have to satisfy new criteria based on the promise of eradicating mass poverty in the foreseeable future (Frankel, 1977: 550).

The growing emphasis on poverty alleviation programmes has also been interpreted as a sign of 'élites competing for political power', seeking to exploit the situation created by the decline of the Congress party as a relatively stable and dominant party based on a measure of political consensus. This decline was in part a reflection of the process of disintegration set in motion by commercialization, by the rapid growth of wage labour, and by vertical dependency in a traditional society on which the Congress party organization and patronage networks had depended heavily. Reinforcing the decline of the Congress party was a parallel process of politicization of the lower rural classes, which also contributed to the weakening of traditional vertical ties. Indira Gandhi's *Garibi hatao* slogan can thus be interpreted as an integral part of the strategy for mobilizing mass support in this changed situation (Kohli, 1987: 84–8).

Indira Gandhi's resounding success in the 1972 elections helped place poverty alleviation programmes firmly on the national agenda. From the Fourth Five Year Plan onwards, the scope and scale of these programmes were progressively extended.

The Fourth Plan declared that area development plans were to be prepared for tribal areas, whilst SCs were to be better integrated within the rural communities in which they lived. The planners found it necessary, however, to emphasize that in both cases the special programmes were meant to supplement the allocations under the normal Plan. Evidently there was a fear that the States might divert resources allocated for SC/ST welfare, provided under the regular programme, to other uses.

Initially, a distinction was made between marginal farmers and landless labourers, on the one hand, and potentially viable small farmers, on the other. For the former, the emphasis was to be on encouraging activities such as animal husbandry and on expanding employment opportunities on a sustained basis through area development programmes by improving local infrastructure and developing natural resources. For small farmers it was felt that measures (including subsidized credit) to enable them to gain better access to irrigation and other yield-augmenting inputs would be more appropriate. In addition, there were to be special programmes for problem areas—notably drought-prone districts, desert regions, and hill areas.

Faced with criticisms of the Emergency, Indira Gandhi went to the electorate with a 20-point programme which was no more than a refurbished version of the *Garibi hatao* package. The opposition was highly critical of the centralization of power and the authoritarianism of her regime, and argued for a different approach to development, based on decentralization of power, participation of the people, full employment, and poverty alleviation as the central themes.

The Janata government set up a committee to provide a blueprint for decentralization (Government of India, 1978). The new Planning Commission pleaded for active state support to organize the poor (Government of India, 1978) in order to enable them to press their interests more effectively in the political process. These ideas were not pursued seriously. However, the revised Fifth Plan, prepared under the Janata government, provided a substantially larger allocation for rural development and employment programmes. After Indira Gandhi's return to power, the scale of poverty alleviation programmes was enlarged, even though they had paid her such poor political dividends earlier. The Sixth Plan (1980–5) nearly trebled the allocation for rural development and employment programmes. All this underlines the enduring political appeal of these programmes and their becoming an accepted part of the discourse and agenda of practically all political parties and consequently of the Five Year Plans.

Indira Gandhi's second tenure witnessed some changes in her government's approach to poverty alleviation programmes. The minimum needs programme continued, but there was still no attempt to ensure that the allocation of resources between State and districts, for various basic needs, was designed to create a minimum standard of facilities everywhere. The creation of national technology missions for rural water supply and immunization constituted the only major departure from the earlier pattern. The three major area-based programmes (namely, drought-prone, desert, and hill areas) were also continued on an enlarged scale without much change in concept. Others went through significant changes in organization and process along with an expansion of the scale of outlays.

The small farmer and the marginal farmer/agricultural labour programmes were discontinued and replaced by the Integrated Rural Development Programme (IRDP) from 1977 onwards. Two types of employment-generating programmes were launched. The National Rural Employment Programme (NREP), which commenced in 1980, was designed to utilize the accumulated food stocks to create employment, and improve the nutrition of the rural poor. Subsidized supply of foodgrains was an essential part of this scheme. It was distinguished from earlier food-for-work programmes by its emphasis on creating specified durable community assets, with detailed guidelines as to the category of people who were to be benefited and also as to the kinds of assets which were to be created. Three years later, another programme, the Rural Labour Employment Guarantee Programme (RLEGP), was aimed primarily at providing additional employment to landless labourers, without laying any special emphasis on the creation of durable assets.

In 1982, a programme for the Development of Women and Children in rural areas was undertaken to help women increase their income opportunities and provide support services (including child care facilities) to enable them to take up economically viable activities. The scheme of Training Rural Youth for Self Employment (TRYSEM) was started in 1977. It was intended to help young people from poor rural families to set up production units of their own in both agricultural and non-agricultural sectors. The target group for all these programmes was to be families with incomes below a specified level; quotas were also specified for sub-categories among the poor, especially SCs, STs, and women. The total allocation for IRDP and employment programmes in the Sixth Plan (1980–5) exceeded Rs 40 billion—compared to Rs 15 billion in the Fifth Plan.

Significantly, the political initiative, as well as the funding of targeted poverty alleviation programmes, came for the most part from the Centre. There were, however, important exceptions. The most notable of these was the Employment Guarantee Scheme (EGS) of Maharashtra; the Mid-day Meals Programme of Tamil Nadu, and some other States; Andhra Pradesh's Rice for the Poor at Rs 2 per kilogram; and a variety of social security programmes.

Maharashtra's EGS, started in 1972 and given statutory form in 1979, undertook to provide gainful and productive employment on approved works for all unskilled persons in rural areas who were prepared to take up manual labour. The guarantee was limited to manual, unskilled work in the vicinity of the worker's residence, with payments consistent with prescribed minimum wages. The original intention was to take up only schemes which created durable community assets; but this has been gradually relaxed to include works benefiting individuals. The programme was started entirely at the initiative of Maharashtra State and has not lost its identity among the numerous anti-poverty and employment programmes sponsored by the Centre. The cost of the EGS has also risen rapidly (from approximately Rs 210 million in 1975–6 to Rs 3180 million in 1988–9) and is several times larger than the cost of the centrally sponsored NREP. Significantly, despite the fact that the EGS was a highly innovative programme conceived in a Congress-ruled State, it was not accepted as a model for the centrally-sponsored national employment programmes. It is only recently, in the context of discussions about the 'Right to Work', that the EGS model has been considered seriously as an alternative to a constitutional guarantee.

The central government has sponsored a special programme for supplementary feeding of children below six years of age and for mothers. In addition, a number of States have introduced special nutrition programmes. The most important of these is the mid-day meals programme introduced in Tamil Nadu, Gujarat, Maharashtra and West Bengal, with substantial financial support. The Tamil Nadu programme, which seeks to provide one meal a day throughout the year to all pre-school children registered with *balwadis*, and all school-going children up to the age of ten years, was estimated as benefiting 8 million children (as at 1984) at a cost of Rs 2240 millions (met entirely by the State from the proceeds of special levies). This outlay accounted for nearly 10 per cent of the Tamil Nadu budget (Harriss, 1991). Arguments about whether the State can afford such a large subsidy have not made much impact because of its evident popularity among the mass of the people and the beneficial effects on childrens' health and school attendance.

Similarly, the 'rice at Rs 2 per kilogram' scheme of Andhra Pradesh provides direct consumption subsidies to the poor. This scheme also carries an extremely high financial burden, borne entirely out of the State's own resources (Radhakrishna and Indrakant, 1987). Expensive as it is, none of the successive governments have found it politically expedient to curtail the programme or the magnitude of subsidies to it.

Several States have introduced social security schemes of widely varying design. Commencing with the introduction of old age pensions, the tendency has been to extend such schemes to other categories, e.g. accident relief, maternity assistance, support for destitute and disabled people. One of the most comprehensive schemes is in operation in Tamil Nadu. The annual cost, currently, is approximately Rs 700 million, again borne entirely by the State budget (Guhan, 1989).

Most States do not have a strong enough resource base to sustain such programmes on a significant scale. The few which do, e.g. Maharashtra, Tamil Nadu, and Andhra Pradesh, are in a relatively strong fiscal position and can afford to ignore the Planning Commission's disapproval. The Planning Commission has, indeed, been forced to accept the necessity of allocating substantial resources for targeted poverty alleviation programmes. It has also acquiesced, however reluctantly, in a variety of subsidies for the benefit of the poor. However, the Commission has sought to discourage States from setting up their own subsidy programmes, partly because they divert resources from productive investment, and also because there is a strongly held belief at the Centre that the enhancement of asset base and employment opportunity is a far more effective way of ensuring a sustained improvement in living standards. Hence, the emphasis is on programmes targeted towards regions with special problems.

The scale and range of such programmes has expanded considerably and currently account for perhaps as much as a sixth of the public sector plan outlay. The allocation of the outlay is a matter for central decision-making. The States resent this and complain about excessive centralization and rigidity in the design of the programmes. Nevertheless, they have seldom opposed, or rejected, the expansion of the centrally sponsored anti-poverty programmes. Apart from their obvious political appeal, these programmes provide a means for the States to receive additional central funds. Moreover, because monitoring and accountability are weak, the programmes encourage elaborate and extensive networks of patronage, reaching out to every corner of the country. It is a network in which State functionaries (political and bureaucratic) can participate, as well as ruling party workers at grass roots level. The role of professional intermediaries who mediate between the village and the State has also expanded considerably. At the same time, criticisms about duplication, waste and corruption have simultaneously intensified.

Organizational problems facing poverty alleviation programmes

A number of evaluation studies (carried out by the government as well as by independent researchers) have drawn attention to the many deficiencies in the working of these programmes (*see*, e.g. Goverment of India, 1985a; Godbole, 1990).[2] Among the main points repeatedly made, the following are the most common:

1. The benefits of the programmes are not reaching the intended beneficiaries because it is easy for the better-off, by virtue of their superior connections with local bureaucracy, to appropriate a sizeable part of the benefits.

2. Even when the beneficiaries are part of the target groups, their selection is often biased, and intermediaries tend to siphon off a substantial part of the resources.

3. The intended beneficiaries lack knowledge/information about the programmes under which they could seek assistance, or about its form and terms.

4. Intended beneficiaries are not organized to press their claims, to correct biased selection and other defects.

[2] For an interesting micro-study of the relation between politics at the grassroots level and the working of IRDP, *see* Gebert (1989).

5. The fragmentation of effort and lack of integration of various anti-poverty programmes cause much waste.

The pros and cons and organizational problems of different types of anti-poverty programme have also been discussed extensively.

Such deficiencies as there are have evoked two different strands of response: one emphasizing the need for decentralization of power and the other focusing on improving administration. The former derives from the view that the root of the problem lies in the centralization of resources and power. Indira Gandhi's (Mark I) regime, which culminated in the Emergency, carried the process of concentration of power into the hands of the central government—and that too into the hands of a small group of individuals—to an extreme limit. A major plank of Jayaprakash Narayan's Nav Nirman movement was the need to reverse this process and take active steps to devolve more power not only from the Centre to the States, but also from the States down to the representative level government institutions below. One of the first actions of the Janata Government after assuming office was to set up a Committee on Panchayati Raj institutions (PRIs). The Resolution appointing that Committee declared

> The Government accords the highest priority to rural development ... The Government considers that the maximum degree of decentralization both in planning and in implementation is necessary.... It has accordingly decided, in consultation with the State governments and Union Territories, to set up a committee to inquire into the working of Panchayati Raj institutions and to suggest measures to strengthen them and enable a decentralized system of planning and development to be effective. [Government of India 1978b.]

The Committee found the activities of the PRIs to be unsatisfactory. By way of explanation, the Committee referred to the frequent tendency of some States to override one or other of the tiers of the system; the failure to hold elections as provided by the law; the lukewarm attitude of the high level political élite—including MLAs and MPs—to the need to strengthen the democratic process at grassroot level, and the perception of the Panchayati Raj leadership as a threat to their own position; factionalism, corruption, and political interference in the functioning of PRIs; and the tendency of economically and socially privileged segments to dominate decision-making.

The Committee reiterated the view of the Balwantrai Mehta team that

> establishment of democratic bodies below the state level is imperative from the political and socio-developmental perspectives. Democratic institutions with periodic elections at all levels will provide a forum for the assertion of their strength in large number by the weaker segments. With ... adequate opportunities to exercise power at various levels

national energy in political recrimination will yield to constructive competition and worthwhile cooperation. [Government of India, 1978b.]

They recommended a three-tier system of Panchayati Raj with both development and some regulatory functions (through Nyaya Panchayats); periodic elections supervised by the State election officers (with reservation for SCs/STs); devolution of substantial powers from States to local bodies; constitution of a State level Finance Commission to determine the devolution of resources from the States to the PRIs; and the establishment of development staff under the Chief Executive Officer of the district, answerable to the elected Zilla Parishad.

This report was neither placed formally on the table of the Parliament, nor discussed with the States. But some States, notably Karnataka and West Bengal, made a serious effort to resuscitate PRIs as instruments of decentralized development. In West Bengal, the Left Front government has shown a consistently positive attitude to decentralization since the late '70s. Elected representatives of the PRIs have been actively involved in decision-making and implementation of local development programmes; and, increasingly, the responsibility for local planning, along with the allocation of resources, is being devolved upon them. This process has been helped by several circumstances. Thus, for example, the same coalition has been voted back to power in four successive general elections; the implementation of a substantial land reform programme which has secured the rights of a large number of *bargadars* (small tenants) over land they cultivate; and the ruling party has been elected to a large majority of PRIs in three successive elections.

Likewise, Karnataka passed a Panchayati Raj bill—substantially along the lines recommended by the Asoka Mehta committee—soon after the Janata government was elected. But, because of a delay in the grant of presidential assent, the law became effective only in 1985. Under the Janata and later under the Janata Dal, a serious effort was made to devolve substantial powers and resources to the *zilla parishads* and PRIs, and to confront the many difficult political and administrative issues which the process entails. But, unlike in West Bengal, the party in power at the State level has changed during the last 10 years: inter- and intra-party conflicts, and conflicts between the PRIs and the bureaucracy have impeded the progress of the reforms. The recent action of the State government seeking to restrict the freedom of PRIs and, more importantly, the decision to postpone elections to these bodies have been decisive setbacks.

Kerala also has gone through a similar experience; the *Kerala District Councils Act* took nearly 10 years to pass, reflecting the reluctance of the State legislature to devolve power. Even after the law came into effect the State government has been slow to transfer resources to the PRIs, leaving them in an economically difficult situation.

The other approach focuses on administration. The Planning Commission issued a series of guidelines on district planning as early as 1969, in which the need for building up capabilities in terms of organization and expertise was emphasized. In 1977, a committee on Block Level Planning made a number of suggestions for strengthening the machinery and processes of local development planning. But progress was slow, and the problems of fragmentation and waste increased in the wake of the proliferation of rural development and anti-poverty programmes. A comprehensive review of the problem was undertaken by another committee set up by the Planning Commission in 1984. This committee recognized that

> For decentralized planning to be making headway, [the] institutional mechanism has to be made broad-based with active involvement of local representatives, and endowed with greater degree of autonomy in local decision-making. But a full fledged transfer of power to decide all sectoral allocation can be achieved only 'at some distant point of time'. [Government of India, 1984.]

Various initiatives and improvisations (necessary to bring about much needed technical and administrative changes and changes in attitudes) should be introduced cautiously. Adequate time should be allowed for assimilating them.

While District Panchayat Authorities should have adequate representation for elected representatives, representatives of economic/occupational and social groups and voluntary agencies, it should be coordinated by the District Collector. The development officials at district level and below should be deputed and answerable to the District Panchayat Authorities. There is a great deal of emphasis on upgrading technical expertise, training of planning personnel, and building professional planning organizations for expert scrutiny and evaluation of projects. The role of elected PRIs was thus relegated decisively in to the background. The committee suggested that a certain portion of local development funds may be given to District Panchayat Authorities to be spent at their discretion.

Underlying this view is a deep scepticism about the efficacy of PRIs. Criticisms have been made along the following lines:

1. There is no strong grass roots pressure for decentralization.

2. Under the existing social structure and property relations, the rural élite dominate these institutions and appropriate much of the benefits of development.

3. Weaker segments are sceptical about decentralization without 'basic changes in property relations'.

4. State level leaders, involved in a struggle for their political survival, use the opportunity provided by the rapid growth of plan outlay for consolidating their own power by controlling the allocation of funds for schools, hospitals, roads, and other essentially local works (Rao, 1989).

In the meantime, the trend towards the expansion of poverty alleviation programmes, largely sponsored by the Centre, has continued, indeed gathered further momentum. Rajiv Gandhi saw them as a means of mobilizing the electoral support of the poor. At the same time, several changes were made in the design and organization of these programmes, in an attempt to ameliorate deficiencies. Interest in political and administrative decentralization revived during the '80s, partly because of the strong plea made by the Rao Committee (which investigated the organization and management of poverty alleviation programmes) and the Economic Advisory Council to the Prime Minister; and partly because of the popularity and apparent success of the West Bengal and Karnataka experiments.

The Seventh Five Year Plan recognized the need to combine administrative improvement with political decentralization, whilst leaning heavily towards the former. It proposed a three pronged strategy whereby 'poverty alleviation programmes will be formulated and implemented in a decentralized manner with the participation of local people at the grass roots level through village *panchayat*s, *panchayat samiti*s' in a gradual fashion; the existing administrative arrangements would be restructured in order to ensure that rural development and poverty alleviation are planned in an integrated way and implemented effectively; and emphasis would be placed on group-oriented activities for beneficiaries.

Recent shifts in the debate on decentralization

In 1987–8 Rajiv Gandhi organized a series of workshops to discuss with the District Magistrates from all over India ways of creating 'responsive administration'. On the basis of the deliberation of these workshops, the Chief Secretaries of the governments of various States produced an overall report in 1988. Its thrust was to make the District Magistrate the focal point of district planning. Besides being in charge of law and order, (s)he was to be the chief executive of the *zilla parishad* with elected representatives playing a distinctly secondary, if not passive, role. The distrust of the political process is made quite clear in the following passage from the Chief Secretaries' report:

> The Presidentship of Zilla Parishads proving such a glittering prize, furious contests will develop with money and communal/caste influences overriding considerations of competence and commitment to social justice among the candidates leading to the capture of the top position by undesirable elements. This may be contrasted with the existing position where the bureaucracy may be sluggish but tries to be fair in its dealings as a rule. And making the collector CEO of ZP will bring knowledge, experience and influence of the Collector and enhance its effectiveness.

The entire exercise of drawing a blueprint for a restructuring of administration in areas which constitutionally fall in the States' domain was undertaken without consulting or involving the Chief Ministers of States. Even the participation of District Magistrates was not cleared through the State governments. In 1989, the Centre decided to deal with the Panchayats in implementing the Jawahar Rozgar Yojana (JRY) by itself. The funds available under the JRY were transferred directly to *panchayats*, which were empowered to utilize them (subject to certain guidelines on the categories of works/beneficiaries) for local development. This move was widely seen as an attempt on the part of the Centre to bypass the States and establish control over the district administration and even *panchayats* by passing on substantial funds directly to them.

Around the same time, the Centre introduced a Constitutional Amendment Bill on Panchayati Raj. The bill reflected the general consensus among political parties on giving constitutional status to Panchayati Raj, making periodic elections to their bodies mandatory, as in the case of the States and the Centre. Other provisions included reservations in order to ensure adequate representation for SCs, STs, and women on local government councils; the creation of a State level Finance Commission with a view to determining the devolution of resources to the PRIs. However, the bill was far too detailed in spelling out the modalities of distribution of the powers between States and the PRIs, leaving little flexibility to the States to work out the specific allocation of functions, procedures, etc. in the light of their own particular circumstances.

The feeling that the bill encroached widely on the power of the States was reinforced by the parallel moves, described above, for the Centre to establish direct channels through the district administrations and the *panchayats*. That the bill was vague on the role of the district administration and its relation to PRIs deepened such suspicion. In the event, a general election was called. The Parliament was dissolved, and the amendment lapsed.

The Janata Dal government introduced a fresh amendment which recognized the States' power to legislate details within the broad principles outlined in the Constitution. The bill tabled in Parliament in 1991 follows broadly the same approach. It is now before a Select Committee.[3]

The introduction of an amendment to the Constitution, and the declarations of support for its objectives from virtually all parties, reflects a general recognition that democratically elected local government institutions have not been given a fair chance; that the system has become far too centralized to cater to the diverse needs of different regions; and that the various anti-poverty programmes implemented by the state bureaucracy, besides being wasteful, are not effective in reaching the target groups.

At the same time, the leisurely way in which the amendment is being processed is indicative of continued hesitation and doubts about its implications. The view that elected PRIs are inherently incapable of using funds wisely for the benefit of the poor is still prevalent, not only among officials but also in non-official circles. PRIs, according to this view, will be dominated by the upper castes and substantial landholders, who are unlikely to work in the interests of the weaker segments or the unemployed. The higher levels of the state and its functionaries must, therefore, ensure that the resources for local development meant for the poor do, in fact, reach them. An extreme version of this view sees the District Collector as a better custodian of the interests of the poor than elected representatives.

It is altogether naïve to suppose that under the existing dispensation the District Collector acts autonomously and is not influenced by pressure from the upper castes and classes. Despite central control and implementation through the bureaucracy, inefficiency, wastage and corruption are the rule rather than the exception. Restructuring the developmental bureaucracy at the district level, itself by no means an easy task as experience has shown, is not enough. Making people better aware of their rights/entitlements under various programmes; insisting on greater information about the benefits of specific programmes in a particular locality; allowing the village assembly to decide who are the eligible beneficiaries on the basis of well-specified criteria; and monitoring the distribution of benefits are all features that are essential in helping generate *political* pressure in order to ensure more effective implementation of the programmes.

If political pressures must be brought into the process, the case for an institutionalized framework based on representative bodies (subject to the discipline of mandatory periodic elections) becomes very

[3] When this volume was going to Press, the Constitutional amendments were passed.

strong. The experience of Karnataka and West Bengal would suggest that the effectiveness of political decentralization is contingent upon the willingess of the State and the Central government to transfer resources and responsibilities to the PRIs. At a rough estimate, 30–40 per cent of the State plan outlays comprise programmes which, by their nature, have to be planned and implemented locally. The proportion may be even higher if the employment programmes were included.

An essential precondition for the success of decentralization is that these programmes should be transferred to the elected local bodies and the allocation used without interference from the State government. Once this is done, local self-government will acquire the necessary political credibility and become truly effective. The real obstacles lie in the reluctance of State level politicians to shed control over the dispensation of any substantial part of plan resources and in the opposition of the bureaucracy to decentralized administration.

These changes will clearly not take place without considerable political effort. But the passing of the constitutional amendment would be a big step forward. Once PRIs, periodic elections and statutory mechanisms for the devolution of resources are made mandatory, those interested in promoting greater decentralization (including local leaders) will have a better chance of succeeding in their struggles.

REFERENCES

CHANDRASEKAR, B.K., 1989: 'Panchayati Raj Bill: The Real Flaw', *Economic and Political Weekly* 24 (1 July): 26, pp.1433–5.

DAS GUPTA, B., 1977: *New Agrarian Technology and India* (New Delhi: Macmillan).

FRANKEL, F., 1971: *India's Green Revolution* (Princeton: Princeton University Press).

————, 1977: *India's Political Economy 1947–1977* (Delhi: Oxford University Press).

GEBERT, R., 1989: 'Poverty Alleviation and Village Politics in Tamil Nadu —whose interests first?' *Economic and Political Weekly* 24 (28 Jan.): 4, pp.197–202.

GODBOLE, M., 1990: *Rural Employment Strategy: A Quest in the Wilderness* (Bombay: Himalaya Publishing House).

GOVERNMENT OF INDIA, 1953: *First Five Year Plan* (New Delhi: Planning Commission, Government of India).

————, 1961: *Third Five Year Plan* (New Delhi: Planning Commission, Government of India).

————, 1964: 'Notes on Perspective of Development, 1960–61 to 1975–76: Implications for Planning for a Minimum Level of Living' (Planning Commission, Government of India, New Delhi). (mimeo.).

————, 1972: *Towards an Approach to the Fourth Five Year Plan* (New Delhi: Planning Commission, Government of India).

————, 1978: *Draft Five Year Plan 1978–83* (New Delhi: Planning Commission, Government of India).

————, 1978a: *Report of Working Group on Block Level Planning* (New Delhi: Planning Commission, Government of India).

————, 1978b: *Report of the Committee on Panchayati Raj Institutions* (New Delhi: Ministry of Agriculture and Irrigation, Government of India).

————, 1984: *Report of the Working Group on District Planning* (New Delhi: Planning Commission, Government of India).

————, 1985: *Sixth Five Year Plan 1985–90, Vol.II* (New Delhi: Planning Commission, Government of India).

————, 1985a: *Evaluation Report on IRDP* (New Delhi: Planning Commission, Government of India).

GUHAN, S., 1989: 'Social Security Initiatives in Tamil Nadu: 1989' (Working paper 96) (Madras Institute of Development Studies, Madras) (mimeo.).

HARRISS, B., 1991: *Child Nutrition and Poverty in South India—Noon Meals in Tamil Nadu* (New Delhi: Concept).

KOHLI, A., 1987: *The State and Poverty in India: The Politics of Reform* (Cambridge: Cambridge University Press).

MATHEW, G., (Ed.), 1986: *Panchayati Raj in Karnataka Today: Its National Dimensions* (New Delhi: Concept).

MINHAS, B.S., 1974: *Planning and the Poor* (New Delhi: S.Chand).

RADHAKRISHNA, R. and S. INDRAKANT, 1987: 'Effect of Rice Market Intervention Policies in India: The Case of Andhra Pradesh' (Hyderabad: Centre for Economic and Social Studies) (mimeo.).

RAO, C.H.H., 1989: 'Decentralised Planning: An Overview of Experience and Prospects', *Economic and Political Weekly* 24 (25 Feb.): 8, pp.411–16.

RESERVE BANK OF INDIA, 1969: *Report of the All India Rural Credit Survey Committee* (Bombay: Reserve Bank of India).

Index